Photonics

Georg A. Reider

Photonics

An Introduction

 Springer

Georg A. Reider
Photonics Institute
TU Vienna
Vienna, Austria

ISBN 978-3-319-26074-7   ISBN 978-3-319-26076-1   (eBook)
DOI 10.1007/978-3-319-26076-1

Library of Congress Control Number: 2016930556

Springer Cham Heidelberg New York Dordrecht London
Translation from the German language edition: Photonik by Georg Reider, © Springer-Verlag Wien 2012.
All rights reserved

Printed on acid-free paper

Springer International Publishing AG Switzerland is part of Springer Science+Business Media (www.springer.com)

To the memory of Hermann A. Haus

Preface

Photonics deals with the generation, propagation, manipulation, and detection of (usually coherent) light waves. This book provides a comprehensive introduction into this important field, from the electrodynamic and quantum mechanic fundamentals to the level of photonic components and building blocks such as lasers, amplifiers, modulators, waveguides, and detectors.

The book is intended for senior level and graduate students of applied physics and electrical engineering as well as engineers in fields such as laser technology, optical communications, laser materials processing, and medical laser applications who wish to gain an in-depth understanding of photonics.

I have to thank many friends, colleagues, and students for improving, with their comments and questions, the contents and didactic line of the book. I am particularly indebted to Martin Hofer, who has produced many of the illustrations and has contributed indispensable advice. The students who have helped to improve the book over the past years are too numerous to be mentioned by name; I wish to express special thanks to Christian Hartl and Florian Höller for their extraordinary support, however.

As a scientific writer, I am aware of standing on the shoulders of others, and I wish to thank the authors of seminal books and articles that have fostered my understanding of photonics, in particular H. Haus, R.B. Boyd, T.F. Heinz, H. Kogelnik, A.E. Siegman, O. Svelto, and A. Yariv.

Finally, I sincerely wish to thank Silvia Schilgerius and Kay Stoll (Springer), and Fathima Rizwana (SPi Global) for their great support and patience.

Vienna, Austria
Georg A. Reider
June 2015

How to Use This Book

This book is supposed to be self-contained; practically all results are derived from basic principles such as the equations of Maxwell and Schrödinger, respectively, and fundamental mathematical concepts such as Fourier transformation and linear systems. The mathematics used is senior undergraduate level throughout; the steps of derivation and the approximations used are carefully commented. Sections marked with an asterisk ($*$) are intended for a more specialized readership and can be omitted without loss of understanding of the remaining text.

The book does not contain explicit "exercises"; instead, the reader is encouraged to follow the derivations in writing. Important equations are set in shaded boxes and, unless they are actually definitions, should be derived by heart after reading. The book includes many representative examples; most figures, in particular, have been produced using the theory and formalisms presented in the text, and the interested reader is encouraged to reproduce them and vary the input parameters; the general availability of computers and (public domain) mathematical and graphical software (such as Gnuplot and Matlab) renders such simulations a relatively easy yet extremely instructive exercise.

At the end of each chapter, a summary points out the central issues tackled; its purpose is to put the contents of the chapter into a broader context. It is followed by a set of problems that are intended to deepen the understanding of the material. Some of them are quite easy, others more demanding. Many make reference to previous sections of the book, with the goal to provide a "global" understanding of the subject.

A bibliography containing references and suggestions for further reading concludes each chapter; the selected references are not only sources but (with the exception of data bases, of course) highly recommended reading.

Contents

1   Electrodynamic Theory of Light 1
  1.1   The Electromagnetic Field 2
  1.2   Wave Equation 5
      1.2.1   Complex Wave Functions and Amplitudes 6
      1.2.2   Plane Waves 8
  1.3   Propagation Velocities 9
      1.3.1   Phase Velocity 9
      1.3.2   Group Velocity 10
      1.3.3   Beam Velocity* 11
  1.4   Energy Transport 12
      1.4.1   Average Energy Flux Density 13
      1.4.2   Energy Exchange Field/Matter 14
      1.4.3   Energy Transport: Plane Waves 15
  1.5   Polarization States 18
      1.5.1   Jones Formalism 19
      1.5.2   Polarization Optics 21
      1.5.3   Transformation of Jones Vectors and Matrices 27
      1.5.4   Elliptically Polarized States 30
      1.5.5   Poincaré Sphere* 32
  1.6   Inhomogeneous Waves 35
  1.7   Summary 36
  1.8   Problems 37
  References and Suggested Reading 37

2   Wave Propagation in Matter 39
  2.1   Transition Between Different Media 40
      2.1.1   Phase Matching at a Boundary 40
      2.1.2   Reflection and Transmission Coefficients 45
      2.1.3   Total Reflection 52
  2.2   Optical Properties of Isotropic Media 56
      2.2.1   Linear Oscillator Model 56
      2.2.2   Absorption and Reflection 59
      2.2.3   Free Electron Gas Model of Metals 62
      2.2.4   Kramers–Kronig Relations 66

2.3 Wave Propagation in Anisotropic Media............................. 69
  2.3.1 Symmetry Properties of Crystals............................. 70
  2.3.2 Propagation Along the Principal Axes........................ 74
  2.3.3 Propagation in Arbitrary Directions* 75
  2.3.4 Electro-Optic Devices 86
  2.3.5 Liquid Crystal Devices 88
2.4 Other Propagation Effects 90
  2.4.1 Optical Activity 90
  2.4.2 Magneto-Optic Faraday Effect 92
  2.4.3 Wave Propagation in Moving Media* 95
2.5 Summary 98
2.6 Problems.. 99
References and Suggested Reading 100

3 Optical Beams and Pulses... 101
3.1 Beam Propagation... 101
  3.1.1 Paraxial Wave Equation 101
  3.1.2 Gaussian Beams 102
  3.1.3 Optical Components and Gaussian Beams 110
  3.1.4 ABCD-Transformation of Gaussian Beams 117
  3.1.5 Hermite–Gaussian Beams 126
  3.1.6 Fourier Optical Treatment of Beam Propagation.............. 128
3.2 Pulse Propagation 136
  3.2.1 Dispersive Propagation Effects................................ 137
  3.2.2 Nonlinear Propagation Effects 148
3.3 Summary 154
3.4 Problems.. 155
References and Suggested Reading 156

4 Optical Interference 157
4.1 Two Field Interference.. 157
  4.1.1 Michelson Interferometer 158
  4.1.2 Mach–Zehnder and Sagnac Interferometers.................. 162
  4.1.3 S-Matrix.. 162
  4.1.4 Young's Double Slit 165
4.2 Multiple Wave Interference... 167
  4.2.1 Optical Gratings 168
  4.2.2 Dielectric Multilayer Systems............................... 170
  4.2.3 Fabry–Perot Interferometer 177
4.3 Resonators 180
  4.3.1 Spherical Mirror Resonators 182
  4.3.2 3D Resonators 187
4.4 Coherence* 189
  4.4.1 Temporal Coherence... 189
  4.4.2 Spatial Coherence... 193

4.5   Summary 194
4.6   Problems... 195
References and Suggested Reading 196

5   **Dielectric Waveguides** 197
5.1   Planar Waveguides 197
       5.1.1   Eigenmodes 199
       5.1.2   Transverse Mode Profile...................................... 203
       5.1.3   Waveguide Dispersion 204
5.2   Fiber Waveguides 205
       5.2.1   Step Index Fibers 206
       5.2.2   Fiber Losses and Dispersion 212
       5.2.3   Gradient Index Fibers.. 216
5.3   Integrated Optics 218
       5.3.1   Waveguide Couplers... 219
       5.3.2   Splitters and Switches 222
       5.3.3   Waveguide Gratings 227
       5.3.4   Waveguide-Interferometers and Modulators 237
       5.3.5   Active Waveguide Components 240
       5.3.6   Photonic Band Gap Fibers.................................... 241
5.4   Summary 242
5.5   Problems... 243
References and Suggested Reading 244

6   **Light–Matter Interaction** 245
6.1   Optical Interactions with Two Level Systems 245
       6.1.1   Perturbations 247
       6.1.2   Absorption and Stimulated Emission 253
       6.1.3   Spontaneous Emission.. 256
       6.1.4   Line Broadening 259
       6.1.5   Saturation of Absorption 264
6.2   Light Amplification by Stimulated Emission 267
       6.2.1   Four-Level Amplifier 269
       6.2.2   Three-Level Amplifier.. 271
       6.2.3   Pulse Amplification and Absorption 272
6.3   Optical Interactions with Semiconductors 275
       6.3.1   Electronic States in Semiconductors.......................... 275
       6.3.2   Optical Transitions in Semiconductors 283
       6.3.3   Optical Gain Condition 287
       6.3.4   Low Dimensional Semiconductors 288
       6.3.5   Carrier Induced Refractive Index Change 293
6.4   Summary 294
6.5   Problems... 295
References and Suggested Reading 296

7 Optical Oscillators 297
  7.1 Stationary Performance 297
    7.1.1 Rate Equations, Four-Level System 297
    7.1.2 Laser Output Characteristic 300
    7.1.3 Three-Level Laser 304
  7.2 Frequency and Time Behavior of Lasers............................. 305
    7.2.1 Multi-Line vs. Single Line Operation......................... 305
    7.2.2 Mode Selection 307
    7.2.3 Laser Line Width 309
    7.2.4 Relaxation Oscillations and Gain Modulation................ 309
  7.3 Pulsed Lasers... 313
    7.3.1 Q-Switching.. 314
    7.3.2 Mode Locking 318
    7.3.3 Carrier Envelope Phase, CEP 322
  7.4 Atomic and Molecular Lasers 323
    7.4.1 Atomic Solid State Lasers..................................... 325
    7.4.2 Gas Lasers... 328
  7.5 Semiconductor Lasers 331
    7.5.1 Heterostructure Lasers.. 334
    7.5.2 Quantum Well Lasers... 336
    7.5.3 Performance and Technology 336
  7.6 Free Electron Lasers* 340
    7.6.1 "Spontaneous" Emission 342
    7.6.2 Light-Electron Coupling and Amplification.................. 344
  7.7 Summary 348
  7.8 Problems.. 349
  References and Suggested Reading 350

8 Nonlinear Optics and Acousto-Optics 351
  8.1 Nonlinear Susceptibility 351
    8.1.1 Frequency Mixing 353
    8.1.2 Anharmonic Oscillator 358
  8.2 Second Order Processes 360
    8.2.1 Second Harmonic Generation 360
    8.2.2 Phase Matching 365
    8.2.3 Optical Parametric Amplification 369
    8.2.4 Parametric Frequency Conversion* 376
    8.2.5 Second Order Autocorrelation 377
  8.3 Third Order Processes 379
    8.3.1 Third Harmonic Generation 379
    8.3.2 Optical Kerr Effect.. 380
    8.3.3 Third Order Parametric Amplification 384
    8.3.4 Two-Photon Absorption 387
    8.3.5 Raman Amplification 388

	8.3.6	Brillouin Amplification	392
	8.3.7	Phase Conjugation*	395
8.4	Electro-Optic Effects		399
	8.4.1	Linear Electro-Optic Effect	400
	8.4.2	Quadratic Electro-Optic Effect	402
	8.4.3	Field Induced Second Harmonic Generation*	403
8.5	Acousto-Optics		403
	8.5.1	Light Scattering at Sound Waves	403
	8.5.2	Acousto-Optic Modulators	409
8.6	Summary		410
8.7	Problems		411
References and Suggested Reading			412

9   Photodetection 413
9.1	Photoelectric Detectors		414
	9.1.1	Photoelectron Multiplier Tubes	414
	9.1.2	Semiconductor Photodetectors	416
	9.1.3	Detector Arrays	423
	9.1.4	Photoresistors	424
9.2	Characteristic Parameters of Detectors		425
9.3	Photon Statistics		426
9.4	Photometry and Colorimetry		430
	9.4.1	Photometry	430
	9.4.2	Colorimetry	431
9.5	Summary		434
9.6	Problems		435
References and Suggested Reading			436

Index 437

Electrodynamic Theory of Light

1

Electrodynamics describes light as electromagnetic radiation in the frequency range of approximately 10^{15} Hz; in this theory, matter is treated as continuous, with the primary material response being the electric polarization. As in any other frequency range, the electromagnetic field and its interaction with matter is described by Maxwell's equations.

These equations do not imply any natural time- or length scale; they do, however, imply a relation between the two scales in the form of a dispersion relation that relies on c_0, the vacuum speed of light. Electrodynamic phenomena therefore can be scaled arbitrarily if the ratio between time and length scale is conserved. The electric and magnetic properties of matter, however, depend very strongly on the frequency. Magnetization, for example, is practically negligible at optical frequencies and is usually not taken into account in optics.

Many important optical phenomena can be understood only within a quantum mechanical treatment of matter, because they reflect its atomic structure. The concept of electrons, for example, is not implied in electrodynamics (where charge is continuous); emission and detection of light are among the most obvious quantum mechanical effects. A small set of physical constants, most notably the charge and the mass of the electron and Planck's constant, are responsible for the optical properties of matter and, in particular, their spectral dependence. A remarkable consequence of the value of these constants is the existence of a spectral window where a wide range of (condensed) materials is highly transparent and, at the same time, has a very noticeable impact on the phase propagation of electromagnetic waves. This window is what we call the visible spectral range, complemented by adjacent spectral bands in the so-called near infrared and near ultraviolet. Outside this window, condensed matter is either strongly absorbing or does not interact with electromagnetic radiation at all. This is the reason why photonics—the technology of electromagnetic radiation in condensed matter—is staged in the visible and the near infrared.

© Springer International Publishing Switzerland 2016
G.A. Reider, *Photonics*, DOI 10.1007/978-3-319-26076-1_1

Because of the central role the electron and its atomic environment is playing in photonics, it is very convenient to adopt the atomic energy scale of electronvolts ($1\,\mathrm{eV} = 1.602 \times 10^{-19}\,\mathrm{J}$). As we shall see, the energy exchange between light and matter is essentially an energy exchange between individual electrons or atoms and the electromagnetic field and involves a discrete energy quantum of $E = \hbar\omega$, where ω is the angular frequency of the field and $\hbar = h/2\pi$, $h = 6.626 \times 10^{-34}\,\mathrm{Js}$ is Planck's constant; such a quantum of electromagnetic energy is often called a photon, which does not, however, imply that a photon is a "particle"; at this level, it is just a consequence of the atomic (discrete) structure of matter. Expressed in units of eV, the frequency range of main stream photonics lies between several $100\,\mathrm{meV}$ and several eV, corresponding to a wavelength range between several $\mu\mathrm{m}$ and $100\,\mathrm{nm}$. Advanced fields of photonics also operate in the Terahertz and XUV range.

Another frequency range of interest is that of thermal radiation, which can be a major source of noise; expressed in units of eV, it ranges up to several $k_B T$, where $k_B = 1.381 \times 10^{-23}\,\mathrm{JK}^{-1}$ is Boltzmann's constant and T is the absolute temperature in units of Kelvin. At room temperature, $k_B T \approx 26\,\mathrm{meV}$, which is significantly less than typical photonic energies. Thermal noise is therefore usually not a critical issue in photonics, another reason for the enormous success of this technology.

A deeper analysis of electromagnetic radiation shows that it requires a treatment similar to the quantum mechanics of matter. In the framework of *quantum* electrodynamics, electromagnetic radiation is shown to behave in many respects similar to quantum mechanical oscillators and the term "photon" assumes a meaning that is far beyond the aforementioned "token" of energy exchange. One of the most obvious consequences of the quantized nature of electromagnetic radiation is spontaneous emission of light by atoms, a phenomenon that cannot be explained by a semiclassical theory that treats matter quantum mechanically and light electrodynamically. Another fascinating consequence of the quantization of light is the existence of "entangled" photons, the basis of quantum cryptography. Much of the theoretical background of photonics, however, can be treated within a semiclassical treatment, which is also employed throughout this book.

1.1 The Electromagnetic Field

Maxwell's equations, relating the electric field $\mathbf{E}$ [Vm^{-1}] and the magnetic field $\mathbf{H}$ [Am^{-1}] in a medium with polarization density $\mathbf{P}$ [Asm^{-2}], magnetization density $\mathbf{M}$ [Am^{-1}], density of free charges ρ [Asm^{-3}], and current density $\mathbf{j}$ [Am^{-2}], have the form

$$\nabla \times \mathbf{E} = -\mu_0 \frac{\partial \mathbf{H}}{\partial t} - \mu_0 \frac{\partial \mathbf{M}}{\partial t} \tag{1.1}$$

$$\nabla \times \mathbf{H} = \varepsilon_0 \frac{\partial \mathbf{E}}{\partial t} + \frac{\partial \mathbf{P}}{\partial t} + \mathbf{j} \tag{1.2}$$

$$\nabla \cdot (\varepsilon_0 \mathbf{E}) = -\nabla \cdot \mathbf{P} + \rho \qquad (1.3)$$

$$\nabla \cdot (\mu_0 \mathbf{H}) = -\nabla \cdot (\mu_0 \mathbf{M}), \qquad (1.4)$$

where $\varepsilon_0 = 8.854 \times 10^{-12}\,\mathrm{AsV^{-1}\,m^{-1}}$ is the vacuum permittivity and $\mu_0 = 4\pi 10^{-7}\,\mathrm{VsA^{-1}\,m^{-1}}$ the magnetic constant (also called vacuum permeability). In cartesian coordinates, the differential operator ∇ is given by

$$\nabla = \begin{bmatrix} \partial/\partial x \\ \partial/\partial y \\ \partial/\partial z \end{bmatrix} \qquad (1.5)$$

or

$$\nabla = [\partial/\partial x, \partial/\partial y, \partial/\partial z], \qquad (1.6)$$

depending on the vector operation. $\mathbf{P}$ is the response of the medium to the electric field and, for moderate optical fields, a linear function of $\mathbf{E}$,

$$\mathbf{P} = \varepsilon_0 \chi \mathbf{E}; \qquad (1.7)$$

χ is the (dimensionless) electric susceptibility and represents the dielectric properties of the medium. It is common to introduce the electric displacement density $\mathbf{D}$ [$\mathrm{Asm^{-2}}$]

$$\mathbf{D} := \varepsilon_0 \mathbf{E} + \mathbf{P} \qquad (1.8)$$

that combines the "vacuum displacement density" $\varepsilon_0 \mathbf{E}$ with the material polarization density. With Eq. (1.7), we obtain

$$\mathbf{D} = \varepsilon_0 (1 + \chi) \mathbf{E} := \varepsilon_0 \varepsilon \mathbf{E}, \qquad (1.9)$$

where

$$\varepsilon = 1 + \chi \qquad (1.10)$$

is known as relative electric permittivity.[1] In similar fashion, $\mu_0\mathbf{H}$ and $\mu_0\mathbf{M}$ are combined to $\mathbf{B}$ [Vsm^{-2}]

$$\mathbf{B} = \mu_0(\mathbf{H} + \mathbf{M}). \tag{1.11}$$

At optical frequencies, the magnetization $\mathbf{M}$ is usually negligible, so that

$$\mathbf{B} = \mu_0\mathbf{H}. \tag{1.12}$$

In the absence of free charges and currents ($\rho = 0, \mathbf{j} = 0$), the fields are therefore described by

$$\nabla \times \mathbf{E} = -\mu_0 \frac{\partial \mathbf{H}}{\partial t} \tag{1.13}$$

$$\nabla \times \mathbf{H} = \frac{\partial \mathbf{D}}{\partial t}; \tag{1.14}$$

Eqs. (1.3) and (1.4) are implied in Eqs. (1.13) and (1.14) since the divergence of the rotation of a vector field is zero, $\nabla \cdot (\nabla \times \mathbf{a}) = 0$. From Eqs. (1.13) and (1.14) follows, using Stokes's theorem, the continuity of the tangential component of $\mathbf{E}$ and $\mathbf{H}$ at an interface between different media. The continuity of the normal component of $\mathbf{D}$ and $\mathbf{H}$ is implied for solutions of Eqs. (1.13) and (1.14).

In the optical spectral range, the susceptibility χ represents the fundamental response of matter to electromagnetic radiation; it is, however, not a "material constant" but rather a (tensorial) response function giving rise to a wide range of photonic phenomena:

– A medium does not respond instantaneously to the electric field, which implies that χ is frequency dependent; consequences are phase- and group velocity dispersion as well as light absorption by a medium (Sect. 2.2);
– In anisotropic media, the polarization vector is generally not parallel to electric field vector, resulting in effects such as birefringence (Sect. 2.3);
– At sufficiently high electric fields, the relation between electric field and polarization is not linear any more, giving rise to a variety of nonlinear optical effects such as the electro-optic effect (Sect. 2.3.4), self-focusing of optical beams (Sect. 3.1.3), soliton propagation (Sect. 3.2.2.2), and frequency mixing and multiplication (Chap. 8).
– The polarization may be nonlocal in the sense that the polarization at a certain point in space is determined not only by the electric field in this point, but also by the field in the vicinity of the point; a manifestation is optical activity (Sect. 2.4.1).

---

[1]In the following, we will refer to ε simply as "permittivity."

1.2 Wave Equation

We can eliminate the magnetic field from Eqs. (1.13) and (1.14) to obtain a single wave equation for the electric field: taking the rotation of Eq. (1.13) and substituting the time derivative of Eq. (1.14), we obtain

$$\nabla \times (\nabla \times \mathbf{E}) + \mu_0 \frac{\partial^2 \mathbf{D}}{\partial t^2} = \mathbf{0}. \qquad (1.15)$$

In isotropic media, the relation between $\mathbf{P}$ and $\mathbf{E}$ is expressed by a scalar suscepti-bility χ, and $\varepsilon = 1 + \chi$. From Eq. (1.3) in the form $\nabla \cdot \mathbf{D} = \nabla \cdot \varepsilon \varepsilon_0 \mathbf{E} = 0$ follows, for homogeneous media, $\nabla \cdot \mathbf{E} = 0$. With the identity

$$\nabla \times (\nabla \times \mathbf{a}) = \nabla(\nabla \cdot \mathbf{a}) - \nabla^2 \mathbf{a}, \qquad (1.16)$$

we can formulate Eq. (1.15) as

$$-\nabla^2 \mathbf{E} + \mu_0 \frac{\partial^2 \mathbf{D}}{\partial t^2} = \mathbf{0}, \qquad (1.17)$$

where the Laplace operator ∇^2, in cartesian coordinates, is given by

$$\nabla^2 = \frac{\partial^2}{\partial x^2} + \frac{\partial^2}{\partial y^2} + \frac{\partial^2}{\partial z^2}. \qquad (1.18)$$

With

$$c_0 := \frac{1}{\sqrt{\varepsilon_0 \mu_0}}, \qquad (1.19)$$

Eq. (1.17) assumes the form

$$\nabla^2 \mathbf{E}(\mathbf{x}, t) - \frac{\varepsilon}{c_0^2} \frac{\partial^2 \mathbf{E}(\mathbf{x}, t)}{\partial t^2} = \mathbf{0}; \qquad (1.20)$$

for reasons that will become obvious, $c_0 = 2.998 \times 10^8 \text{ ms}^{-1}$ is called vacuum speed of light. Equation (1.20) is the wave equation for the electric field in isotropic, linear, and local media.

1.2.1   Complex Wave Functions and Amplitudes

The structure of Eq. (1.20) allows us to factorize its solutions $\mathbf{E}(\mathbf{x}, t)$ into a spatial and a temporal part. For the temporal part, we choose harmonically oscillating functions: not only do they describe the output of a single mode laser very well, they also represent the base for the Fourier decomposition of more general time varying signals. The ansatz

$$\mathbf{E}(\mathbf{x}, t) = \mathrm{Re}\left[\tilde{\mathbf{E}}(\mathbf{x},\boldsymbol{\omega})\mathrm{e}^{\mathrm{j}\omega t}\right] = \tfrac{1}{2}\left[\tilde{\mathbf{E}}(\mathbf{x}, \omega)\mathrm{e}^{\mathrm{j}\omega t} + c.c.\right], \qquad (1.21)$$

where ω is the angular frequency and $c.c.$ stands for "complex conjugate," is a solution of Eq. (1.20), if $\tilde{\mathbf{E}}(\mathbf{x}, \omega)$ is a solution of the Helmholtz equation

$$\nabla^2 \tilde{\mathbf{E}}(\mathbf{x}, \omega) + \frac{\omega^2 \varepsilon}{c_0^2} \tilde{\mathbf{E}}(\mathbf{x}, \omega) = \mathbf{0}. \qquad (1.22)$$

In cartesian coordinates, each component of $\tilde{E}_i$ must be a solution of the scalar Helmholtz equation

$$\left[\frac{\partial^2}{\partial x^2} + \frac{\partial^2}{\partial y^2} + \frac{\partial^2}{\partial z^2} + \frac{\omega^2 \varepsilon}{c_0^2}\right] \tilde{E}_i(\mathbf{x}, \omega) = 0. \qquad (1.23)$$

A particularly simple solution is the harmonically oscillating function

$$\tilde{\mathbf{E}}(\mathbf{x}, \omega) = \tilde{\mathbf{E}}(\mathbf{k}, \omega)\mathrm{e}^{-\mathrm{j}\mathbf{k}\cdot\mathbf{x}} \qquad (1.24)$$

where $\mathbf{k}$ is known as wave vector and its absolute value k as angular wave number[2] or propagation constant. The complete electric wave function is then

$$\mathbf{E}(\mathbf{x}, t) = \mathrm{Re}\left[\tilde{\mathbf{E}}(\mathbf{x}, t)\right], \qquad (1.25)$$

where

$$\tilde{\mathbf{E}}(\mathbf{x}, t) = \tilde{\mathbf{E}}(\mathbf{k}, \omega)\mathrm{e}^{-\mathrm{j}(\mathbf{k}\cdot\mathbf{x}-\omega t)} \qquad (1.26)$$

is the so-called complex wave function and $\tilde{\mathbf{E}}(\mathbf{k}, \omega)$ the complex amplitude; the imaginary part of the argument of the exponential function is called phase. Inserting Eq. (1.26) into the Helmholtz equation Eq. (1.22) establishes a fundamental relation

---

[2]In spectroscopy, the term "wave number" usually refers to $k/2\pi$.

between the angular frequency ω and the wave number k

$$k^2 = \frac{\omega^2}{c_0^2}\varepsilon; \tag{1.27}$$

this dispersion relation is usually written in the form

$$k = \frac{\omega}{c_0}n, \tag{1.28}$$

where

$$n := \sqrt{\varepsilon} \tag{1.29}$$

is the so-called propagation- or refractive index; in vacuum, the dispersion relation is

$$k = k_0 := \frac{\omega}{c_0}. \tag{1.30}$$

As observable quantity, the electromagnetic field is always real valued; its representation as real part of a complex function offers a number of formal advantages, however. In particular, a phase offset of a component can be incorporated in the complex amplitude: the amplitude

$$\tilde{\mathbf{E}}(\mathbf{k}, \omega) = \tilde{\mathbf{E}}_0 = \begin{bmatrix} E_{0,x}e^{j\phi_{(x)}} \\ E_{0,y}e^{j\phi_{(y)}} \\ E_{0,z}e^{j\phi_{(z)}} \end{bmatrix}, \tag{1.31}$$

for example, represents the wave

$$E_x(\mathbf{x}, t) = E_{0,x}\cos[\omega t - (k_x x + k_y y + k_z z) + \phi_{(x)}]$$
$$E_y(\mathbf{x}, t) = E_{0,y}\cos[\omega t - (k_x x + k_y y + k_z z) + \phi_{(y)}]$$
$$E_z(\mathbf{x}, t) = E_{0,z}\cos[\omega t - (k_x x + k_y y + k_z z) + \phi_{(z)}]. \tag{1.32}$$

Another advantage of the complex representation is that the action of differential operators on Eq. (1.26) can be replaced by simple (vector) operations

$$\frac{\partial}{\partial t} \rightarrow j\omega$$
$$\nabla \rightarrow -j\mathbf{k}. \tag{1.33}$$

Fig. 1.1 Surfaces of
constant phase of a plane
wave

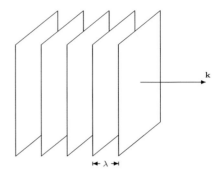

In the following, we will deal with complex wave functions instead of their real part
and take the real part only if necessary; care has to be taken if nonlinear operations
come into play, such as multiplication of fields in nonlinear optics (Chap. 8), or if
the power of the electric field is evaluated; such calculations are based on the real
part of the field.

1.2.2   Plane Waves

Surfaces of constant phase of Eq. (1.26), $\mathbf{k} \cdot \mathbf{x} - \omega t = $ const., are planes normal to the
wave vector $\mathbf{k}$ (Fig. 1.1); these waves therefore are called plane waves; the distance
between planes of equal phase are separated by integer multiples of the so-called
wavelength

$$\lambda := \frac{2\pi}{|\mathbf{k}|}. \tag{1.34}$$

The number $k/2\pi$ is equal to the number of spatial periods per unit length, measured
in the direction of $\mathbf{k}$; k is therefore also called spatial (angular) frequency. In vacuum,

$$\lambda_0 = \frac{2\pi}{k_0} = 2\pi \frac{c_0}{\omega}; \tag{1.35}$$

the vacuum wavelength in the optical region of the electromagnetic spectrum is
of the order of $1\,\mu\text{m}$. The corresponding temporal oscillation period, $2\pi/\omega$, is
about 3×10^{-15} s, or 3 femtoseconds (fs).

Similar to harmonically oscillating *temporal* functions that allow "synthesizing"
arbitrary temporal functions, plane waves can be used to synthesize arbitrary *spatial*
wave functions via a Fourier integral over all possible wave vectors (Sect. 3.1.6).

In practice, there are different conventions to specify the frequency of a wave:
the temporal frequency $\nu = \omega/2\pi$, the quantum energy $\hbar\omega$, the spatial vacuum

Table 1.1 Relations between different parameters characterizing the frequency of an optical wave: frequency v [THz], wave number $(k/2\pi)$ [cm^{-1}], quantum energy $\hbar\omega$ [meV], and vacuum wave length λ_0 [μm], expressed in the appropriate units

	v [THz]	$(k/2\pi)$ [cm^{-1}]	$\hbar\omega$ [meV]	λ_0 [μm]
v [THz]		$0.0300(k/2\pi)$	$0.242\hbar\omega$	$300/\lambda_0$
$(k/2\pi)$ [cm^{-1}]	$33.4v$		$8.07\hbar\omega$	$10^4/\lambda_0$
$\hbar\omega$ [meV]	$4.14v$	$0.124(k/2\pi)$		$1240/\lambda_0$
λ_0 [μm]	$300/v$	$10^4/(k/2\pi)$	$1240/\hbar\omega$	

frequency (spectroscopic wave number) $k/2\pi = 1/\lambda_0$, or the vacuum wave length λ_0. Table 1.1 summarizes the relations between the different parameters.

1.3 Propagation Velocities

1.3.1 Phase Velocity

To determine the phase velocity of the wave function Eq. (1.26), we choose a certain value of the phase

$$\mathbf{k} \cdot \mathbf{x} - \omega t = \text{const.,} \tag{1.36}$$

and calculate the speed at which it propagates through space by taking the spatial derivative

$$\mathbf{k} \cdot \frac{d\mathbf{x}}{dt} - \omega = 0. \tag{1.37}$$

The phase velocity in the direction of the wave vector is then, using Eq. (1.28),

$$v_{\text{ph}} = \left| \frac{d\mathbf{x}}{dt} \right| = \frac{\omega}{k} = \frac{c_0}{n}. \tag{1.38}$$

In vacuum, $n = \sqrt{\varepsilon} = 1$ and the phase velocity is equal to the vacuum velocity of light, c_0. In the visible, this is also a good approximation for the phase velocity in gasses at moderate pressure; the propagation index of transparent condensed media ranges between 1 and about 3, the corresponding phase velocity between c_0 and $c_0/3$. The phase velocity is the relevant velocity for the description of interference effects (Chap. 4). The propagation of optical pulses is governed by the so-called group velocity, which we discuss in the following.

1.3.2 Group Velocity

According to the Fourier theorem, wave packets can be understood as superpositions of monochromatic waves. Since the permittivity ε and thus the phase velocity of the individual Fourier components is frequency dependent, the propagation velocity of the wave packet may be difficult to define and evaluate. We will treat this problem in some detail in Sect. 3.2. As a first approach, we superimpose two waves with slightly different frequencies and determine the propagation velocity of the resulting "beating" envelope. The two frequencies $\omega_0 \pm \Delta\omega$ correspond to two wave numbers $k^0 \pm \Delta k$; assuming equal amplitudes of the two waves, the total field is given by

$$
\mathbf{E}(z, t) = \mathrm{Re}\left[\mathbf{E}_0 e^{-j[(k^0+\Delta k)z - (\omega_0+\Delta\omega)t]} + \mathbf{E}_0 e^{-j[(k^0-\Delta k)z - (\omega_0-\Delta\omega)t]} \right]
$$

$$
= 2\mathbf{E}_0 \cos(z\Delta k - t\Delta\omega)\cos(\omega_0 t - k^0 z), \tag{1.39}
$$

which is an amplitude modulated wave with the carrier frequency ω_0 and the envelope $\cos(z\Delta k - t\Delta\omega)$ (Fig. 1.2). The envelope propagates at the velocity $\mathrm{d}z/\mathrm{d}t = \Delta\omega/\Delta k$, while the phase fronts move at the phase velocity of the carrier,

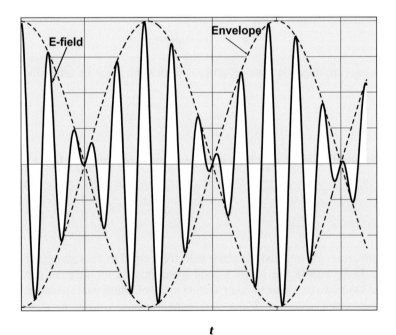

t

Fig. 1.2 Superposition of two monochromatic waves of slightly different frequencies, resulting in a beat signal

$v_{ph} = \omega_0/k^0$. For $\Delta\omega \to 0$, the Taylor expansion

$$\Delta\omega = \frac{d\omega}{dk}\Delta k + \dots, \qquad (1.40)$$

allows us to identify $\Delta\omega/\Delta k$ with $d\omega/dk$, so that the group velocity can be defined as

$$v_g := \frac{d\omega}{dk}. \qquad (1.41)$$

The group delay time $1/v_g$ (the time that the envelope needs to propagate over the unit length) is related to the propagation index $n(\omega)$ by

$$\frac{1}{v_g} = \frac{dk}{d\omega} = \frac{d}{d\omega}\left(\frac{n\omega}{c_0}\right) = \frac{1}{c_0}\left(n + \omega\frac{dn}{d\omega}\right). \qquad (1.42)$$

Therefore,

$$v_g = \frac{c_0}{n + \omega(dn/d\omega)} = \frac{c_0}{n - \lambda_0(dn/d\lambda_0)}; \qquad (1.43)$$

the second relation follows from $d\omega/\omega = -d\lambda_0/\lambda_0$ [see Eq. (1.35)]; since the propagation index is usually tabulated as a function of λ_0, this relation is of particular practical importance.

Depending on the frequency, the derivative $dn/d\lambda_0$ may be positive or negative for a given material, so that the group velocity can be larger or smaller than the phase velocity. Within the transparency range of a medium, the derivative is usually negative $dn/d\lambda_0 < 0$, so that $v_g < v_{ph}$. These spectral ranges of "normal" dispersion alternate with frequency bands of "anomalous" dispersion ($dn/d\lambda_0 > 0$).[3] The mechanism behind the dispersion of a medium will be discussed in Sect. 2.2; its impact on pulse propagation and broadening will be treated in Sect. 3.2.

1.3.3 Beam Velocity*

As already mentioned, wave packets can be tailored in time *and* space to form an optical (pulsed) beam (Sect. 3.1.6). The wave vectors of the Fourier components of such a beam are grouped around a central wave vector $\mathbf{k}^0$ that defines the direction

---

[3]In spectral ranges of very high anomalous dispersion, the group velocity can exceed c_0; this does not contradict special relativity, however, which refers to the *signal* velocity; for details, see Brillouin (1960) and Jackson (1999).

of the beam: $\mathbf{k} = \mathbf{k}^0 + \Delta\mathbf{k}$; each wave vector is related to a frequency $\omega = \omega_0 + \Delta\omega$ according to the dispersion relation $\omega(\mathbf{k})$ that can be expanded as

$$\omega(\mathbf{k}) = \omega_0 + \frac{\partial\omega}{\partial\mathbf{k}}\Delta\mathbf{k} + \ldots \qquad (1.44)$$

The wave packet can be written as three-dimensional integral over $\Delta\mathbf{k}$,

$$\tilde{\mathbf{E}}(\mathbf{x}, t) = \int \tilde{\mathbf{E}}(\Delta\mathbf{k}) e^{-j[(\mathbf{k}^0 + \Delta\mathbf{k})\cdot\mathbf{x} - \omega(\mathbf{k})t]} \, d^3\Delta\mathbf{k}$$

$$= e^{-j(\mathbf{k}^0\cdot\mathbf{x} - \omega_0 t)} \int \tilde{\mathbf{E}}(\Delta\mathbf{k}) e^{-j\Delta\mathbf{k}\cdot[\mathbf{x} - (\partial\omega/\partial\mathbf{k})t]} \, d^3\Delta\mathbf{k}, \qquad (1.45)$$

where $\tilde{\mathbf{E}}(\Delta\mathbf{k})$ is the amplitude corresponding to the wave vector $\mathbf{k}^0 + \Delta\mathbf{k}$. The result is a plane carrier wave $\exp[-j(\mathbf{k}^0 \cdot \mathbf{x} - \omega_0 t)]$ with a spatial-temporal envelope represented by the integral; the vectorial group velocity $\mathbf{v}_{\text{ray}}$ is obtained by choosing a certain value of the envelope phase $\Delta\mathbf{k} \cdot [\mathbf{x} - (\partial\omega/\partial\mathbf{k})t] = \text{const.}$ and extracting $\dot{\mathbf{x}}$ from the temporal derivative

$$\mathbf{v}_{\text{ray}} = \dot{\mathbf{x}} = \begin{bmatrix} \partial\omega/\partial k_x \\ \partial\omega/\partial k_y \\ \partial\omega/\partial k_z \end{bmatrix} = \nabla_{\mathbf{k}}\omega(\mathbf{k}). \qquad (1.46)$$

In isotropic media, the dispersion relation Eq. (1.28) does not depend on the direction of the wave vector,

$$\omega = \frac{c_0}{n}\sqrt{k_x^2 + k_y^2 + k_z^2} \qquad (1.47)$$

and the group velocity is parallel to $\mathbf{k}^0$

$$\mathbf{v}_{\text{ray}} = v_g \frac{\mathbf{k}^0}{|\mathbf{k}^0|}. \qquad (1.48)$$

In anisotropic media, however, the direction of the beam velocity generally differs from $\mathbf{k}^0$, as we shall see in Sect. 2.3.

1.4 Energy Transport

The energy transport of electromagnetic waves is described by Poynting's theorem; to derive it, we multiply Eq. (1.13) with $\mathbf{H}$, and (1.14) with $\mathbf{E}$

$$\mathbf{E} \cdot (\nabla \times \mathbf{H}) = \mathbf{E} \cdot \frac{\partial}{\partial t}(\varepsilon_0\mathbf{E} + \mathbf{P}) \qquad (1.49)$$

$$\mathbf{H} \cdot (\nabla \times \mathbf{E}) = -\mathbf{H} \cdot \left(\mu_0 \frac{\partial \mathbf{H}}{\partial t} \right); \tag{1.50}$$

after subtraction and using $2\mathbf{a} \cdot (\partial \mathbf{a}/\partial t) = \partial(\mathbf{a} \cdot \mathbf{a})/\partial t$, we obtain the equation

$$\mathbf{E} \cdot (\nabla \times \mathbf{H}) - \mathbf{H} \cdot (\nabla \times \mathbf{E}) = \frac{\partial}{\partial t} \left(\varepsilon_0 \frac{\mathbf{E} \cdot \mathbf{E}}{2} + \mu_0 \frac{\mathbf{H} \cdot \mathbf{H}}{2} \right) + \mathbf{E} \cdot \frac{\partial \mathbf{P}}{\partial t} \tag{1.51}$$

which, using the identity $\mathbf{b} \cdot (\nabla \times \mathbf{a}) - \mathbf{a} \cdot (\nabla \times \mathbf{b}) = \nabla \cdot (\mathbf{a} \times \mathbf{b})$, we convert into Poynting's theorem in its differential form

$$- \nabla \cdot (\mathbf{E} \times \mathbf{H}) = \frac{\partial}{\partial t} \left(\varepsilon_0 \frac{\mathbf{E} \cdot \mathbf{E}}{2} + \mu_0 \frac{\mathbf{H} \cdot \mathbf{H}}{2} \right) + \mathbf{E} \cdot \frac{\partial \mathbf{P}}{\partial t}. \tag{1.52}$$

For the interpretation of the individual terms, we employ the divergence-theorem

$$\int_V (\nabla \cdot \mathbf{u}) \, dV = \int_A \mathbf{u} \cdot \mathbf{n} \, dA, \tag{1.53}$$

where A is the surface of the volume V, $\mathbf{n}$ is the outward pointing unit normal vector of a surface element, and dV, dA are differential volume and surface elements, respectively, to transform Eq. (1.52) into

$$\int_A [(\mathbf{E} \times \mathbf{H}) \cdot \mathbf{n}] \, dA = - \int_V \left[\frac{\partial}{\partial t} \left(\varepsilon_0 \frac{\mathbf{E} \cdot \mathbf{E}}{2} + \mu_0 \frac{\mathbf{H} \cdot \mathbf{H}}{2} \right) + \mathbf{E} \cdot \frac{\partial \mathbf{P}}{\partial t} \right] dV. \tag{1.54}$$

The terms $\varepsilon_0 \mathbf{E} \cdot \mathbf{E}/2$ and $\mu_0 \mathbf{H} \cdot \mathbf{H}/2$ represent the electric and magnetic contributions to the vacuum-energy density of the field, while $\mathbf{E} \cdot (\partial \mathbf{P}/\partial t)$ is the power density that is exchanged between the field and the medium. Thus, the right-hand side of Eq. (1.54) is equal to the temporal change of the energy stored in volume V. The left-hand side can therefore be interpreted as energy flux through the surface A, and the Poynting vector

$$\mathbf{S} = \mathbf{E} \times \mathbf{H} \tag{1.55}$$

as energy flux density [W m^{-2}] of the electromagnetic field.

1.4.1 Average Energy Flux Density

Due to the high frequency (10^{14} to 10^{15} Hz) of optical fields, most detectors can only measure the time average of the energy flux density and related quantities. We denote the average of a real valued function $a(t)$ by brackets $\langle a(t) \rangle$ and define it as

$$\langle a(t) \rangle = \lim_{T \to \infty} \frac{1}{T} \int_{-T/2}^{T/2} a(t) \, dt. \tag{1.56}$$

In the complex notation that we have introduced, the average can be expressed very conveniently: assuming two real vectors $\mathbf{a}(t)$, $\mathbf{b}(t)$ with the complex representation

$$\mathbf{a}(t) = \tfrac{1}{2} \left[\tilde{\mathbf{a}}(\omega) e^{j\omega t} + \tilde{\mathbf{a}}^*(\omega) e^{-j\omega t} \right]$$

$$\mathbf{b}(t) = \tfrac{1}{2} \left[\tilde{\mathbf{b}}(\omega) e^{j\omega t} + \tilde{\mathbf{b}}^*(\omega) e^{-j\omega t} \right], \tag{1.57}$$

we obtain

$$\langle \mathbf{a}(t) \times \mathbf{b}(t) \rangle = \tfrac{1}{2} \mathrm{Re} \left[\tilde{\mathbf{a}}(\omega) \times \tilde{\mathbf{b}}^*(\omega) \right] \tag{1.58}$$

and

$$\langle \mathbf{a}(t) \cdot \mathbf{b}(t) \rangle = \tfrac{1}{2} \mathrm{Re} \left[\tilde{\mathbf{a}}(\omega) \cdot \tilde{\mathbf{b}}^*(\omega) \right], \tag{1.59}$$

because $\langle e^{\pm j2\omega t} \rangle = 0$. The average energy flux density can therefore be expressed as

$$\langle \mathbf{S} \rangle = \tfrac{1}{2} \mathrm{Re} \left[\tilde{\mathbf{E}}(\omega) \times \tilde{\mathbf{H}}^*(\omega) \right]. \tag{1.60}$$

1.4.2   Energy Exchange Field/Matter

The average electric vacuum field energy density of a stationary field is constant, since

$$\left\langle \frac{\partial}{\partial t} \frac{\varepsilon_0 \mathbf{E} \cdot \mathbf{E}}{2} \right\rangle = \left\langle \varepsilon_0 \mathbf{E} \cdot \frac{\partial \mathbf{E}}{\partial t} \right\rangle = \tfrac{1}{2} \mathrm{Re} \left[-j\omega \varepsilon_0 \tilde{\mathbf{E}}(\omega) \cdot \tilde{\mathbf{E}}^*(\omega) \right] = 0; \tag{1.61}$$

the same applies to the magnetic vacuum field energy, so that, according to Eq. (1.54) (and not surprisingly), the average energy flux through a closed surface vanishes in vacuum. In the presence of a polarizable medium, this is generally not the case because of the last term in Eq. (1.54), which is the product of the polarization current density $\partial \mathbf{P}/\partial t$ and the electric field. In complex notation, $\mathbf{P}(t) = \mathrm{Re}[\tilde{\mathbf{P}}(t)]$ with

$$\tilde{\mathbf{P}}(t) = \tilde{\mathbf{P}}(\omega)\mathrm{e}^{\mathrm{j}\omega t}; \qquad (1.62)$$

according to Eq. (1.7),

$$\tilde{\mathbf{P}}(\omega) = \varepsilon_0 \tilde{\chi} \tilde{\mathbf{E}}(\omega), \qquad (1.63)$$

where we assume the susceptibility to be scalar but complex, $\tilde{\chi} = \chi' + \mathrm{j}\chi''$ (implying a phase shift between the electric field and the polarization); the polarization current density is therefore

$$\frac{\partial \tilde{\mathbf{P}}(t)}{\partial t} = \mathrm{j}\omega\varepsilon_0 \tilde{\chi} \tilde{\mathbf{E}}(t) \qquad (1.64)$$

and

$$\left\langle \mathbf{E} \cdot \frac{\partial \mathbf{P}}{\partial t} \right\rangle = \tfrac{1}{2}\mathrm{Re}\left[\tilde{\mathbf{E}}(\omega) \cdot \left[\mathrm{j}\omega\varepsilon_0 \tilde{\chi} \tilde{\mathbf{E}}(\omega) \right]^* \right] = -\chi'' \frac{\omega\varepsilon_0 \tilde{\mathbf{E}}(\omega) \cdot \tilde{\mathbf{E}}^*(\omega)}{2};$$

$$(1.65)$$

this is the power density that is transferred from the field to the medium; it is proportional to the imaginary part of the susceptibility and vanishes only if the polarization is exactly in phase with the electric field (compare Sect. 2.2). The complex polarization current density $\partial \tilde{\mathbf{P}}/\partial t = \mathrm{j}\omega \tilde{\mathbf{P}}$ is then $\pi/2$ out of phase with the electric field and the power exchange is purely reactive, which means that the energy deposited in the medium in one half cycle is returned to the field in the consecutive one. Thus, only the quadrature component of the polarization gives rise to a net energy exchange (Fig. 1.3).

1.4.3 Energy Transport: Plane Waves

The results obtained so far are applicable to spatially arbitrary, harmonically oscillating waves; for plane waves Eq. (1.26), Maxwell's equations Eqs. (1.13) and (1.14) simplify to

$$\mathbf{k} \times \mathbf{E} = \mu_0 \omega \mathbf{H} \qquad (1.66)$$
$$\mathbf{k} \times \mathbf{H} = -\omega \mathbf{D}. \qquad (1.67)$$

Fig. 1.3 Relative phases of $\tilde{\mathbf{E}}$, $\tilde{\mathbf{P}}$, and $\partial \tilde{\mathbf{P}} / \partial t$ in a medium with complex susceptibility, shown in the complex plane

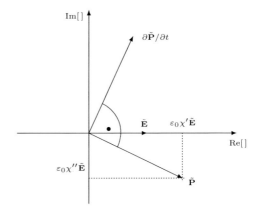

Fig. 1.4 Geometric relation of the vectors **E**, **D**, **H**, **k**, and **S** of a plane wave in an isotropic medium

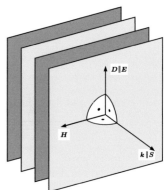

The vectors **D**, **H**, and **k** are mutually orthogonal (Fig. 1.4); in isotropic media, the additional relation $\mathbf{E} \| \mathbf{D}$ applies. Since $|k| = n\omega / c_0 = n\omega \sqrt{\varepsilon_0 \mu_0}$, the absolute values of **H** and **E** are then related by

$$|\mathbf{H}| = n \sqrt{\frac{\varepsilon_0}{\mu_0}} |\mathbf{E}| =: \frac{n}{Z_0} |\mathbf{E}|, \qquad (1.68)$$

where

$$Z_0 = \sqrt{\frac{\mu_0}{\varepsilon_0}} \simeq 377\,\Omega \qquad (1.69)$$

is called vacuum impedance.

The Poynting vector $\mathbf{S} = \mathbf{E} \times \mathbf{H}$ is orthogonal to $\mathbf{E}$ and $\mathbf{H}$; in isotropic media it is also parallel to $\mathbf{k}$,

$$\mathbf{S} = \mathbf{E} \times \mathbf{H} = \frac{1}{\mu_0 \omega} \mathbf{E} \times (\mathbf{k} \times \mathbf{E}) = \frac{\mathbf{E} \cdot \mathbf{E}}{\mu_0 \omega} \mathbf{k} \tag{1.70}$$

(the second equality follows from $\mathbf{a} \times (\mathbf{b} \times \mathbf{c}) = \mathbf{b}(\mathbf{a} \cdot \mathbf{c}) - \mathbf{c}(\mathbf{a} \cdot \mathbf{b})$ and is generally valid only in isotropic media where $\mathbf{E} \cdot \mathbf{k} = 0$). Using Eq. (1.59), its time average can be expressed as

$$I = |\langle \mathbf{S} \rangle| = n \frac{\tilde{\mathbf{E}}(\omega) \cdot \tilde{\mathbf{E}}^*(\omega)}{2Z_0} = n \frac{\mathbf{E}_0^2}{2Z_0} \tag{1.71}$$

and is called irradiance or intensity I.[4] From this equation follows the useful relation

$$|\mathbf{E}_0| = \sqrt{\frac{2Z_0 I}{n}} \tag{1.72}$$

that allows us to calculate the electric field amplitude from a given energy flow density. The electric field of a 1 mW laser with a beam cross section of 1 mm^2, for example, is about 10^3 V/m.

If the area illuminated by the wave is not normal to the wave vector, the intensity is given by the normal projection

$$\mathbf{n} \cdot \langle \mathbf{S} \rangle = \frac{n \mathbf{E}_0^2}{2Z_0} \cos \theta, \tag{1.73}$$

where θ is the angle of incidence, measured in reference to the surface normal (Fig. 1.5).

Fig. 1.5 Electromagnetic energy flux density at oblique incidence on a surface

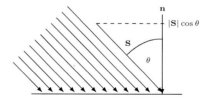

---

[4]In photonics, the term *intensity* is generally used instead of *irradiance*.

1.5   Polarization States

An important property of an optical wave is its polarization state, i.e., the orientation
of the electric field vector in space; it influences, among other things, the reflection
and transmission behavior at interfaces between different media. According to
Eq. (1.67), the electric field vector lies in a plane normal to **k** and has two degrees
of freedom.[5] In general, the electric field vector of a harmonically oscillating wave
describes an ellipse in this plane, rotating with a period of $2\pi/\omega$; depending on the
sense of rotation, this state is called left or right elliptically polarized. If the ellipse
degenerates to a line, the state is linearly polarized; another special case is circularly
polarized light.

It is convenient to describe these states in a cartesian coordinate system whose
z-axis is chosen to be parallel to **k**. The electric field can then be represented by a
two-dimensional vector; the general case Eq. (1.32) is given by

$$E_x(z,t) = E_{0,x}\cos(\omega t - kz)$$
$$E_y(z,t) = E_{0,y}\cos(\omega t - kz + \Delta\phi); \qquad (1.74)$$

for convenience, the origin of the time coordinate is chosen such that $\phi_{(x)} = 0$ and
$\phi_{(y)} = \Delta\phi$.

If the two field components are in phase ($\Delta\phi = 0$), **E** oscillates along a line
oriented under the angle $\varphi = \arctan(E_{0,y}/E_{0,x})$ in respect to the x-axis; such a field
is called linearly polarized.

If the phase difference is $\Delta\phi = \pm\pi/2$ and $E_{0,x} = E_{0,y} = E_0$, then the field
vector in a given plane $z = 0$ describes a circle

$$E_x(t) = E_0\cos\omega t$$
$$E_y(t) = \mp E_0\sin\omega t \qquad (1.75)$$

and the wave is called circularly polarized. For an observer facing the light wave,
the temporal rotation is clockwise (cw) for $\Delta\phi = \pi/2$ and counterclockwise (ccw)
for $\Delta\phi = -\pi/2$, respectively; the two states are called right (cw) or left (ccw)
circularly polarized and denoted by the symbols σ^+, σ^-. A snapshot ($t = 0$) of
the spatial trace of the field vector of right (left) polarized light shows a right (left)-
handed helix

$$E_x(z) = E_0\cos kz$$
$$E_y(z) = \pm E_0\sin kz \qquad (1.76)$$

with a pitch length of $\lambda = 2\pi/k$ (Fig. 1.6).

---

[5]This statement is generally valid only in isotropic media; in anisotropic media, the electric field
can have a longitudinal component and this and the following statements refer to the transverse
component of **E**.

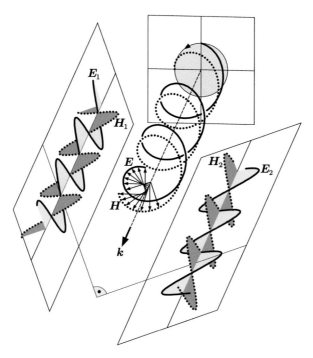

Fig. 1.6 Snapshot of left circularly polarized light and its decomposition into two linearly polarized waves

The general case, with arbitrary ratio $E_{0,y}/E_{0,x}$ and arbitrary phase difference $\Delta\phi$ will be discussed in detail in Sect. 1.5.4.

1.5.1 Jones Formalism

Since the absolute value of the amplitude is irrelevant for the polarization state, states can be represented by unit vectors, called Jones vectors **J** (Table 1.2).

Linearly polarized light is represented by

$$\begin{bmatrix} \cos\varphi \\ \sin\varphi \end{bmatrix}, \tag{1.77}$$

circularly polarized fields by

$$\sigma^{\pm} = \frac{1}{\sqrt{2}} \begin{bmatrix} 1 \\ e^{\pm j\pi/2} \end{bmatrix} = \frac{1}{\sqrt{2}} \begin{bmatrix} 1 \\ \pm j \end{bmatrix}. \tag{1.78}$$

Table 1.2 Jones vectors of selected polarization states; φ denotes the angle between polarization and x-axis

Polarization state	Jones vector
Linear	$\begin{bmatrix} \cos\varphi \\ \sin\varphi \end{bmatrix}$
Right circular σ^+	$\frac{1}{\sqrt{2}} \begin{bmatrix} 1 \\ j \end{bmatrix}$
Left circular σ^-	$\frac{1}{\sqrt{2}} \begin{bmatrix} 1 \\ -j \end{bmatrix}$
General (elliptical)	$\begin{bmatrix} \cos\alpha \\ \sin\alpha\, e^{j\Delta\phi} \end{bmatrix}$

For the interpretation of α and $\Delta\phi$, see Sect. 1.5.4

The general, elliptically polarized state Eq. (1.74) corresponds to the Jones vector

$$\frac{1}{\sqrt{E_{0,x}^2 + E_{0,y}^2}} \begin{bmatrix} E_{0,x} \\ E_{0,y} e^{j\Delta\phi} \end{bmatrix} = \begin{bmatrix} \cos\alpha \\ \sin\alpha e^{j\Delta\phi} \end{bmatrix} \tag{1.79}$$

and will be discussed in Sect. 1.5.4.

1.5.1.1 Orthogonal Polarization States

Two Jones vectors are called orthogonal if their scalar product is zero,

$$\mathbf{J}^{(1)} \cdot \mathbf{J}^{(2)*} = 0. \tag{1.80}$$

Examples are two linearly polarized states oriented along φ and $\varphi + \pi/2$, respectively, or left/right circularly polarized states σ^+, σ^-. A state orthogonal to Eq. (1.79) is obviously

$$\begin{bmatrix} \sin\alpha \\ -\cos\alpha e^{j\Delta\phi} \end{bmatrix}. \tag{1.81}$$

A pair of orthogonal states (Jones vectors) establishes a base that allows constructing any other state by appropriate linear combination. In particular, any other orthogonal base can be constructed; for example, the sum and difference, respectively, of σ^+ and σ^- produce a linearly polarized orthogonal base

$$\frac{1}{\sqrt{2}}(\sigma^+ + \sigma^-) = \begin{bmatrix} 1 \\ 0 \end{bmatrix}$$

$$\frac{1}{j\sqrt{2}}(\sigma^+ - \sigma^-) = \begin{bmatrix} 0 \\ 1 \end{bmatrix}, \tag{1.82}$$

and a circularly polarized base can be obtained from a linearly polarized base by a complex-valued combination

$$\sigma^\pm = \frac{1}{\sqrt{2}} \begin{bmatrix} 1 \\ 0 \end{bmatrix} + \frac{1}{\sqrt{2}} e^{\pm j\pi/2} \begin{bmatrix} 0 \\ 1 \end{bmatrix}. \tag{1.83}$$

These relations are not only mathematical transformations, but also represent physical reality, since linearly polarized light, for example, can be synthesized by two superimposed circularly polarized waves and vice versa.

The polarization state can change during propagation; as we will see, however, for a given propagation system there are always so-called eigenstates that are conserved during propagation (Sect. 1.5.2.5). In lossless media, these states can be shown to be orthogonal to each other and represent a "natural base" for the description of wave propagation in the respective system.

1.5.2 Polarization Optics

1.5.2.1 Wave Plates

In Sect. 2.3, we will encounter various optical components that can alter the polarization state; their operation can be represented by a specific Jones matrix T, that relates an arbitrary input state $\mathbf{J}_{\text{in}}$ to the corresponding output state $\mathbf{J}_{\text{out}}$

$$\mathbf{J}_{\text{out}} = T\mathbf{J}_{\text{in}}; \tag{1.84}$$

Table 1.3 summarizes Jones matrices of important components. Many of these elements rely on the dependence of the phase velocity on the polarization state. In birefringent materials (Sect. 2.3), for example, there are two orthogonal, linearly polarized eigenstates $\mathbf{J}_{\text{f,s}}$ with different phase velocities, denoted as "fast" and "slow"; the corresponding propagation indices are n_{f} and n_{s}. An incoming field of arbitrary polarization is decomposed in two waves $\propto \mathbf{J}_{\text{f,s}} e^{-j(n_{\text{f,s}}\mathbf{k}_0 \cdot \mathbf{x} - \omega t)}$ that develop, during propagation, a phase difference of

$$\Delta\phi_{\text{V}} = (n_{\text{s}} - n_{\text{f}})k_0 d, \tag{1.85}$$

where d is the thickness of the medium; such plates are called retarders or wave plates (Fig. 1.7). In a coordinate system with the x-axis parallel to $\mathbf{J}_{\text{f}}$, the Jones

Table 1.3 Jones matrices of important polarization optical elements

Optical element	Orientation	Jones matrix
Linear polarizer	$\parallel x$-axis	$\begin{bmatrix} 1 & 0 \\ 0 & 0 \end{bmatrix}$
Linear polarizer	φ to x-axis	$\begin{bmatrix} \cos^2\varphi & \sin\varphi\cos\varphi \\ \sin\varphi\cos\varphi & \sin^2\varphi \end{bmatrix}$
Polarization rotator		$\begin{bmatrix} \cos\varphi & -\sin\varphi \\ \sin\varphi & \cos\varphi \end{bmatrix}$
$\lambda/2$-Wave plate	f$\parallel x$-axis	$\begin{bmatrix} 1 & 0 \\ 0 & -1 \end{bmatrix}$
$\lambda/4$-Wave plate	f$\parallel x$-axis	$\begin{bmatrix} 1 & 0 \\ 0 & -j \end{bmatrix}$
$\lambda/4$-Wave plate	f $\pm 45°$ to x-axis	$\frac{1}{\sqrt{2}}\begin{bmatrix} 1 & \mp j \\ \mp j & 1 \end{bmatrix}$
$\Delta\phi_V$-Wave plate	f$\parallel x$-axis	$\begin{bmatrix} 1 & 0 \\ 0 & e^{-j\Delta\phi_V} \end{bmatrix}$
Right circular polarizer		$\begin{bmatrix} 1 & -j \\ j & 1 \end{bmatrix}$
Left circular polarizer		$\begin{bmatrix} 1 & j \\ -j & 1 \end{bmatrix}$
Mirror, normal incidence		$\begin{bmatrix} 1 & 0 \\ 0 & -1 \end{bmatrix}$

matrix has the form

$$T = \begin{bmatrix} 1 & 0 \\ 0 & e^{-j\Delta\phi_V} \end{bmatrix}. \tag{1.86}$$

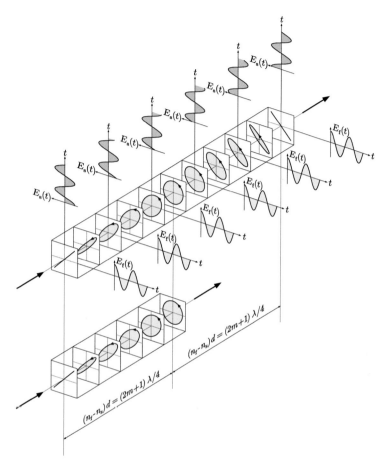

Fig. 1.7 Evolution of linearly polarized light ($\varphi = 45°$) in a $\lambda/2$-wave plate: in the middle of the plate, the light is circularly polarized, the state at the output is the mirror image of the input state

A so-called half-wave plate produces a phase shift of π (corresponding to $\lambda/2$), and is represented by

$$T = \begin{bmatrix} 1 & 0 \\ 0 & -1 \end{bmatrix}, \tag{1.87}$$

which is equivalent to a mirror operation about the x-axis. Linearly polarized light emerges linearly polarized from such a component, but its polarization direction is flipped from φ to $-\varphi$; for $\varphi = 45°$, the output is actually orthogonal to the input state (note that this change of the polarization direction is *not* due to a rotation; a rotator would rotate all states by the same angle). Circularly polarized light changes its sense of rotation. A mirror has the same effect, resulting, however, from the inversion of the propagation direction.

A quarter wave plate ($\Delta\phi_V = \pi/2$) with the Jones matrix

$$T = \begin{bmatrix} 1 & 0 \\ 0 & -j \end{bmatrix},$$ (1.88)

converts circularly polarized into linearly polarized light and vice versa.

1.5.2.2 Polarization Rotators

Optically active and magneto-optic materials (Sects. 2.4.1 and 2.4.2) have circularly polarized eigenstates with the propagation indices $n^{\pm}$; they act as circular retarders, inducing a phase shift of

$$\Delta\phi_V = (n^- - n^+)k_0 d$$ (1.89)

between the two circularly polarized states. In a circularly polarized base [Eq. (1.78)], the Jones matrix has the form

$$T_c = \begin{bmatrix} 1 & 0 \\ 0 & e^{-j\Delta\phi_V} \end{bmatrix}_c,$$ (1.90)

which, as we shall see from an inspection of Eq. (1.124), corresponds to a polarization rotator that rotates an incoming state by an angle of

$$\varphi = -\Delta\phi_V/2 = (n^+ - n^-)k_0 d/2.$$ (1.91)

1.5.2.3 Polarizers

Polarizers (also called polarization filters) are components that transmit one particular polarization state only; an incoming state is decomposed into the transmitted eigenstate and its orthogonal complement, which is absorbed or directed into a different direction; in other words, a polarizer projects the input state onto the transmitted eigenstate. The matrix of a polarizer for x-polarized light is therefore

$$T = \begin{bmatrix} 1 & 0 \\ 0 & 0 \end{bmatrix}.$$ (1.92)

1.5.2.4 Composite Systems

A series of polarization optical elements with Jones matrices T_i can be represented by a single system Jones matrix that is the product of the individual matrices in the exact sequence of transmission

Fig. 1.8 Effect of a series of polarizers: (**a**) two crossed polarizers block the transmission completely; (**b**) addition of a third polarizer, rotated by 45°, results in a transmission of up to 25 % (depending on the input state)

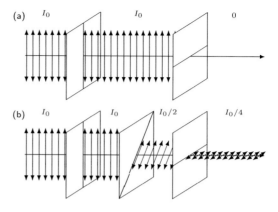

$$T_{\text{total}} = \dots T_3 T_2 T_1. \tag{1.93}$$

A pair of two mutually orthogonal linear polarizers (first y, then x polarized, Fig. 1.8a), for example, has the system matrix

$$T = \begin{bmatrix} 1 & 0 \\ 0 & 0 \end{bmatrix} \begin{bmatrix} 0 & 0 \\ 0 & 1 \end{bmatrix} = \begin{bmatrix} 0 & 0 \\ 0 & 0 \end{bmatrix}, \tag{1.94}$$

and transmits no light at all. Adding a third polarizer, oriented under 45°, between the two polarizers (Fig. 1.8b), results in the matrix

$$T = \frac{1}{2} \begin{bmatrix} 1 & 0 \\ 0 & 0 \end{bmatrix} \begin{bmatrix} 1 & 1 \\ 1 & 1 \end{bmatrix} \begin{bmatrix} 0 & 0 \\ 0 & 1 \end{bmatrix} = \frac{1}{2} \begin{bmatrix} 0 & 1 \\ 0 & 0 \end{bmatrix}, \tag{1.95}$$

which is equivalent to an x-polarizer with 50 % attenuation. The same 45° polarizer inserted before or after the original pair of crossed polarizers would, of course, not alter the zero transmission; this is an instructive example demonstrating the non-commutativity of polarization optics.

1.5.2.5 Polarization Eigenstates

An eigenvector or eigenstate (or eigenmode) of a matrix T is a vector that, if multiplied with T, remains unchanged apart from a (complex) factor, called eigenvalue. The eigenvectors of a Jones matrix are the polarization eigenstates of the corresponding optical element; to determine these states, we have to solve the equation

$$TJ = \lambda_T J \tag{1.96}$$

or

$$(T - \lambda_T 1)J = 0, \tag{1.97}$$

where 1 is the unit matrix

$$1 := \begin{bmatrix} 1 & 0 \\ 0 & 1 \end{bmatrix}; \tag{1.98}$$

explicitly,

$$\begin{bmatrix} T_{11} - \lambda_T & T_{12} \\ T_{21} & T_{22} - \lambda_T \end{bmatrix} \begin{bmatrix} J_1 \\ J_2 \end{bmatrix} = 0. \tag{1.99}$$

For this system to have non-trivial (i.e., non-zero) solutions, the determinant $\det(T - \lambda_T 1)$ must be zero. Thus, the characteristic equation

$$(T_{11} - \lambda_T)(T_{22} - \lambda_T) - T_{21}T_{12} = 0 \tag{1.100}$$

has to be solved, yielding two eigenvalues $\lambda_T^{(1)}$ and $\lambda_T^{(2)}$. Corresponding eigenvectors $J^{(1,2)}$ are found by inserting the values λ_T into one of the equation of Eq. (1.99); the length of the eigenvectors is not defined, since any multiple aJ of an eigenvector J is also an eigenvector. It is, however, convenient to normalize the eigenvectors to unit length.

Once the set of eigenvectors is given, any arbitrary state can be written as a linear combination of these eigenvectors,

$$J = a_1 J^{(1)} + a_2 J^{(2)}; \tag{1.101}$$

the output state of the optical element represented by T is then

$$TJ = a_1 \lambda_T^{(1)} J^{(1)} + a_2 \lambda_T^{(2)} J^{(2)}. \tag{1.102}$$

1.5.2.6 Lossless Systems

Since Jones vectors represent electric fields, Eq. (1.71) implies

$$J_{out} J^*_{out} = J_{in} J^*_{in}, \tag{1.103}$$

provided the system is lossless and the input and output propagation indices are equal. Following the arguments regarding the scattering matrix of lossless systems [Eq. (4.19)], we find that the Jones matrix of a lossless elements is unitary

$$\left[T^*\right]^{\mathrm{T}} = T^{-1}. \tag{1.104}$$

Such matrices can be shown to have orthogonal, generally elliptic eigenvectors $\mathbf{J}_{1,2}$ with eigenvalues of unit modulus, i.e., $\lambda_T^{(1,2)} = e^{-j\phi_{1,2}}$. In the eigenvector base "b" formed by $\mathbf{J}_{1,2}$, the Jones matrix is diagonal and can be written as

$$T = \begin{bmatrix} 1 & 0 \\ 0 & e^{-j\Delta\phi_V} \end{bmatrix}_b, \tag{1.105}$$

where $\Delta\phi_V = \phi_2 - \phi_1$, and an irrelevant common factor has been dropped. In the eigenbase, the action of the optical element is to introduce a phase shift of $\Delta\phi_V$ between the eigenstates. Thus, any lossless polarization optical system can be understood as retarder acting on its eigenstates.

1.5.3 Transformation of Jones Vectors and Matrices

Jones vectors and matrices are usually represented in a certain *cartesian* coordinate system. The corresponding base vectors are linearly polarized states along the coordinate axes. It is often useful to express them in a different, for example, circularly polarized base or in a linearly polarized base that is rotated in respect to the original; this can be achieved by simple linear transformations.

1.5.3.1 Rotated Cartesian Base

We assume a Jones vector that is represented, in the original cartesian system, by $\mathbf{J}_1 = \begin{bmatrix} J_{1,1} \\ J_{1,2} \end{bmatrix}$ and try to find the components $\mathbf{J}_{1'} = \begin{bmatrix} J_{1',1} \\ J_{1',2} \end{bmatrix}$ of the same vector in a system rotated by φ_R. In the original system, the components can be expressed by $J_{1,1} = r\cos\varphi$, $J_{1,2} = r\sin\varphi$ (Fig. 1.9). Rotating the coordinate system by φ_R changes φ to $\varphi - \varphi_R$. Using $\cos(\varphi - \varphi_R) = \cos\varphi\cos\varphi_R + \sin\varphi\sin\varphi_R$ and $\sin(\varphi - \varphi_R) = -\cos\varphi\sin\varphi_R + \sin\varphi\cos\varphi_R$ we obtain

$$J_{1',1} = r\cos(\varphi - \varphi_R) = J_{1,1}\cos\varphi_R + J_{1,2}\sin\varphi_R$$

$$J_{1',2} = r\sin(\varphi - \varphi_R) = -J_{1,1}\sin\varphi_R + J_{1,2}\cos\varphi_R. \tag{1.106}$$

Fig. 1.9 Coordinates of a vector in two different cartesian reference frames, rotated by φ_R in respect to each other

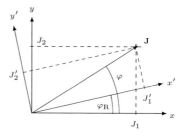

This can be expressed by

$$\mathbf{J}_{l'} = A_{l \to l'} \mathbf{J}_l, \qquad (1.107)$$

where $A_{l \to l'}$ is the rotation matrix

$$A_{l \to l'} = A_{\varphi_R} = \begin{bmatrix} \cos \varphi_R & \sin \varphi_R \\ -\sin \varphi_R & \cos \varphi_R \end{bmatrix}. \qquad (1.108)$$

The inverse transformation is obtained by multiplication of $\mathbf{J}_{l'}$ with the inverse matrix

$$\mathbf{J}_l = A_{l' \to l} \mathbf{J}_{l'} = A_{l \to l'}^{-1} \mathbf{J}_{l'}, \qquad (1.109)$$

where

$$A_{l \to l'}^{-1} = A_{-\varphi_R}. \qquad (1.110)$$

To transform a Jones *matrix* into the rotated system, we multiply Eq. (1.84) by $A_{l \to l'}$ and obtain, with Eqs. (1.107) and (1.109),

$$\mathbf{J}_{l',\text{out}} = A_{l \to l'} \mathbf{J}_{l,\text{out}} = A_{l \to l'} T \mathbf{J}_{l,\text{in}} = A_{l \to l'} T A_{l \to l'}^{-1} \mathbf{J}_{l',\text{in}}, \qquad (1.111)$$

so that

$$\mathbf{J}_{l',\text{out}} = T_{l'} \mathbf{J}_{l',\text{in}} \qquad (1.112)$$

with the transformed Jones matrix

$$T_{l'} = A_{l \to l'} T A_{l \to l'}^{-1}. \qquad (1.113)$$

Equations (1.107) and (1.113) are not restricted to rotations, but constitute the general coordinate transformation rules for Jones vectors and matrices.

1.5.3.2 Physical Rotation of Polarization States and Optical Components

In practice, one frequently knows the matrix T of a polarization optical component in a certain orientation and needs to know the matrix T^{φ} of the component in a different orientation. Rotating the element by φ is mathematically (but not physically) equivalent to a rotation of the reference frame by $-\varphi$. The matrix of the rotated element is therefore

$$T^\varphi = A_{-\varphi} T A_\varphi.$$ (1.114)

The Jones matrix, for example, of a linear polarizer rotated by φ in respect to the x-axis is

$$T = A_{-\varphi} \begin{bmatrix} 1 & 0 \\ 0 & 0 \end{bmatrix} A_\varphi = \begin{bmatrix} \cos^2 \varphi & \cos \varphi \sin \varphi \\ \cos \varphi \sin \varphi & \sin^2 \varphi \end{bmatrix}.$$ (1.115)

By the same argument, the components of a Jones vector that has been physically rotated (by a polarization rotator) by an angle φ are equal to that of the original vector in a reference frame rotated by $-\varphi$. The Jones matrix of a polarization rotator is therefore

$$T = A_{-\varphi_R} = \begin{bmatrix} \cos \varphi & -\sin \varphi \\ \sin \varphi & \cos \varphi \end{bmatrix}.$$ (1.116)

1.5.3.3 Transformation to a Circularly Polarized Base

Next, we analyze the transformation between a linearly and a circularly polarized base; we indicate the reference base by a subscript l and c, respectively. In a circularly polarized base, the states σ^+, σ^- [Eq. (1.78)] are represented by

$$\sigma^+ = \begin{bmatrix} 1 \\ 0 \end{bmatrix}_c, \quad \sigma^- = \begin{bmatrix} 0 \\ 1 \end{bmatrix}_c.$$ (1.117)

The transformation matrix $A_{c \to l}$ must be such that

$$\frac{1}{\sqrt{2}} \begin{bmatrix} 1 \\ j \end{bmatrix}_l = \begin{bmatrix} A_{11} & A_{12} \\ A_{21} & A_{22} \end{bmatrix}_{c \to l} \begin{bmatrix} 1 \\ 0 \end{bmatrix}_c$$ (1.118)

$$\frac{1}{\sqrt{2}} \begin{bmatrix} 1 \\ -j \end{bmatrix}_l = \begin{bmatrix} A_{11} & A_{12} \\ A_{21} & A_{22} \end{bmatrix}_{c \to l} \begin{bmatrix} 0 \\ 1 \end{bmatrix}_c.$$ (1.119)

Obviously, the columns of $A_{c \to l}$ are given by the representation of $\sigma^\pm$ in the linear base

$$A_{c \to l} = \frac{1}{\sqrt{2}} \begin{bmatrix} 1 & 1 \\ j & -j \end{bmatrix},$$ (1.120)

and the inverse matrix is

$$A_{1 \to c} = A_{c \to 1}^{-1} = \frac{1}{\sqrt{2}} \begin{bmatrix} 1 & -j \\ 1 & j \end{bmatrix}. \qquad (1.121)$$

The transformation of Jones matrices follows Eq. (1.113)

$$T_1 = A_{c \to 1} T_c A_{c \to 1}^{-1}; \qquad (1.122)$$

the matrix Eq. (1.90) of a circular retarder,

$$T_c = \begin{bmatrix} 1 & 0 \\ 0 & e^{-j\Delta\phi_V} \end{bmatrix}_c \qquad (1.123)$$

in particular, is transformed to

$$\begin{bmatrix} \cos(\Delta\phi_V/2) & -\sin(\Delta\phi_V/2) \\ \sin(\Delta\phi_V/2) & \cos(\Delta\phi_V/2) \end{bmatrix}_1 \qquad (1.124)$$

in the linear base. Comparison with Eq. (1.116) shows that a circular retarder with retardation $\Delta\phi_V$ acts as polarization rotator that rotates an arbitrary input state by $\varphi = -\Delta\phi_V/2$.

1.5.3.4 Eigenbase
Of particular interest is a base consisting of the (normalized) eigenvectors of a polarization optical device. In its eigenbase, the Jones matrix of the device has the form

$$\begin{bmatrix} \lambda_T^{(1)} & 0 \\ 0 & \lambda_T^{(2)} \end{bmatrix}_b \propto \begin{bmatrix} 1 & 0 \\ 0 & \frac{\lambda_T^{(2)}}{\lambda_T^{(1)}} \end{bmatrix}_b, \qquad (1.125)$$

as follows from Eq. (1.102); for lossless systems with $\lambda_T^{(1,2)} = e^{-j\phi_{1,2}}$, this is the matrix of a retarder [Eq. (1.105)]; as stated above, any lossless polarization optical element simply acts as a retarder on its eigenvectors.

1.5.4   Elliptically Polarized States

In a circular base, a general state

$$E^+ \sigma^+ + E^- e^{j\Delta\phi_c} \sigma^+ \qquad (1.126)$$

is represented by

$$
\begin{bmatrix} E^+ \\ E^- e^{j\Delta\phi_c} \end{bmatrix}_c
$$

(1.127)

where $E^\pm$ are assumed to be non-negative and real. For the special case $\Delta\phi_c = 0$, the representation in the linear base is

$$
\frac{1}{\sqrt{2}} \begin{bmatrix} 1 & 1 \\ j & -j \end{bmatrix} \begin{bmatrix} E^+ \\ E^- \end{bmatrix}_c = \frac{1}{\sqrt{2}} \begin{bmatrix} E^+ + E^- \\ j(E^+ - E^-) \end{bmatrix}_l;
$$

(1.128)

this corresponds, at $z = 0$, to the wave Eq. (1.74)

$$
\mathbf{E}(t) = \mathrm{Re}\left[\tilde{\mathbf{E}}e^{j\omega t}\right] = \frac{1}{\sqrt{2}} \begin{bmatrix} (E^+ + E^-)\cos\omega t \\ (E^- + E^-)\sin(\omega t) \end{bmatrix}_l;
$$

(1.129)

the locus of the electric field vector is obtained by elimination of t

$$
\frac{E_x^2}{(E^+ + E^-)^2} + \frac{E_y^2}{(E^+ - E^-)^2} = 1,
$$

(1.130)

which is an ellipse with major axis $(E^+ + E^-)$ and minor axis $|E^+ - E^-|$; the sign of $E^+ - E^-$ determines the handedness of the corresponding elliptically polarized wave. According to Eq. (1.124), the phase shift $\Delta\phi_c$ is equivalent to a rotation by $\Delta\phi_c/2$, so that the general state Eq. (1.127) is elliptically polarized, with an ellipticity of

$$
\tan\epsilon = \frac{E^+ - E^-}{E^+ + E^-},
$$

(1.131)

rotated by the angle

$$
\varphi = \Delta\phi_c/2
$$

(1.132)

in respect to the x-axis.

Because of these simple relations, the easiest way to determine the ellipticity and orientation of an arbitrary state, given in an arbitrary base, is to transform it into the circular base where it can be expressed in the form

$$
\widetilde{E}_0 \begin{bmatrix} \cos\alpha_c \\ \sin\alpha_c e^{j\Delta\phi_c} \end{bmatrix}_c.
$$

(1.133)

Since the scalar (complex) multiplier $\widetilde{E}_0$ does not influence the polarization state, the polarization ellipse can then be determined using Eqs. (1.131) and (1.132). For the general state in a linear base,

$$\begin{bmatrix} \cos\alpha_l \\ \sin\alpha_l e^{j\Delta\phi_l} \end{bmatrix}_l, \tag{1.134}$$

in particular, we obtain after a somewhat lengthy, but simple calculation (see, e.g., Kliger et al. 1990)

$$\sin 2\epsilon = \sin 2\alpha_l \sin \Delta\phi_l \tag{1.135}$$

$$\tan 2\varphi = \tan 2\alpha_l \cos \Delta\phi_l. \tag{1.136}$$

In an analogue fashion, the inverse relations

$$\cos 2\alpha_l = \cos 2\epsilon \cos 2\varphi \tag{1.137}$$

$$\cot \Delta\phi_l = \cot 2\epsilon \sin 2\varphi \tag{1.138}$$

are obtained. An arbitrary state (Fig. 1.10) can thus be characterized alternatively by the parameter pair (ϵ, φ) or $(\alpha_l, \Delta\phi_l)$, that refer to the circularly or linearly polarized base, respectively.

1.5.5 Poincaré Sphere*

The identification of a polarization state by two angular parameters allows associating it with a point on a sphere. If, in particular, 2ϵ is associated with the "geographic latitude," and 2φ with the "geographic longitude," one obtains a spherical map of all possible polarization states, known as Poincaré sphere (Fig. 1.11). North- and south-pole, respectively, correspond to right and left circularly polarized states. The equator, with $\epsilon = 0$, comprises all linearly polarized states, and its intersection with

Fig. 1.10 The locus of the electric field vector of elliptically polarized light is an ellipse with ellipticity $\tan \epsilon$, oriented at the azimuthal angle φ

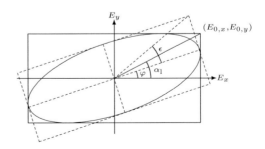

Fig. 1.11 On the Poincaré sphere, every polarization state is represented by a point P with the latitude 2ϵ and the longitude 2φ. Relations (1.35) to (1.138) follow from spherical geometry by identifying $HP = 2\alpha_1$ and $\sphericalangle XHP = \Delta\phi_1$: in a right-angled spherical triangle, the hypothenuse c and the two legs a, b are related by $\cos c = \cos a \cos b$, and the angel A opposite a is given by $\cot A = \cot a \sin b$, for example (see Fig. 1.10 for the meaning of ϵ, φ, and α_1)

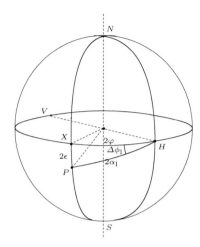

the 0-meridian ($\varphi = 0$) denotes horizontal polarization. Mutually orthogonal states occupy antipodal positions: using Eq. (1.81), we obtain for the state orthogonal to $(2\epsilon, 2\varphi)$ the coordinates $2\epsilon' = -2\epsilon$, $2\varphi' = 2\varphi + \pi$.

The parameter set $(2\epsilon, 2\varphi)$ relates to the circular base. As we have seen above, a circular retarder Eq. (1.123) with phase shift $\Delta\phi_V$ rotates any state by the angle $\varphi = -\Delta\phi_V/2$. On the Poincaré sphere, this means that a state is moved zonally (with constant 2ϵ) by the angle $-\Delta\phi_V$. The poles, as eigenstates of the circular rotator, are not affected, and define the rotation axis.

Mathematically, the circular base is just one out of an infinite set of possible orthogonal bases. Its practical significance results from the fact that the parameter set $(2\epsilon, 2\varphi)$ allows for an intuitive geometric interpretation. As we have seen, however, *any* lossless polarization optical device generates an eigenbase, that corresponds to two antipodal points on the Poincaré sphere; the action of such a device is that of a retarder that changes the phase shift $\Delta\phi_b$ of the input state (represented in the eigenbase)

$$\begin{bmatrix} E_{b,1} \\ E_{b,2}\mathrm{e}^{\mathrm{j}\Delta\phi_b} \end{bmatrix}_b , \tag{1.139}$$

to $\Delta\phi_b - \Delta\phi_V$ while keeping the ratio $E_{b,2}/E_{b,1}$ unchanged. An input state is therefore rotated on a circle around the axis constituted by the two eigenstates by an angle of $-\Delta\phi_V$ (Fig. 1.12). Varying $\Delta\phi_V$, any state on the circle can be reached. If the circle intersects the equator or a pole, for example, the retarder can convert the input state into linearly or circularly polarized light, respectively.

Rotating the retarder itself (around an axis parallel to the propagation direction) by an angle φ_R moves its eigenstates by an angle of $2\varphi_R$ on a zonal circle of the Poincaré sphere. Rotation of a linear retarder (with eigenstates on the equator) allows positioning its antipodal eigenstates anywhere on the equator. A linear retarder (and only a linear retarder) of variable phase shift and orientation can

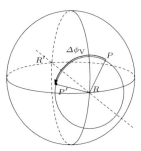

Fig. 1.12 Wave plates can be represented on the Poincaré sphere by an axis through their eigenstates R, R'. The transformation of a state P by the wave plate is equivalent to the rotation of P around the axis by an angle equal to the phase shift of the wave plate

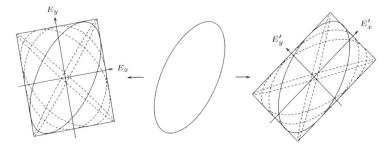

Fig. 1.13 One and the same polarization state occupies different "boxes" in differently oriented cartesian coordinate systems. A birefringent wave plate with axes parallel to the coordinate axes can transform a state into any other state within the box by appropriate choice of its phase shift

convert any given input state into any desired output state: for this purpose, one has to construct the symmetry plane between the input state and the target state and position the linear retarder at the intersection between this plane and the equator. Note that a circular retarder, by comparison, can only convert states of identical ellipticity into each other.

The practical importance of birefringent wave plates (i.e., of linear retarders) merits a few further remarks on its action. The two linearly polarized eigenstates of such a wave plate generate a cartesian coordinate system in which a given state has the coordinates (Fig. 1.13)

$$\frac{1}{\sqrt{E_{0,x}^2 + E_{0,y}^2}} \begin{bmatrix} E_{0,x} \\ E_{0,y} e^{j\Delta\phi_1} \end{bmatrix} = \begin{bmatrix} \cos\alpha_1 \\ \sin\alpha_1 e^{j\Delta\phi_1} \end{bmatrix}; \qquad (1.140)$$

the retarder changes the value of $\Delta\phi_1$, but does not affect the aspect ratio $\tan\alpha_1 = E_{0,y}/E_{0,x}$. If the phase shift of the retarder is adjustable, any state within a box defined by the corner points $(\pm E_{0,x}, \pm E_{0,y})$ can be accessed. Rotating the wave plate generates a new cartesian coordinate system in which the same input state occupies a different box; again, any state within this box can be reached by appropriate choice of the phase shift. To transform a given input state to a selected output state, one has to find the box that contains both states; the orientation of this box indicates the required orientation of the wave plate.

A linear retarder of variable phase shift can be realized by stacking two wedges of a birefringent material, one of them movable in respect to the other, on top of each other to form a parallel plate of variable thickness and corresponding phase shift. If the setup is mounted on a rotation stage, one obtains full control over orientation and phase shift of the retarder. Such a Babinet–Soleil compensator can, as already stated, transform any polarization state into any other. An alternative scheme is to use a rotatable electro-optic Pockels cell (Sect. 2.3.4) whose birefringence can be controlled by an external electric field.

1.6   Inhomogeneous Waves

Plane waves as defined in Eq. (1.26) are spatially homogeneous in the sense that the (complex) amplitude $\tilde{\mathbf{E}}(\mathbf{k}, \omega)$ does not depend on $\mathbf{x}$. Many relations derived in this chapter refer to such waves. We now want to discuss a superposition of two such waves; we will encounter superpositions of plane waves in the discussion of interference, in the theory of planar optical waveguides, in the Fourier optical treatment of optical beams and other phenomena.

We choose two waves with orthogonal waves vectors, parallel to the x, z plane

$$\mathbf{k}_{1,2} = k_0 \frac{\sqrt{2}}{2} [1, 0, \pm 1]. \tag{1.141}$$

Particularly instructing is the situation where both waves are linearly polarized and coplanar to the wave vectors

$$\tilde{\mathbf{E}}_{1,2}(\mathbf{x}, t) = \tilde{E}_0 \frac{\sqrt{2}}{2} \begin{bmatrix} \mp 1 \\ 0 \\ 1 \end{bmatrix} e^{-j(\mathbf{k}_{1,2} \cdot \mathbf{x} - \omega t)}. \tag{1.142}$$

Figure 1.14 shows the resulting total field; the surfaces of constant phase are planes normal to $\mathbf{k}_1 + \mathbf{k}_2$, moving at the phase velocity $\sqrt{2}c_0 > c_0$. There are planes of purely longitudinal electric field, normal and parallel, respectively, to the phase fronts, and other planes, with purely transverse electric field. Between these planes, the electric field has transverse as well as longitudinal components and the electric field vector actually rotates parallel to the plane spanned by the wave vectors. The Poynting vector vanishes in the planes of purely longitudinal electric field, because, as can be easily shown, the magnetic field vanishes in these planes; it reaches its maximum value in the planes of purely transverse electric field, where it is directed normal to the phase fronts. Between these planes, it exhibits transverse components that cancel when averaged over an oscillation period. This example shows that

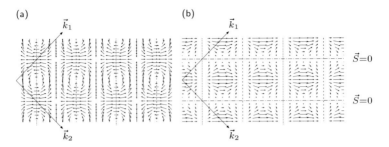

Fig. 1.14 Superposition of two plane waves with orthogonal wave vectors $\mathbf{k}_{1,2}$, linearly polarized coplanar with the wave vectors: (**a**) snapshot of the electric field, (**b**) snapshot of the Poynting vector

superpositions of plane waves have properties that are quite different from that of homogeneous waves and have to be carefully considered, especially if the angle between the wave vectors is large. This is the case, for example, in tightly focused laser beams or in waveguides with strong guiding.

1.7   Summary

Photonics is predominantly concerned with electromagnetic fields that have a well-defined phase, frequency, and propagation direction. Plane monochromatic waves are therefore very useful and popular elements to describe photonic processes. In this Chapter, the properties of such fields are discussed in detail. An important parameter of light waves is the propagation velocity; in addition to the phase velocity, which is defined for a monochromatic plane wave, the concept of group- and ray velocity is introduced which applies to "packages" of plane waves, i.e., to light pulses and beams, and is discussed in more detail in the respective sections of Chap. 3.

Particular emphasis is laid in this Chapter on the polarization state of light waves, a property that is exploited in many photonic devices. The analysis of polarization optical devices in terms of eigenstates, eigenvalues, and eigenbases in the framework of the Jones vector formalism has, in addition to its practical importance, a didactic purpose as it familiarizes the reader with the mathematical concept of the Hilbert space, where wave functions are treated as vectors. This concept is fundamental not only for the understanding of quantum mechanics (Chap. 6) but also for the coupling of modes in waveguides and other processes; even though it is not explicitly elaborated on in this book, it pervades large parts of it.

1.8   Problems

1. A laser beam is focussed onto an area of $5\,\mu m^2$; what beam power is needed to reach an electric field strength equal to that experienced by an electron in an H-atom ($E_{at} = e/4\pi\epsilon_0 a_0^2$, $a_0 = 5.3 \times 10^{-11}$ m)? With a pulsed laser (5 fs pulses), what pulse energy is needed for that?

2. Reproduce Fig. 1.2, assuming $v_g = 0.9\,v_{ph}$ with two frequencies that differ by 10 % and plot the wave as function of time and of distance, respectively; vary z in steps of λ and observe what happens.

3. Plot a snapshot of the vector field of the Poynting vector of a plane wave at $t = 0$ in a fashion similar to Fig. 1.14; compare linearly and circularly polarized light.

4. Reproduce Fig. 1.14 (a) for $\tilde{\mathbf{E}}_{1,2}$ coplanar with $\mathbf{k}_{1,2}$, (b) for $\tilde{\mathbf{E}}_{1,2} \perp \mathbf{k}_{1,2}$, (c) $\tilde{\mathbf{E}}_1$ coplanar with $\mathbf{k}_{1,2}$ and $\tilde{\mathbf{E}}_2 \perp \mathbf{k}_{1,2}$; step up the time in increments of $T/8$ and observe what happens.

5. Visualize the general polarization state Eq. (1.79) in a plot similar to Fig. 1.10; vary $\Delta\phi$ and α and observe what happens.

6. Assume a stack of two $\lambda/4$ wave plates, rotated by $45°$ in respect to each other. What are the polarization eigenstates of this system? Express linearly and circularly polarized light, respectively, in this eigenbase.

7. Derive the transmittance of an electro-optic modulator that consists of two polarizers oriented along the x-axis and an electrically controlled variable retarder that is rotated by $45°$ in respect to the x-axis as a function of the phase delay $\Delta\phi_V$.

References and Suggested Reading

Born, M., & Wolf, E. (1999). *Principles of optics*. New York: Cambridge University Press.
Brillouin, L. (1960). *Wave propagation and group velocity*. New York: Academic Press.
Jackson, J. D. (1999). *Classical electrodynamics*. New York: Wiley.
Klein, M. V., & Furtak, T. E. (1986). *Optics*. New York: John Wiley.
Kliger, D. S., Lewis, J. W., & Randall, D. A. (1990). *Polarized light in optics and spectroscopy*. New York: Academic Press.
Lipson, S. G., & Lipson. H. (1969). *Optical physics*. London: Cambridge University Press.
Saleh, B. E., & Teich, M. C. (2007). *Fundamentals of photonics*. New York: Wiley.

Wave Propagation in Matter

2

While the focus of Chap. 1 was on the properties of electromagnetic waves, we now turn to the optical properties of matter and their impact on wave propagation. To emphasize this perspective, we use Eq. (1.8) to write Eq. (1.17) in the form

$$\nabla^2 \mathbf{E} - \varepsilon_0 \mu_0 \frac{\partial^2 \mathbf{E}}{\partial t^2} = \mu_0 \frac{\partial^2 \mathbf{P}}{\partial t^2}. \tag{2.1}$$

The right-hand side of this differential equation can be understood as a source term driving a new field that is superimposed on the original field.

Consider a very thin sheet of an isotropic, lossless material with susceptibility χ, suspended in vacuum and irradiated by a plane wave under normal incidence. The polarization induced in the sheet gives rise to a new wave which, for symmetry reasons, consists of a forward and a backward propagating plane wave of equal amplitude. The forward propagating component is added to the driving field; since it is out of phase by $\pi/2$ [compare Eq. (8.40)], however, the resulting forward wave is slightly retarded in comparison to the original field. If we introduce a second sheet at a distance of a quarter wavelength after the first one, essentially the same happens: the total forward propagating field is further retarded, and a second backward propagating wave is generated. In reference to the first one, however, this wave is delayed by half a wave length (or π) and will cancel the contribution from the first sheet.

If we fill up the entire space with such sheets, the forward propagating wave is continuously retarded and propagates at a phase velocity smaller than c_0; the degree of retardation increases with the susceptibility of the medium, since the amplitude of the partial waves is proportional to χ. There will, however, be no backward propagating field because for any chosen sheet there is another one that produces a cancelling wave. The situation changes if the medium is inhomogeneous, or if, for example, only a half space is filled with the medium: the backward contributions

© Springer International Publishing Switzerland 2016
G.A. Reider, *Photonics*, DOI 10.1007/978-3-319-26076-1_2

from a quarter wavelength thick front layer of the medium are not compensated and add up to a reflected wave.

The mathematical treatment of his process is rather complicated and is the subject of the Ewald–Oseen extinction theorem[1]; the amplitudes of the reflected and transmitted waves can be calculated quite easily using boundary conditions, however, as we will show in the following.

2.1 Transition Between Different Media

2.1.1 Phase Matching at a Boundary

2.1.1.1 Reflection and Refraction

Before we evaluate the amplitudes of the reflected and transmitted waves, respectively, we first want to find their propagation directions, i.e., their respective wave vectors. Assume a plane wave

$$\mathbf{E}^i = \mathbf{E}_0^i e^{-j(\mathbf{k}^i \cdot \mathbf{x} - \omega t)} \tag{2.2}$$

incident on a plane interface between two dielectric media (i) and (t) with the propagation indices n_i and n_t, respectively. Both media are supposed to be lossless and isotropic; depending on the relative magnitude of n_i and n_t, a medium is called optically denser or thinner than the other.

The reflected and transmitted waves are also expected to be plane waves

$$\mathbf{E}^{r,t} = \mathbf{E}_0^{r,t} e^{-j(\mathbf{k}^{r,t} \cdot \mathbf{x} - \omega t)}; \tag{2.3}$$

each wave vector can be decomposed into a component $\mathbf{k}_\perp^{i,r,t}$ normal to the interface, and a tangential component $\mathbf{k}_\parallel^{i,r,t}$. Right at the interface, there is no contribution of $\mathbf{k}_\perp^{i,r,t}$ to the wave functions

$$\mathbf{E}^{i,r,t}\big|_{IF} = \mathbf{E}_0^{i,r,t} e^{-j(\mathbf{k}_\parallel^{i,r,t} \cdot \mathbf{x} - \omega t)}. \tag{2.4}$$

Because of the translational invariance of the planar interface, the ratios $e^{-j(\mathbf{k}_\parallel^{r,t} \cdot \mathbf{x} - \omega t)} / e^{-j(\mathbf{k}_\parallel^i \cdot \mathbf{x} - \omega t)}$ must be independent of $\mathbf{x}$ and the tangential components of the participating wave vectors must consequently be equal

$$\mathbf{k}_\parallel^i = \mathbf{k}_\parallel^r = \mathbf{k}_\parallel^t. \tag{2.5}$$

---

[1]See, e.g., Born and Wolf (1999).

Fig. 2.1 Wave vectors of incident, transmitted and reflected waves at a planar shift invariant interface: the tangential components must satisfy the phase matching condition $\mathbf{k}_\parallel^i = \mathbf{k}_\parallel^r = \mathbf{k}_\parallel^t$, the length of the respective vectors is given by the dispersion relation $|\mathbf{k}^{i,r,t}| = n_{i,r,t} k_0$

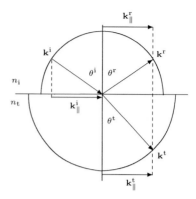

From this phase matching condition follows immediately that the three wave vectors are parallel to the plane of incidence, defined by the incident wave vector and the surface normal. The normal components of the wave vectors follows from the dispersion relation (1.28) $|\mathbf{k}^{i,r,t}| = k^{i,r,t} = n_{i,r,t} k_0$ to be

$$(k_\perp^{i,r,t})^2 = (n_{i,r,t} k_0)^2 - |\mathbf{k}_\parallel^{i,r,t}|^2, \tag{2.6}$$

where $n_r = n_i$, so that the wave vectors are fully determined (Fig. 2.1). Introducing the angles $\theta^{i,r,t}$ between the surface normal and the respective wave vectors, we obtain the relations

$$\sin \theta^r = \sin \theta^i \tag{2.7}$$

for the reflected wave, and Snell's law

$$n_t \sin \theta^t = n_i \sin \theta^i \tag{2.8}$$

for the transmitted wave; the change of the propagation direction of the transmitted wave is known as refraction.

2.1.1.2 Total Reflection

Inspection of Fig. 2.2 shows that if medium (i) is optically denser than medium (t) ($n_i > n_t$), then for sufficiently large angles of incidence, the tangential component of the wave vector in medium (t) is larger than its length. In this case, there exists no refracted wave and the energy contained in the incident wave is completely transferred to the reflected wave, a process called total internal reflection. The

Fig. 2.2 Same as Fig. 2.1, showing the situation $|\mathbf{k}_\parallel^t| > |\mathbf{k}^t|$, resulting in total reflection

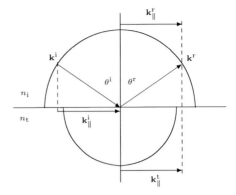

condition for total reflection is

$$|\mathbf{k}_\parallel^i| = n_i k_0 \sin \theta^i > n_t k_0, \qquad (2.9)$$

and can be expressed as

$$\theta^i > \theta_{\text{crit}} = \arcsin \frac{n_t}{n_i}, \qquad (2.10)$$

where θ_{crit} is called critical angle of total reflection.

2.1.1.3 Diffraction

If the boundary between the two media is not invariant under translation, the above argument Eq. (2.5) does not hold and the fields radiated in the forward or backward direction are, in general, diffusely scattered waves. An important exception is that of a spatially periodic boundary; such structures are called line gratings and play an important role in photonics. The ratio between incident and transmitted or reflected wave, evaluated at the surface, can then be written as a Fourier series

$$e^{-j(\mathbf{k}_\parallel^{r,t}\cdot\mathbf{x}-\omega t)}/e^{-j(\mathbf{k}_\parallel^i\cdot\mathbf{x}-\omega t)} \propto \sum_{m=-\infty}^{\infty} F_m e^{-jm\mathbf{K}_g\cdot\mathbf{x}}, \qquad (2.11)$$

where $\mathbf{K}_g$ is a vector parallel to the interface and normal to the lines of the grating, with an absolute value of $2\pi/\Lambda$, Λ is the spatial period, and F_m are the Fourier components. Consequently, the wave vectors of the emitted waves have the tangential components (Fig. 2.3)

$$\mathbf{k}_\parallel^{r,t} = \mathbf{k}_\parallel^i + m\mathbf{K}_g \qquad (2.12)$$

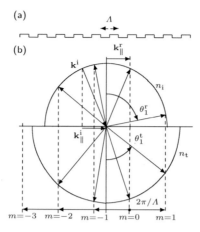

Fig. 2.3 Geometric relations of the diffracted wave vectors: (**a**) surface profile, (**b**) diffracted wave vectors

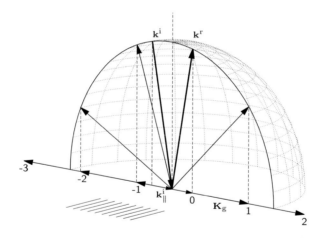

Fig. 2.4 Diffraction at a periodically modulated interface (line grating) with lines oriented normal to the plane of incidence; the integers denote the order of diffraction

with integer m. Provided that $|\mathbf{k}_{\parallel}^i + m\mathbf{K_g}| < n_{i,t}k_0$, so-called diffracted waves of order m are radiated in addition to the "ordinary" reflected and transmitted waves ($m = 0$).

If $\mathbf{K_g} \| \mathbf{k}_{\parallel}^i$ (i.e., if the lines of the grating are normal to the plane of incidence, Fig. 2.4), the wave vectors of the diffracted waves lie in the plane of incidence; the diffraction angles $\theta_m^{r,t}$ are obtained from Eq. (2.6)

$$n_{r,t} \sin \theta_m^{r,t} = n_i \sin \theta^i + m\lambda_0/\Lambda, \qquad (2.13)$$

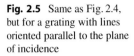

Fig. 2.5 Same as Fig. 2.4, but for a grating with lines oriented parallel to the plane of incidence

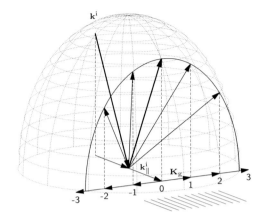

as illustrated in Fig. 2.3; obviously, m is restricted to values $|n_i \sin \theta^i + m\lambda_0/\Lambda| \leq n_{r,t}$.[2] The number of diffracted waves increases with the ratio Λ/λ_0; if $\Lambda < \lambda_0/2n_{i,t}$, there exist no propagating diffracted waves and the boundary, in regard to the scattering of light waves, behaves like a translationally invariant interface.

The applicability of Eq. (2.11) is not limited to line gratings oriented normal to the plane of incidence. A grating with lines parallel to the plane of incidence, for example, produces diffracted waves whose wave vectors are constructed according to Fig. 2.5 and lie on a cone with a half top angle equal to $90° - \theta^i$; a similar construction allows us to calculate the diffracted wave vectors for gratings of arbitrary orientation.

Under conditions of total internal reflection ($\theta^i > \arcsin \frac{n_t}{n_i}$), a periodically modulated boundary can mediate the radiation of transmitted waves, as illustrated in Fig. 2.6; coupling of waves with the aid of gratings is a frequently employed scheme in photonics.

The diffraction angles Eq. (2.12) depend on the wavelength of the incident light; line gratings are therefore important components for spectral filtering and analysis; in Sects. 3.1.6 and 4.2.1, we will discuss these applications in more detail. Another important application is the temporal compression of light pulses (Sect. 3.2.1.7).

Note that all findings of Sect. 2.1.1 are independent of the nature of the waves; they apply to plane electromagnetic, acoustic, as well as quantum mechanical DeBroglie waves. The *amplitudes* of the respective waves can be calculated from specific boundary conditions, as we shall see in the following section for the reflected and transmitted electromagnetic waves. For the amplitudes of *diffracted* electromagnetic waves see, e.g., Petit (1980).

---

[2]Higher diffraction orders exist as so-called evanescent waves, compare Sect. 2.1.3.

Fig. 2.6 An interface grating
allows producing transmitted
waves under conditions of
total internal reflection, in the
illustrated case diffracted
waves of order $m = -1, -2$

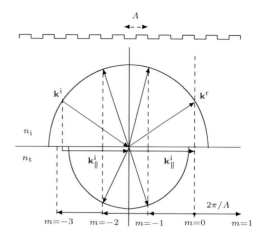

2.1.2   Reflection and Transmission Coefficients

From Maxwell's equations follows the continuity of the tangential components of
E and **H** at a boundary. The amplitudes of the transmitted and reflected waves
must be such that the sum of the incident and reflected fields on the one hand,
and the transmitted on the other have equal tangential components. To simplify
the treatment of this problem, we decompose the incident field into two linearly
polarized components, one of which is polarized normal to the plane of incidence
(σ-polarized—not to be confused with $\sigma^{\pm}$ polarized light), the other one parallel
to the plane of incidence (π-polarized).[3] It follows immediately from the boundary
conditions that the polarization state of these two components is conserved during
reflection and transmission, so that we are dealing with polarization eigenstates of
reflection and transmission at a plane surface.

In our coordinate system Fig. 2.7 (with the xz-plane as plane of incidence), the
wave vectors are

$$\mathbf{k}^{\mathrm{i,r,t}} = \left[k_x^{\mathrm{i,r,t}}, 0, k_z^{\mathrm{i,r,t}} \right]. \tag{2.14}$$

The boundary conditions for σ-polarized light are

$$E_y^{\mathrm{i}} + E_y^{\mathrm{r}} = E_y^{\mathrm{t}} \tag{2.15}$$

$$H_x^{\mathrm{i}} + H_x^{\mathrm{r}} = H_x^{\mathrm{t}}. \tag{2.16}$$

---

[3]The terms σ- and π-polarized are derived from the German terms *s*enkrecht and *p*arallel.

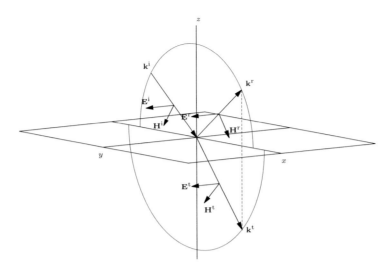

Fig. 2.7 Coordinate convention for incident, reflected, and transmitted σ-polarized waves; the actual magnitudes and signs of the fields are obtained after multiplying the base vectors with the reflection and transmission coefficients r_σ and t_σ, respectively. Note that the orientation of the coordinate systems is a matter of convention—each of them could also be rotated by $180°$ around the wave vector, resulting in different signs for the coefficients

With Eq. (1.66), Eq. (2.16) can be expressed as

$$(\mathbf{k}^i \times \mathbf{E}^i)_x + (\mathbf{k}^r \times \mathbf{E}^r)_x = (\mathbf{k}^t \times \mathbf{E}^t)_x, \tag{2.17}$$

so that

$$k_z^i E_y^i + k_z^r E_y^r = k_z^t E_y^t. \tag{2.18}$$

Using Eq. (2.15), $k_z^r = -k_z^i$ and $E_y^{i,r,t} = E^{i,r,t}$, we obtain the relations

$$E^r = \frac{1 - k_z^t/k_z^i}{1 + k_z^t/k_z^i} E^i =: r_\sigma E^i \tag{2.19}$$

$$E^t = \frac{2}{1 + k_z^t/k_z^i} E^i =: t_\sigma E^i. \tag{2.20}$$

Substituting $k_z^{i,t} = n_{i,t} k_0 \cos \theta^{i,t}$, these equations can be cast in the form

$$r_\sigma = \frac{n_i \cos \theta^i - n_t \cos \theta^t}{n_i \cos \theta^i + n_t \cos \theta^t} \tag{2.21}$$

$$t_\sigma = \frac{2n_i \cos \theta^i}{n_i \cos \theta^i + n_t \cos \theta^t}, \tag{2.22}$$

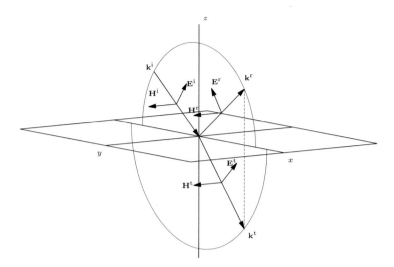

Fig. 2.8 Same as Fig. 2.7 for π-polarized light

where $\cos \theta^t$ follows from Eq. (2.8)

$$\cos \theta^t = \sqrt{1 - (n_i/n_t)^2 \sin^2 \theta^i}. \tag{2.23}$$

Equations (2.21) and (2.22) are known as Fresnel coefficients for the reflected and transmitted electric field (see the summary Table 2.1 and Figs. 2.9 and 2.10).

For π-polarized light we choose the coordinate system Fig. 2.8 and follow the above calculations with the roles of **E** and **H** interchanged: the boundary conditions are

$$H_y^i + H_y^r = H_y^t \tag{2.24}$$

$$E_x^i + E_x^r = E_x^t. \tag{2.25}$$

With Eq. (1.67) we convert Eq. (2.25) to

$$\frac{1}{\varepsilon_i} \left(k_z^i H_y^i + k_z^r H_y^r \right) = \frac{1}{\varepsilon_t} k_z^t H_y^t; \tag{2.26}$$

from Eq. (2.24) and $H_y^{i,r,t} = H^{i,r,t}$ follows

$$\frac{H^r}{H^i} = \frac{\varepsilon_t k_z^i - \varepsilon_i k_z^t}{\varepsilon_t k_z^i + \varepsilon_i k_z^t}$$

$$\frac{H^t}{H^i} = \frac{2\varepsilon_t k_z^i}{\varepsilon_t k_z^i + \varepsilon_i k_z^t}; \tag{2.27}$$

according to Eq. (1.68), $E^{\mathrm{r,t}}/E^{\mathrm{i}} = \sqrt{\varepsilon_{\mathrm{i}}/\varepsilon_{\mathrm{i,t}}}(H^{\mathrm{r,t}}/H^{\mathrm{i}})$, so that we finally obtain the Fresnel coefficients for π-polarized fields

$$r_\pi = \frac{1 - \varepsilon_{\mathrm{i}}k_z^{\mathrm{t}}/\varepsilon_{\mathrm{t}}k_z^{\mathrm{i}}}{1 + \varepsilon_{\mathrm{i}}k_z^{\mathrm{t}}/\varepsilon_{\mathrm{t}}k_z^{\mathrm{i}}} \tag{2.28}$$

$$t_\pi = \frac{2\sqrt{\varepsilon_{\mathrm{i}}/\varepsilon_{\mathrm{t}}}}{1 + \varepsilon_{\mathrm{i}}k_z^{\mathrm{t}}/\varepsilon_{\mathrm{t}}k_z^{\mathrm{i}}}, \tag{2.29}$$

or, with $k_z^{\mathrm{i,t}} = n_{\mathrm{i,t}}k_0 \cos \theta^{\mathrm{i,t}}$ and $n_{\mathrm{i,t}} = \sqrt{\varepsilon_{\mathrm{i,t}}}$

$$r_\pi = \frac{n_{\mathrm{t}} \cos \theta^{\mathrm{i}} - n_{\mathrm{i}} \cos \theta^{\mathrm{t}}}{n_{\mathrm{t}} \cos \theta^{\mathrm{i}} + n_{\mathrm{i}} \cos \theta^{\mathrm{t}}} \tag{2.30}$$

$$t_\pi = \frac{2n_{\mathrm{i}} \cos \theta^{\mathrm{i}}}{n_{\mathrm{t}} \cos \theta^{\mathrm{i}} + n_{\mathrm{i}} \cos \theta^{\mathrm{t}}}. \tag{2.31}$$

The total field on the input side is the vectorial sum of incident and reflected field; for σ-polarized light, the two contributing electric fields have only components normal to the plane of incidence, so that the total electric field is also normal to the wave vector. For this reason, σ-polarized light is called TE-(transverse electric). The superposition of incident and reflected electric field for π-polarized light displays a rather complex structure with a spatially varying longitudinal component (compare Sect. 1.6); the magnetic field, however, has only a component normal to the plane of incidence and π polarized light is consequently denoted as TM-(transverse magnetic).

Figure 2.9 shows exemplary ($n_{\mathrm{t}}/n_{\mathrm{i}} = 1.5$) Fresnel coefficients for either polarization as a function of the angle of incidence. At normal incidence ($\theta^{\mathrm{i}} = 0$) the signs of r_σ and r_π are opposite; the reason for this seemingly contradictory result is the sign convention in Figs. 2.8 and 2.7. The phase change of the reflected light at normal incidence depends on the sign of $n_{\mathrm{i}} - n_{\mathrm{t}}$ and is π for $n_{\mathrm{i}} < n_{\mathrm{t}}$ and 0 for $n_{\mathrm{i}} > n_{\mathrm{t}}$. Figure 2.10 shows the Fresnel coefficients at the interface between a dense and a thin medium ($n_{\mathrm{i}}/n_{\mathrm{t}} = 1.5$); above the critical angle of total reflection Eq. (2.10), the absolute value of the reflection coefficient is equal to 1, a situation that will be discussed in more detail in Sect. 2.1.3.

Table 2.1 Fresnel coefficients r and t, reflectance R and transmittance T, and phase shift ϕ for total reflection, at a boundary between media with refractive indices n_{i}, n_{t}; θ^{i} and θ^{t} are related by $\cos \theta^{\mathrm{t}} = \sqrt{1 - (n_{\mathrm{i}}^2/n_{\mathrm{t}}^2) \sin^2 \theta^{\mathrm{i}}}$

	r	t	R	T	$\tan\left(\frac{\phi}{2}\right)$
σ	$\dfrac{n_{\mathrm{i}} \cos \theta^{\mathrm{i}} - n_{\mathrm{t}} \cos \theta^{\mathrm{t}}}{n_{\mathrm{i}} \cos \theta^{\mathrm{i}} + n_{\mathrm{t}} \cos \theta^{\mathrm{t}}}$	$\dfrac{2n_{\mathrm{i}} \cos \theta^{\mathrm{i}}}{n_{\mathrm{i}} \cos \theta^{\mathrm{i}} + n_{\mathrm{t}} \cos \theta^{\mathrm{t}}}$	$\lvert r_\sigma \rvert^2$	$\dfrac{n_{\mathrm{t}} \cos \theta^{\mathrm{t}}}{n_{\mathrm{i}} \cos \theta^{\mathrm{i}}}\lvert t_\sigma \rvert^2$	$\dfrac{\sqrt{n_{\mathrm{i}}^2 \sin^2 \theta^{\mathrm{i}} - n_{\mathrm{t}}^2}}{n_{\mathrm{i}} \cos \theta^{\mathrm{i}}}$
π	$\dfrac{n_{\mathrm{t}} \cos \theta^{\mathrm{i}} - n_{\mathrm{i}} \cos \theta^{\mathrm{t}}}{n_{\mathrm{t}} \cos \theta^{\mathrm{i}} + n_{\mathrm{i}} \cos \theta^{\mathrm{t}}}$	$\dfrac{2n_{\mathrm{i}} \cos \theta^{\mathrm{i}}}{n_{\mathrm{t}} \cos \theta^{\mathrm{i}} + n_{\mathrm{i}} \cos \theta^{\mathrm{t}}}$	$\lvert r_\pi \rvert^2$	$\dfrac{n_{\mathrm{t}} \cos \theta^{\mathrm{t}}}{n_{\mathrm{i}} \cos \theta^{\mathrm{i}}}\lvert t_\pi \rvert^2$	$\dfrac{n_{\mathrm{i}}^2}{n_{\mathrm{t}}^2}\dfrac{\sqrt{n_{\mathrm{i}}^2 \sin^2 \theta^{\mathrm{i}} - n_{\mathrm{t}}}}{n_{\mathrm{i}} \cos \theta^{\mathrm{i}}}$

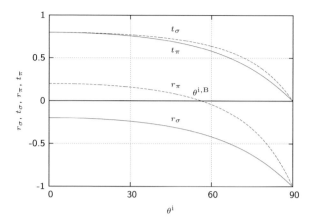

Fig. 2.9 Reflection and transmission coefficients $r_{\sigma,\pi}$ and $t_{\sigma,\pi}$ at a planar interface between two dielectrics as a function of the angle of incidence, shown for $n_t/n_i = 1.5$

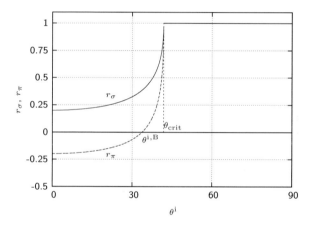

Fig. 2.10 Same as Fig. 2.9 at the boundary between an optically dense and an optically thin medium ($n_i/n_t = 1.5$); for $\theta^i > \theta_{\mathrm{crit}}$, $r_{\sigma,\pi}$ is complex with an absolute value $|r_{\sigma,\pi}| = 1$

2.1.2.1 Reflectance and Transmittance

The Fresnel coefficients refer to the (complex) amplitudes of the reflected and transmitted electric fields. In practice, the incident field is often a collimated light "beam" that can be approximated by a plane wave with finite lateral extension, carrying a certain input power. The fraction of the reflected and transmitted beam power relative to the input power is given by the reflectance R and transmittance T, respectively. To obtain R and T, we choose an area element on the interface and calculate the energy flows on both sides of the interface by projecting the respective Poynting-vectors onto the surface normal [Eq. (1.73)]. Energy conservation requires

$$\frac{n_i|E^i|^2}{2Z_0}\cos\theta^i = \frac{n_i|E^r|^2}{2Z_0}\cos\theta^r + \frac{n_t|E^t|^2}{2Z_0}\cos\theta^t. \tag{2.32}$$

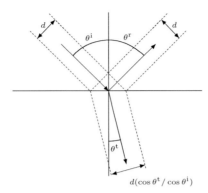

Fig. 2.11 The lateral extension of a transmitted light beam, measured in the plane of incidence, differs from that of the incident and reflected beam by the ratio $\cos\theta^t / \cos\theta^i$

With $E^r = r_{\sigma,\pi} E^i$ and $E^t = t_{\sigma,\pi} E^i$ we obtain

$$1 = |r_{\sigma,\pi}|^2 + \frac{n_t \cos\theta^t}{n_i \cos\theta^i} |t_{\sigma,\pi}|^2. \tag{2.33}$$

The two terms on the right-hand side can be identified with the reflectance and transmittance, respectively,

$$R_{\sigma,\pi} = |r_{\sigma,\pi}|^2 \tag{2.34}$$

$$T_{\sigma,\pi} = \frac{n_t \cos\theta^t}{n_i \cos\theta^i} |t_{\sigma,\pi}|^2. \tag{2.35}$$

The ratio $\cos\theta^t / \cos\theta^i$ takes into account that the lateral extension of the transmitted beam, measured in the plane of incidence, differs from that of the incident beam by this factor; an input beam with circular cross section is refracted into a beam with elliptical cross section (Fig. 2.11).

Inspection of Fig. 2.12 shows that R_σ increases with θ^i and approaches 1 at grazing incidence; R_π follows the same trend for grazing incidence, but vanishes at the so-called Brewster angle $\theta^{i,B}$, which, according to Eq. (2.30), must satisfy

$$n_t \cos\theta^i = n_i \cos\theta^t. \tag{2.36}$$

To solve this equation, we combine it with Snell's law Eq. (2.8) in the form

$$n_t \cos(90° - \theta^t) = n_i \cos(90° - \theta^i) \tag{2.37}$$

to find that the two angles must be complementary, $90° - \theta^t = \theta^i$, i.e., $\sin\theta^t = \cos\theta^i$. The electric field and thus the polarization density **P** in the second medium

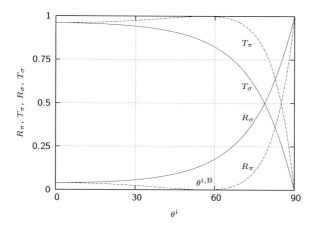

Fig. 2.12 Reflectance R and transmittance T for σ- and π-polarized light as a function of the angle of incidence ($n_t/n_i = 1.5$)

then oscillate exactly parallel to the reflected wave vector; since, as we have argued earlier, the source of the reflected wave is the polarization density (more precisely, the displacement density) in the second medium, and the reflected field must be transverse, the amplitude of the reflected wave is zero. With Snell's law Eq. (2.8), the Brewster-condition $\sin \theta^t = \cos \theta^i$ can be expressed in the form

$$\theta^{i,B} = \arctan \frac{n_t}{n_i}. \tag{2.38}$$

The existence of the Brewster angle is frequently exploited in photonic setups to avoid undesired reflections by appropriate arrangement of optical elements in the light path.

At normal incidence, Eqs. (2.21) and (2.30) yield

$$R = \left(\frac{n_t - n_i}{n_t + n_i}\right)^2 = \left(\frac{n_t/n_i - 1}{n_t/n_i + 1}\right)^2. \tag{2.39}$$

Figure 2.13 shows the reflectance at a boundary between air and different media; R increases with the optical density of the medium. Glasses with typical refractive indices between 1.3 and 1.8 reflect several % of the input power at normal incidence; semiconductors may exhibit much higher refractive indices and reflectance values above 30 %.

2.1.2.2 Reflection and Transmission for Arbitrary Polarization

We now employ the Jones formalism (Sect. 1.5.1) to analyze the reflection and transmission of light with arbitrary polarization. As we have seen, the polarization

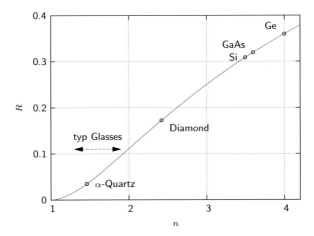

Fig. 2.13 Reflectance at an interface between air ($n_i = 1$) and a medium of refractive index $n_t = n$ at normal incidence

eigenstates of a reflecting interface are π- and σ-polarized waves, respectively. In this eigenbase, the Jones matrices describing reflection and transmission are

$$\boldsymbol{T}^r = \begin{bmatrix} r_\pi & 0 \\ 0 & r_\sigma \end{bmatrix}, \tag{2.40}$$

$$\boldsymbol{T}^t = \begin{bmatrix} t_\pi & 0 \\ 0 & t_\sigma \end{bmatrix}. \tag{2.41}$$

Provided that r, t are real numbers, linearly polarized light remains linearly polarized, but in general changes the plane of polarization. At Brewster's angle,

$$\boldsymbol{T}^r = \begin{bmatrix} 0 & 0 \\ 0 & r_\sigma \end{bmatrix}, \tag{2.42}$$

so that a dielectric surface acts as (lossy) linear reflective polarizer. Since $|r_\pi| \le |r_\sigma|$, "natural" light is predominantly σ-polarized after reflection at a dielectric surface. This effect is used in photography to reduce specular reflections by employing polarization filters of appropriate orientation.

2.1.3 Total Reflection

Under conditions of total reflection [Figs. 2.2 and 2.10, Eq. (2.10)], the normal component of the wave vector turns imaginary

$$k_z^t = k_0 \sqrt{n_t^2 - n_i^2 \sin^2 \theta^i} =: -j\gamma^t. \tag{2.43}$$

In medium (t), for which we assume $z > 0$, the field is then given by

$$E^t = E_0^t e^{-j(\mathbf{k \cdot x} - \omega t)} = E_0^t e^{-\gamma^t z} e^{-j(k_x^i x - \omega t)}. \qquad (2.44)$$

The amplitude of this inhomogeneous, so-called evanescent wave decays exponentially with increasing distance from the interface so that the wave is essentially confined to a layer of thickness $1/\gamma^t$. This penetration depth is on the order of a wavelength unless the angle of incidence is very close to the critical angle, where it grows quickly and approaches infinity at $\theta^i = \theta_{crit}$ [Eq. (2.43)].

According to Eq. (2.19), the reflection coefficient for σ-polarized light under total reflection conditions is

$$r_\sigma = \frac{1 + j(\gamma^t/k_z^i)}{1 - j(\gamma^t/k_z^i)} =: e^{j\phi_\sigma}, \qquad (2.45)$$

while the reflectance is $R = rr^* = 1$; the reflectance of a metallic mirror, for comparison, is usually less than 0.9. According to Eq. (2.45), r_σ is complex and introduces a phase shift of the reflected wave that amounts to

$$\frac{\phi_\sigma}{2} = \arctan \frac{\gamma^t}{k_z^i} = \arctan \frac{\left(n_i^2 \sin^2 \theta^i - n_t^2\right)^{1/2}}{n_i \cos \theta^i} \qquad (2.46)$$

(Fig. 2.14). For π-polarized light, Eq. (2.28) yields

$$r_\pi = \frac{1 + j(n_i/n_t)^2(\gamma^t/k_z^i)}{1 - j(n_i/n_t)^2(\gamma^t/k_z^i)} =: e^{j\phi_\pi} \qquad (2.47)$$

with

$$\frac{\phi_\pi}{2} = \arctan \frac{n_i^2}{n_t^2} \frac{\gamma^t}{k_z^i} = \arctan \frac{n_i^2}{n_t^2} \frac{\left(n_i^2 \sin^2 \theta^i - n_t^2\right)^{1/2}}{n_i \cos \theta^i}. \qquad (2.48)$$

The phase shifts for π- and σ-polarized differ by an amount that depends on θ^i and the ratio n_i/n_t (Fig. 2.14). The Jones matrix for total reflection is

$$T^r = \begin{bmatrix} e^{j\phi_\pi} & 0 \\ 0 & e^{j\phi_\sigma} \end{bmatrix} = e^{j\phi_\pi} \begin{bmatrix} 1 & 0 \\ 0 & e^{j(\phi_\sigma - \phi_\pi)} \end{bmatrix}, \qquad (2.49)$$

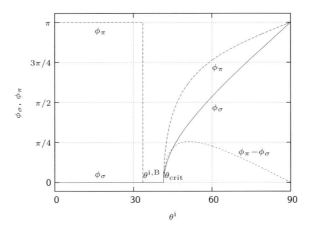

Fig. 2.14 Phase shift of the electric field induced by (total) reflection at an optically thinner medium ($n_t/n_i = 1/1.5$)

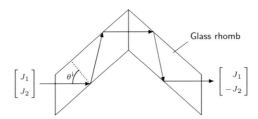

Fig. 2.15 Broadband $\lambda/2$-retarder, based upon the phase difference $\phi_\pi - \phi_\sigma$ at total reflection inside Fresnel rhombs

equivalent to that of a retarder (see Table 1.3). For a sufficiently large ratio n_i/n_t, a phase shift difference of $\phi_\pi - \phi_\sigma = \pi/4$ can be obtained. Multiple internal reflection inside a so-called Fresnel rhomb (Fig. 2.15) allows implementing the equivalent of $\lambda/4$ or $\lambda/2$ wave plates. Since $\phi_\pi - \phi_\sigma$ depends only slightly on the wavelength, such retarders work over a broad spectral range.

Next we want to evaluate the total field in the input half space $z < 0$, which is the superposition of the incident and reflected wave; for σ-polarized light, the latter differs from the incident wave only by the sign of the normal component of its wave vector, $k_z^r = -k_z^i$, and the phase shift Eq. (2.46)

$$
\begin{aligned}
\mathbf{E}^{\text{tot}} &= \mathbf{E}^i + \mathbf{E}^r \\
&= \mathbf{E}_0^i e^{-j(k_x^i x - \omega t)} \left[e^{-jk_z^i z} + e^{j\phi_\sigma} e^{jk_z^i z} \right] \\
&= \mathbf{E}_0^i e^{-j(k_x^i x - \omega t - \phi_\sigma/2)} \left[e^{-j(k_z^i z + \phi_\sigma/2)} + e^{j(k_z^i z + \phi_\sigma/2)} \right] \\
&= \mathbf{E}_0^{\text{tot}} \cos(k_z^i z + \phi_\sigma/2) e^{-j(k_x^i x - \omega t)},
\end{aligned}
\tag{2.50}
$$

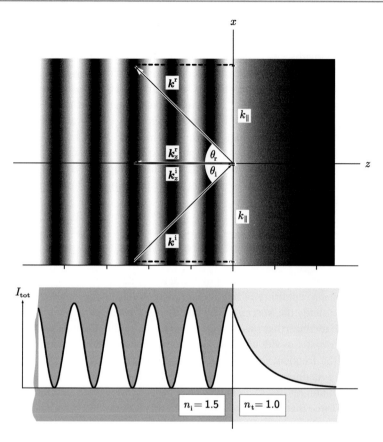

Fig. 2.16 Intensity distribution resulting from total reflection of a σ-polarized wave; the energy flow is parallel to the interface

where $\mathbf{E}_0^{\text{tot}} = 2\mathbf{E}_0^{\text{i}}e^{j\phi_\sigma/2}$. The phase fronts of this inhomogeneous wave are planes normal to the interface and the plane of incidence, and travel at a phase velocity of $\omega/k_x^{\text{i}} = c_0/(n_{\text{i}} \sin \theta^{\text{i}}) < c_0/n_{\text{t}}$. The amplitude is a spatially oscillating function of the distance from the interface with a period of $2\pi/k_z^{\text{i}} = \lambda_0/(n_{\text{i}} \sin \theta^{\text{i}})$ exhibiting nodal planes of zero electric field, as illustrated in Fig. 2.16. The cosine function describing the spatial modulation of the amplitude is shifted by $\phi_\sigma/2$ away from the interface. This is a consequence of the boundary conditions that require the field at the interface to be continuously differentiable.

Total internal reflection is of fundamental importance for the operation of dielectric waveguides; we will return to this matter in Chap. 5.

2.1.3.1 Optical Tunneling Effect

If the optically thinner medium (refraction index n_L) is sandwiched between two optically dense media ($n_{H,1}$, $n_{H,2}$), a light wave can be transmitted through the optically thin medium even under total reflection conditions $|\mathbf{k}_\parallel| > n_L k_0$, provided that the refractive index of the output medium is large enough to support a propagating wave, $n_{H,2} k_0 > |\mathbf{k}_\parallel|$; the transmission coefficient decreases roughly exponentially with distance $\propto e^{-\gamma^t d}$ and the direction of the wave vector of the transmitted wave is given by Snell's law, $n_{H,2} \sin \theta^t = n_i \sin \theta^{H,1}$. This so-called optical tunnel effect is used in various photonic components (for example, high power beam splitters) and is the basis of scanning–tunneling optical microscopy that allows "tapping" the evanescent light scattered from sub-wavelength features of a specimen.

2.2   Optical Properties of Isotropic Media

We have introduced the propagation or refractive index $n = \sqrt{\varepsilon} = \sqrt{1 + \chi}$ as a function of the susceptibility of the medium, which relates the polarization density to the electric field; the susceptibility itself was treated as a phenomenological property of the medium that was considered as a continuum. We now want to present a simple mechanistic model of the susceptibility that qualitatively explains, among other things, the frequency dependence of the refractive index and the absorption coefficient of a medium. The approach of this Drude–Lorentz model is to treat the medium as containing discrete charges (electrons or ions) of a certain mass, held in place by a force that resembles a spring. In this picture, the polarization density of a medium is the vectorial sum over all microscopic dipole moments per unit volume. As we will see, the mass of the oscillating charged particles limits the frequency up to which they can contribute to the polarization; in the visible and near infrared region of the electromagnetic frequency spectrum, only electrons and protons (hydrogen ions) are light enough to contribute.

2.2.1   Linear Oscillator Model

The model assumes charged particles of mass m_e that are elastically tied to their respective equilibrium position by a restoring force $a\mathbf{x}$ proportional to the displacement $\mathbf{x}$; any movement of the particles is damped by a term $b\dot{\mathbf{x}}$ that scales linearly with the velocity. The light field acts on the charged particles via the Coulomb force $-e\mathbf{E}$; the Lorentz force $-e\dot{\mathbf{x}} \times \mathbf{B}$ by the magnetic component of the light field can usually be neglected in comparison to the Coulomb force. The equation of motion for such a particle is

$$m_e \ddot{\mathbf{x}} + b\dot{\mathbf{x}} + a\mathbf{x} = -e\mathbf{E}(t), \qquad (2.51)$$

where $\mathbf{E}$ is the local electric field, which we assume to oscillate at a frequency ω, $\mathbf{E}(t) = \mathbf{E}(\omega)\cos\omega t$. We use complex notation [Eq. (1.21)] for all oscillating quantities, $\mathbf{a}(t) = \text{Re}\left[\tilde{\mathbf{a}}(\omega)e^{j\omega t}\right]$, and assume stationary conditions. The complex displacement amplitude is then

$$\tilde{\mathbf{x}}(\omega) = \frac{-e/m_e}{(\omega_0^2 - \omega^2) + j\omega\Gamma}\tilde{\mathbf{E}}(\omega), \tag{2.52}$$

where

$$\omega_0^2 = a/m_e \tag{2.53}$$

is the resonance frequency of the linear oscillator and $\Gamma = b/m_e$ is the damping coefficient. The corresponding dipole moment $\tilde{\mathbf{p}} = -e\,\tilde{\mathbf{x}}$ is given by

$$\tilde{\mathbf{p}}(\omega) = \frac{e^2/m_e}{(\omega_0^2 - \omega^2) + j\omega\Gamma}\tilde{\mathbf{E}}(\omega), \tag{2.54}$$

and gives rise to a polarization density of

$$\tilde{\mathbf{P}}(\omega) = n_e\frac{e^2/m_e}{(\omega_0^2 - \omega^2) + j\omega\Gamma}\tilde{\mathbf{E}}(\omega), \tag{2.55}$$

where n_e is the particle density. The ratio between induced dipole moment and local field is called polarizability. The local field is usually different from the external field because it is influenced by the dipoles in the immediate environment of the particle; for a coarse description of the relevant processes, we will neglect this difference, however.

Comparison of Eq. (2.55) with Eq. (1.7) yields the (complex) susceptibility

$$\tilde{\chi}(\omega) = \chi' + j\chi'' = \frac{n_e e^2}{\varepsilon_0 m_e}\frac{1}{(\omega_0^2 - \omega^2) + j\omega\Gamma} \tag{2.56}$$

that we can decompose into its real and imaginary part

$$\chi' = \chi_0\frac{1 - (\omega/\omega_0)^2}{[1 - (\omega/\omega_0)^2]^2 + (\omega\Gamma/\omega_0^2)^2} \tag{2.57}$$

$$\chi'' = -\chi_0\frac{(\omega\Gamma/\omega_0^2)}{[1 - (\omega/\omega_0)^2]^2 + (\omega\Gamma/\omega_0^2)^2}, \tag{2.58}$$

where $\chi_0 := n_e e^2/\varepsilon_0 m_e \omega_0^2$ is the (real valued) low-frequency susceptibility.

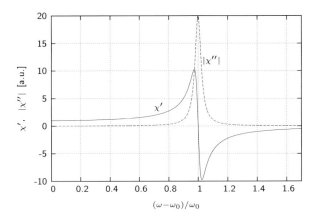

Fig. 2.17 Frequency dependence of χ' and $|\chi''|$ according to the linear oscillator model

The resulting frequency dependence (dispersion) of χ' and χ'' is shown in Fig. 2.17: outside a narrow range in the vicinity of the resonance frequency ω_0, χ' is a monotonically increasing function of ω (so-called normal dispersion, see Sect. 1.3.2). Within the resonance range, χ' decreases with ω (anomalous dispersion). The imaginary part χ'' shows a bell-shaped frequency dependence, centered roughly at ω_0, with a width of $\approx \Gamma$.

In the vicinity of the resonance, $\omega \approx \omega_0$, we can apply the approximations $(\omega_0^2 - \omega^2) = (\omega_0 + \omega)(\omega_0 - \omega) \approx 2\omega_0(\omega_0 - \omega)$ and $\omega\Gamma \approx \omega_0\Gamma$; Eq. (2.56) then simplifies to

$$\tilde{\chi}(\omega) \approx \chi_0 \frac{\omega_0/2}{(\omega_0 - \omega) + \mathrm{j}\Gamma/2} \qquad (2.59)$$

so that

$$\chi' \approx \chi_0 \frac{[1 - (\omega/\omega_0)]/2}{[1 - (\omega/\omega_0)]^2 + (\Gamma/\omega_0)^2/4} \qquad (2.60)$$

$$\chi'' \approx -\chi_0 \frac{(\Gamma/\omega_0)/4}{[1 - (\omega/\omega_0)]^2 + (\Gamma/\omega_0)^2/4}. \qquad (2.61)$$

The resulting functional shape of $|\chi''(\omega)|$ is known as Lorentz line shape, with a peak value of $|\chi''_{max}| = \chi_0\omega_0/\Gamma$ at $\omega = \omega_0$ that scales inversely with the damping coefficient Γ. In this approximation, Γ is equal to the FWHM (full width at half maximum)-line width of $|\chi''|$ and also equal to the width of the range of anomalous dispersion.

2.2.2 Absorption and Reflection

Plane monochromatic waves

$$\tilde{\mathbf{E}}(\mathbf{x}, t) = \tilde{\mathbf{E}}_0 e^{-j(\mathbf{k}\cdot\mathbf{x}-\omega t)} \tag{2.62}$$

are solutions of the wave equation Eq. (1.20), even if the permittivity $\tilde{\varepsilon} = \tilde{\chi} + 1$ is complex. The dispersion relation Eq. (1.27) requires

$$\tilde{\mathbf{k}}^2 = \tilde{\varepsilon}\left(\frac{\omega}{c_0}\right)^2 =: \tilde{n}^2 k_0^2, \tag{2.63}$$

implying a complex propagation index

$$\tilde{n} = \sqrt{\varepsilon' + j\varepsilon''} =: n - j\kappa; \tag{2.64}$$

n and κ are obtained from $\tilde{\varepsilon}$ by setting

$$n^2 - \kappa^2 = \varepsilon', \quad 2n\kappa = -\varepsilon''. \tag{2.65}$$

Elimination of κ yields

$$4n^4 - 4n^2\varepsilon' - \varepsilon''^2 = 0 \tag{2.66}$$

so that

$$n = \sqrt{\frac{(\varepsilon'^2 + \varepsilon''^2)^{1/2} + \varepsilon'}{2}} \tag{2.67}$$

$$\kappa = \sqrt{\frac{(\varepsilon'^2 + \varepsilon''^2)^{1/2} - \varepsilon'}{2}}. \tag{2.68}$$

Figure 2.18 shows the frequency dependence of these two parameters for the $\tilde{\varepsilon}(\omega)$ shown in Fig. 2.17. Equation (2.62) then assumes the form

$$\tilde{\mathbf{E}}(\mathbf{x}, t) = \tilde{\mathbf{E}}_0 e^{-\kappa k_0 z} e^{-j(n k_0 \cdot \mathbf{x} - \omega t)}; \tag{2.69}$$

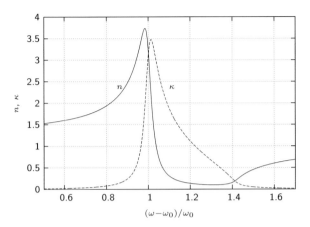

Fig. 2.18 Real and imaginary part of the complex refractive index $n + j\kappa$ of a single resonance dielectric medium as a function of frequency

obviously, the imaginary part κ of the complex propagation index $\tilde{n}$ is responsible for an exponential spatial decay of the field amplitude. According to Eq. (1.71), the intensity is proportional to the absolute square of the amplitude, so that

$$\frac{I(z)}{I(0)} = e^{-2\kappa k_0 z} =: e^{-\alpha z} \tag{2.70}$$

with

$$\alpha = 2\kappa k_0; \tag{2.71}$$

α is known as absorption coefficient and the distance $1/\alpha = 1/2\kappa k_0$ is called absorption length, that is the distance after which the intensity is reduced to a fraction of $1/e$. The phase term $e^{-jnk_0 z}$ determines the phase velocity $c_{\mathrm{ph}} = c_0/n$.

2.2.2.1 Multiple Resonances
The Drude–Lorentz model can be extended to systems with several different resonance frequencies $\omega_{0,i}$ by adding up the individual contributions of the susceptibility, weighted by their respective density $n_{e,i}$, so that Eq. (2.56) is modified to

$$\tilde{\chi}(\omega) = \sum_i \chi_{0,i} \frac{\omega_{0,i}^2}{(\omega_{0,i}^2 - \omega^2) + j\omega \Gamma_i} \tag{2.72}$$

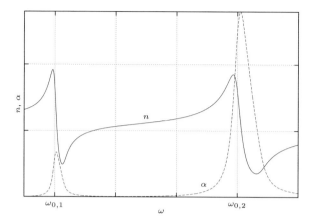

Fig. 2.19 Refractive index n and absorption coefficient α of a medium with two resonances; while significant absorption is confined to narrow bands near the resonance frequencies, the refractive index is influenced by the resonances over the entire spectral range

with $\chi_{0,i} := n_{e,i}e^2/\varepsilon_0 m_e \omega_{0,i}^2$. As can be seen in Fig. 2.19, showing the calculated refractive index and absorption coefficient of a medium with two resonances, there is a spectral range between the resonances where absorption is very low while the impact of the resonances on the dispersion is quite significant. The reason for this is that the imaginary part of the susceptibility, which is, according to Eq. (1.65), responsible for the energy transfer from the field to the medium, decreases with $1/(\omega - \omega_0)^2$, while the real part follows a broader $1/(\omega - \omega_0)$—dependence.

The resonance frequency scales with $1/\sqrt{m_e}$; assuming that the "spring constant" a is similar for different microscopic systems, the resonance frequencies of an electron and a hydrogen ion (with a mass of $1836\,m_e$), respectively, differ be a factor of $\sqrt{1836} \approx 40$; since typical electronic resonance frequencies lie in the near UV spectral range, vibrational resonances of ions and atoms are located in the IR. The large spectral separation between electronic and vibronic resonances is responsible for the unique "spectral window" between IR and UV, i.e., for the transparency of most dielectrics in the visible.

2.2.2.2 Diluted Materials

Many photonic materials consist of a transparent host material with refractive index n_w that is doped with (or contaminated by) absorbing atoms, ions, or molecules in low concentrations. The susceptibility of the composite material is then a sum of the dominating, real valued host susceptibility $\chi_w = n_w^2 - 1$ and the complex dopant contribution $\tilde{\chi}_{dot}$ that is relatively small due to the low concentration; the resulting complex refractive index is given by

$$n - j\kappa = \sqrt{(1 + \chi_w) + \chi'_{dot} + j\chi''_{dot}} = n_w \left(1 + \frac{\chi'_{dot}}{n_w^2} + j\frac{\chi''_{dot}}{n_w^2}\right)^{1/2}. \quad (2.73)$$

Assuming $|\tilde{\chi}_{dot}|/n_w^2 \ll 1$, we can use the approximation $\sqrt{1 + x} \approx 1 + x/2$ to obtain

$$n = n_w + \frac{\chi'_{dot}}{2n_w} \quad (2.74)$$

$$\kappa = -\frac{\chi''_{dot}}{2n_w}, \quad (2.75)$$

and, with Eq. (2.71)

$$\alpha = -\frac{\chi''_{dot}}{n_w}k_0. \quad (2.76)$$

In "diluted" media, the spectral shape of κ therefore follows the Lorentz lineshape of $\chi''_{dot}(\omega)$ [Eq. (2.61)]; this also holds for the absorption coefficient α, provided that the bandwidth of the resonance is small.

2.2.2.3 Reflectance of Strongly Absorbing Media

The Fresnel coefficients (Table 2.1) and related equations are also valid for complex-valued refractive index. At normal incidence, in particular, the reflectance at the interface between air ($n_i = 1$) and an absorber $\tilde{n}_t = n - j\kappa$ is given by

$$R = \frac{(n-1)^2 + \kappa^2}{(n+1)^2 + \kappa^2}. \quad (2.77)$$

The complex refractive index as displayed in Fig. 2.18 yields a reflectance as shown in Fig. 2.20. It is interesting to note that strong resonant absorption results in a spectral band of high reflectance.

2.2.3 Free Electron Gas Model of Metals

The resonant behavior discussed above is due to the restoring force ax in the equation of motion Eq. (2.51) and is characteristic for bound electrons. Many optical and electronic properties of metals, on the other hand, can be described in good

Fig. 2.20 Reflectance R and absorption coefficient α of an absorbing dielectric medium according to the Drude–Lorentz model

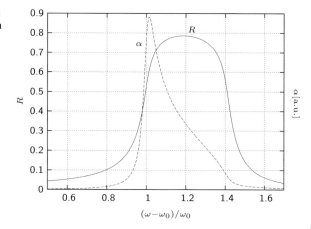

approximation by modeling the conduction electrons as a free electron gas, i.e., by setting the restoring force in the equation of motion equal to zero

$$\ddot{\mathbf{x}} + \Gamma \dot{\mathbf{x}} = -\frac{e}{m_e}\mathbf{E}(t); \qquad (2.78)$$

the complex susceptibility $\tilde{\chi}(\omega)$ according to Eq. (2.56) is then

$$\tilde{\chi}(\omega) = -\left(\frac{n_e e^2}{\varepsilon_0 m_e}\right)\frac{1}{\omega^2 - j\omega\Gamma}. \qquad (2.79)$$

The physical source of the damping term in metals are electron collisions that occur within an average collision time τ_e; to establish a relation between Γ and τ_e, we expose the electrons to a constant electric field; the stationary velocity of the electrons according to Eq. (2.78) is

$$\dot{\mathbf{x}} = -\frac{e}{m_e \Gamma}\mathbf{E}. \qquad (2.80)$$

Assuming that the electron velocity is completely randomized by a collision and the average velocity immediately after a collision is consequently equal to zero, the average velocity of the electrons in the static field is also equal to the acceleration of the electrons, $-(e/m_e)\mathbf{E}(t)$, multiplied with the average time τ_e between consecutive collisions

$$\dot{\mathbf{x}} = -\frac{e}{m_e}\tau_e\mathbf{E}, \qquad (2.81)$$

so that we can set

$$\Gamma = \frac{1}{\tau_e}. \tag{2.82}$$

In order to relate τ_e to a macroscopic observable (namely the conductivity), we multiply $\dot{\mathbf{x}}$ with the density of free electrons n_e and the electron charge $-e$ to obtain the current density

$$\mathbf{j} = -n_e e \dot{\mathbf{x}} = \frac{n_e \tau_e e^2}{m_e} \mathbf{E} = \sigma_e \mathbf{E}, \tag{2.83}$$

where σ_e is the conductivity of the metal, so that we finally obtain

$$\Gamma = \frac{1}{\tau_e} = \frac{n_e e^2}{\sigma_e m_e}. \tag{2.84}$$

Equation (2.79) can now be cast in the form

$$\chi' = -\frac{\omega_p^2 \tau_e^2}{1 + \omega^2 \tau_e^2}, \tag{2.85}$$

$$\chi'' = -\frac{\omega_p^2 \tau_e}{\omega(1 + \omega^2 \tau_e^2)}, \tag{2.86}$$

where

$$\omega_p^2 := \frac{n_e e^2}{\varepsilon_0 m_e} \tag{2.87}$$

is the so-called plasma frequency.

Aluminum, for example, has a conductivity of $\sigma_e = 36 \times 10^6 \ \Omega^{-1} \mathrm{m}^{-1}$ and an electron density of $n_e = 0.18 \times 10^{30} \ \mathrm{m}^{-3}$, so that $\tau_e \approx 7 \times 10^{-15}$ s. The plasma frequency of aluminum, according to Eq. (2.87), is $\omega_p = 24 \times 10^{15} \ \mathrm{s}^{-1}$, corresponding to a photon energy of 15.8 eV or a wavelength of 78 nm (UV).

At frequencies below ω_p, κ is very large (Fig. 2.21), resulting in the high reflectance and short penetration depth that is characteristic for metals (Fig. 2.22). For $\omega > \omega_p$, the assumption $\omega \tau_e \gg 1$ is usually justified, so that one can use the

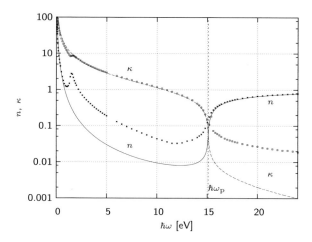

Fig. 2.21 Real and imaginary part of the complex refractive index of aluminum: measured values (*dots*) and theoretical values according to the Drude model; data from Palik (1997)

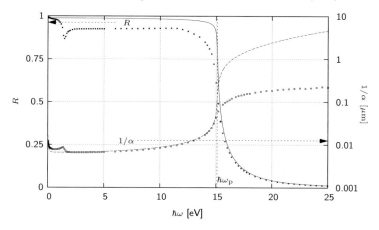

Fig. 2.22 Spectral reflectance R and absorption length $1/\alpha$ of aluminum: measured values (*dots*) and theoretical values according to the Drude model; data from Palik (1997)

approximations

$$\chi' \approx -\frac{\omega_p^2}{\omega^2}, \qquad \chi'' \approx 0 \tag{2.88}$$

and therefore

$$n = \sqrt{\chi + 1} \approx \left(1 - \frac{\omega_p^2}{\omega^2}\right)^{1/2} \approx 1 - \frac{\omega_p^2}{2\omega^2}. \tag{2.89}$$

In the vicinity of the plasma frequency, the refractive index is much smaller than 1, implying a very large value of the phase velocity and of the ratio λ/λ_0; this means that the electron plasma oscillates in phase over a large distance. Above ω_p, the refractive index remains $n < 1$ and $\kappa \approx 0$; metals therefore become transparent in the far UV and provide, at large angles of incidence, total *external* reflection for light incident from the vacuum. The latter effect is exploited for the design of UV- and X-ray reflective optics (refractive optics cannot be realized in this spectral range because of $n \approx 1$).

The detailed optical properties of metals are, of course, more complicated—the excellent agreement of the experimental data for aluminum with the Drude model is rather exceptional. Nonetheless, important features of the optics of metals in general are predicted correctly by the model. The electrons of doped semiconductors can also be modeled as a free electron gas; since the electron density n_e can be controlled by the doping level, the plasma frequency can be tuned over a wide range, allowing one to produce, for example, IR-mirrors with a sharp cutoff at ω_p.

2.2.4   Kramers–Kronig Relations

The polarization $P(t)$ (represented here as a scalar) at a given instant of time is the integrated response of the medium to the electric field up to that instant. Provided that the interaction is linear, we can write

$$P(t) = \int_{-\infty}^{\infty} h(t - t')\varepsilon_0 E(t')\, dt', \tag{2.90}$$

where $h(t)$ represents the "memory function" of the medium. To understand the meaning of $h(t)$, we assume that the incident field is proportional to a Dirac delta-impulse $E(t') \propto \delta(t')$ arriving at $t' = 0$. The resulting time dependent polarization is then proportional to $h(t)$, which is consequently called impulse response function.

If we apply, instead, an oscillating field $E(t) = \mathrm{Re}\left[\tilde{E}(\omega)\mathrm{e}^{\mathrm{j}\omega t}\right]$, then $P(t) = \mathrm{Re}\left[\tilde{P}(\omega)\mathrm{e}^{\mathrm{j}\omega t}\right]$ will oscillate at ω and we obtain, with $t'' := t - t'$

$$\tilde{P}(\omega)\mathrm{e}^{\mathrm{j}\omega t} = \int_{-\infty}^{\infty} h(t - t')\varepsilon_0 \tilde{E}(\omega)\mathrm{e}^{\mathrm{j}\omega t'}\, dt'$$

$$= \varepsilon_0 \tilde{E}(\omega)\mathrm{e}^{\mathrm{j}\omega t} \int_{-\infty}^{\infty} h(t'')\mathrm{e}^{-\mathrm{j}\omega t''}\, dt''. \tag{2.91}$$

Therefore,

$$\tilde{P}(\omega) = \varepsilon_0 \tilde{E}(\omega)\tilde{H}(\omega), \tag{2.92}$$

where

$$\tilde{H}(\omega) = \int_{-\infty}^{\infty} h(t)e^{-j\omega t}\, dt \tag{2.93}$$

is the Fourier transform of the impulse response. A comparison with Eq. (1.7) shows that this so-called transfer function is identical to the susceptibility

$$\tilde{\chi}(\omega) = \chi'(\omega) + j\chi''(\omega) = H(\omega). \tag{2.94}$$

As a response function relating two real valued observables, $h(t)$ must also be real valued. Moreover, $h(t)$ must vanish for negative times, $h(t < 0) = 0$ since in a causal system, the response cannot precede the stimulus. We therefore can write, with Eq. (2.93),

$$\chi'(\omega) = \int_{0}^{\infty} h(t) \cos \omega t\, dt$$

$$\chi''(\omega) = -\int_{0}^{\infty} h(t) \sin \omega t\, dt \tag{2.95}$$

which are the Fourier (Co)Sinus-transforms of $h(t)$; obviously, $\chi'(-\omega) = \chi'(\omega)$ and $\chi''(-\omega) = -\chi''(\omega)$, and therefore $\tilde{\chi}(-\omega) = \tilde{\chi}^*(\omega)$. The inverse transformation

$$h(t) = \frac{1}{2\pi} \int_{-\infty}^{\infty} \tilde{\chi}(\omega)e^{j\omega t}\, d\omega \tag{2.96}$$

can be written as

$$h(t) = \frac{1}{\pi} \int_{0}^{\infty} \left[\chi'(\omega) \cos \omega t - \chi''(\omega) \sin \omega t \right] d\omega. \tag{2.97}$$

For $t \geq 0$, $h(-t) = 0$ and $h(t) = h(t) \pm h(-t)$ and we obtain

$$h(t) = \frac{2}{\pi} \int_{0}^{\infty} \chi'(\omega) \cos \omega t\, d\omega =$$

$$= -\frac{2}{\pi} \int_{0}^{\infty} \chi''(\omega) \sin \omega t\, d\omega. \tag{2.98}$$

As a consequence of causality, there is a one-to-one relation between the real and imaginary parts of the transfer function. To obtain a more explicit result, we substitute Eq. (2.98) in Eq. (2.95)

$$\chi'(\omega) = -\frac{2}{\pi} \int_{0}^{\infty} \cos \omega t \int_{0}^{\infty} \chi''(\omega') \sin \omega' t\, d\omega'\, dt$$

$$\chi''(\omega) = -\frac{2}{\pi} \int_0^\infty \sin \omega t \int_0^\infty \chi'(\omega') \cos \omega' t \, d\omega' \, dt \tag{2.99}$$

and obtain

$$\chi'(\omega) = -\frac{2}{\pi} \int_0^\infty \chi''(\omega') \int_0^\infty \cos \omega t \sin \omega' t \, dt \, d\omega'$$

$$\chi''(\omega) = -\frac{2}{\pi} \int_0^\infty \chi'(\omega') \int_0^\infty \sin \omega t \cos \omega' t \, dt \, d\omega'. \tag{2.100}$$

Applying

$$\int_0^\infty \cos \omega t \sin \omega' t \, dt = \frac{\omega'}{\omega'^2 - \omega^2}$$

$$\int_0^\infty \sin \omega t \cos \omega' t \, dt = \frac{\omega}{\omega^2 - \omega'^2} \tag{2.101}$$

we finally obtain the so-called Kramers–Kronig relations

$$\chi'(\omega) = -\frac{2}{\pi} \mathcal{P} \int_0^\infty \frac{\omega' \chi''(\omega')}{\omega'^2 - \omega^2} \, d\omega'$$

$$\chi''(\omega) = -\frac{2}{\pi} \mathcal{P} \int_0^\infty \frac{\omega \chi'(\omega')}{\omega^2 - \omega'^2} \, d\omega', \tag{2.102}$$

where $\mathcal{P}$ denotes the Cauchy principal value. These relations show that any absorption mechanism inevitably produces dispersion. Moreover, it allows us to calculate, for example, the dispersion from experimentally obtained absorption data.

The Kramers–Kronig relations also apply to other complex material properties (Hodgson 1970; Lucarini et al. 2005; Toll 1956) such as $\tilde{n} = n - j\kappa$. To warrant the convergence of the integrals, one uses $(n - 1)$ as real part and obtains

$$n(\omega) - 1 = \frac{2}{\pi} \mathcal{P} \int_0^\infty \frac{\omega' \kappa(\omega')}{\omega'^2 - \omega^2} \, d\omega' \tag{2.103}$$

$$\kappa(\omega) = \frac{2}{\pi} \mathcal{P} \int_0^\infty \frac{\omega [n(\omega') - 1]}{\omega^2 - \omega'^2} \, d\omega'. \tag{2.104}$$

Similar relations hold for the complex reflection coefficient $r = (\tilde{n} - 1)/(\tilde{n} + 1) = |r| e^{j\phi}$ at the surface of an absorptive medium with complex refractive index $\tilde{n}$; for

$\ln r = \ln |r| + j\phi$ one obtains

$$\phi(\omega) = -\frac{2}{\pi} \mathcal{P} \int_0^\infty \frac{\omega \ln |r(\omega')|}{\omega^2 - \omega'^2} \, d\omega', \qquad (2.105)$$

which is an important relation because $|r|$ can be measured relatively easily, allowing one to determine $\tilde{n}$.

2.3 Wave Propagation in Anisotropic Media

We now extend our treatment of wave propagation to optically anisotropic media (usually crystals), where the relation between **P** and **E** (and therefore **D** and **E**) depends on the direction of **E** within the medium. In the framework of the linear oscillator model, the reason for this is the anisotropy of the restoring force.

One consequence of optical anisotropy is the dependence of the propagation index on the direction of the wave vector and the polarization state of the wave. As we shall see, for a given direction of the wave vector there exist two linear polarization states with well defined, generally different propagation indices. At a border between an anisotropic medium and another one, the two states are refracted in different directions—this is the reason why anisotropic media are also called birefringent.

In an anisotropic, linear medium, the vectors **P** and **E** are generally not collinear, but related by the more general linear equation

$$P_1 = \varepsilon_0 \chi_{11} E_1 + \varepsilon_0 \chi_{12} E_2 + \varepsilon_0 \chi_{13} E_3$$
$$P_2 = \varepsilon_0 \chi_{21} E_1 + \varepsilon_0 \chi_{22} E_2 + \varepsilon_0 \chi_{23} E_3$$
$$P_3 = \varepsilon_0 \chi_{31} E_1 + \varepsilon_0 \chi_{32} E_2 + \varepsilon_0 \chi_{33} E_3, \qquad (2.106)$$

or

$$P_i = \varepsilon_0 \sum_{j=1}^{3} \chi_{ij} E_j; \qquad (2.107)$$

in the following we will adopt Einstein's convention, according to which the double occurrence of an index in one term implies summation over the values of this index, so that Eq. (2.107) can be written as

$$P_i = \varepsilon_0 \chi_{ij} E_j. \qquad (2.108)$$

In vector notation,

$$\mathbf{P} = \varepsilon_0 \boldsymbol{\chi} \mathbf{E}, \tag{2.109}$$

where $\boldsymbol{\chi}$ is the susceptibility tensor[4] with the components χ_{ij}.

2.3.1 Symmetry Properties of Crystals

2.3.1.1 Transformation of Tensors

Tensors are usually represented in a cartesian coordinate system and are written in matrix form; the number of indices of the tensor components indicates the rank of the tensor—$\mathbf{P}$ and $\mathbf{E}$ are tensors of first rank and $\boldsymbol{\chi}$ is a second rank tensor, for example. The matrix representation of one and the same tensor is, of course, different in different reference systems—we have discussed this fact already in the context of the Jones formalism. The results Eqs. (1.107) and (1.113) for the transformation of two-dimensional Jones vectors and matrices can be immediately extended to three-dimensional vectors and tensors: if the transformation between the two (cartesian) coordinate systems is given by the matrix A_{ij}, $i, j = 1 \ldots 3$, then a vector $\mathbf{a}$ with the original coordinates a_i is transformed to

$$a_i' = A_{ij} a_j; \tag{2.110}$$

the inverse transformation is

$$a_k = A_{kl}^{-1} a_l'. \tag{2.111}$$

A tensor m with components m_{jk} is transformed to

$$m_{il}' = A_{ij} m_{jk} A_{kl}^{-1}. \tag{2.112}$$

The transformations relevant in the present context are rotations, mirror operations including inversion, and combinations thereof; Table 2.2 summarizes the corresponding transformation matrices.

A common property of these transformation matrices is that the inverse matrix is obtained by transposition, $A_{ik}^{-1} = A_{ki}$; the transformation (2.112) can therefore be written as

$$m_{il}' = A_{ij} A_{lk} m_{jk}. \tag{2.113}$$

---

[4]An excellent introduction into tensors can be found in Nye (1985).

Table 2.2 Selected symmetry operations and corresponding transformation matrices

Operation	Transformation matrix
Rotation around z-axis	$\begin{bmatrix} \cos\varphi_R & \sin\varphi_R & 0 \\ -\sin\varphi_R & \cos\varphi_R & 0 \\ 0 & 0 & 1 \end{bmatrix}$
Reflection at xy-plane	$\begin{bmatrix} 1 & 0 & 0 \\ 0 & 1 & 0 \\ 0 & 0 & -1 \end{bmatrix}$
Inversion	$\begin{bmatrix} -1 & 0 & 0 \\ 0 & -1 & 0 \\ 0 & 0 & -1 \end{bmatrix}$

As in the case of Jones matrices, a transformation of the reference system is closely related to a physical transformation of the system under study: the tensor components of a crystal that is rotated by an angle $-\varphi_R$ are equal to the components of the crystal in a reference system rotated by φ_R. By the same token, physical reflection and inversion of the crystal (even if it is not physically possible) is equivalent to reflection and inversion of the reference frame.

Crystalline materials are characterized by their symmetry, i.e., by the invariance of their properties under certain transformations; threefold rotational symmetry, for example, means that the material is indistinguishable from the same material, rotated by 120° around a certain axis. The tensor components of such a medium must therefore be solutions of the equation $\mathbf{m}' = \mathbf{m}$

$$A_{ij}A_{lk}m_{jk} = m_{il}, \qquad (2.114)$$

where A_{ij} is any of the symmetry operations of the material. If the system is invariant under several different transformations, one obtains a set of such equations, forcing certain components to be zero and others to be linearly dependent of each other. Isotropic media, for example, are invariant under arbitrary rotations, and centrosymmetric media are invariant under inversion. As a result, the susceptibility tensor of isotropic, centrosymmetric media has the form $\chi_{ij} = \delta_{ij}\chi$, where χ is a scalar; all off-diagonal components are zero and the diagonal elements are identical.[5]

---

[5] An isotropic medium is not necessarily centrosymmetric; a chiral liquid is an example of an isotropic, yet non-centrosymmetric medium, see Sect. 2.4.1.

In Chap. 8, we will encounter tensors of higher rank such as the quadratic nonlinear susceptibility $\chi^{(2)}$ that links the square of the electric field to the nonlinear polarization,

$$P_i^{(2)} = \varepsilon_0 \chi_{ijk}^{(2)} E_j E_k. \tag{2.115}$$

As can be easily shown, the transformation of such a third rank tensor follows the pattern of Eq. (2.113)

$$m'_{lmn} = A_{li} A_{mj} A_{nk} m_{ijk}; \tag{2.116}$$

for a fourth rank tensor,

$$m'_{rsuv} = A_{ri} A_{sj} A_{uk} A_{vl} m_{ijkl}. \tag{2.117}$$

2.3.1.2 Principal Axes

The vectors **D** and **E** are related by Eq. (1.8)

$$\mathbf{D} = \varepsilon_0 \mathbf{E} + \mathbf{P} = \varepsilon_0 \boldsymbol{\varepsilon} \mathbf{E} = \varepsilon_0 \mathbf{E}(1 + \chi) \tag{2.118}$$

where $\boldsymbol{\varepsilon}$ is the permittivity tensor and

$$\mathbf{1} = \begin{bmatrix} 1 & 0 & 0 \\ 0 & 1 & 0 \\ 0 & 0 & 1 \end{bmatrix} \tag{2.119}$$

is the second rank unit tensor. In tensor notation,

$$\boldsymbol{\varepsilon} = \mathbf{1} + \chi. \tag{2.120}$$

In nonmagnetic, lossless media, the tensors $\boldsymbol{\varepsilon}$ and χ are symmetric (see, e.g., Haus 1984)

$$\varepsilon_{ij} = \varepsilon_{ji} \tag{2.121}$$

so that they can be diagonalized, i.e., a coordinate system can be found where $\boldsymbol{\varepsilon}$ has the form

$$\boldsymbol{\varepsilon} = \begin{bmatrix} \varepsilon_{11} & 0 & 0 \\ 0 & \varepsilon_{22} & 0 \\ 0 & 0 & \varepsilon_{33} \end{bmatrix} =: \begin{bmatrix} \varepsilon_{(x)} & 0 & 0 \\ 0 & \varepsilon_{(y)} & 0 \\ 0 & 0 & \varepsilon_{(z)} \end{bmatrix}. \tag{2.122}$$

Fig. 2.23 If the **E** vector of a wave is parallel to a principal axis (i) of a birefringent medium, the wave is a polarization eigenstate with the propagation index $n_{(i)} = \varepsilon_{(i)}^{1/2}$; the corresponding **k** vector is normal to (i) and has the length $k_0 n_{(i)}$

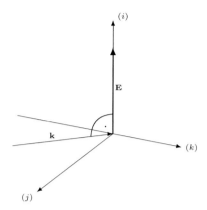

The axes of this reference frame are called principal axes of the medium, and $\varepsilon_{(i)}$ are called principal values of $\boldsymbol{\varepsilon}$.

If **E** is parallel to a principal axis (i), then $\mathbf{D} = \varepsilon_0 \boldsymbol{\varepsilon} \mathbf{E} = \varepsilon_0 \varepsilon_{(i)} \mathbf{E}$. If $\boldsymbol{\varepsilon}$ is not yet diagonal, the principal axes can therefore be found by solving the equation $\boldsymbol{\varepsilon} \mathbf{E} = \varepsilon_{(i)} \mathbf{E}$, i.e., by finding the eigenvectors of the matrix $\boldsymbol{\varepsilon}$; the eigenvectors define the direction of the principal axes and the corresponding eigenvalues are equal to the principal values. Since symmetric matrices have orthogonal eigenvectors, the reference frame generated by the principal axes is orthogonal.

The scalar relation $\mathbf{D} = \varepsilon_0 \varepsilon_{(i)} \mathbf{E}$ that is valid for eigenstates of $\boldsymbol{\varepsilon}$ is formally identical to the relation between **D** and **E** in an isotropic medium. A light wave with an electric field parallel to a principal axis (i) therefore propagates, according to the dispersion relation Eq. (1.28) for isotropic media, with the phase velocity $c_0/n_{(i)}$ where $n_{(i)} = \varepsilon_{(i)}^{1/2}$. The corresponding wave vector is normal to (i) and has the length $k_0 n_{(i)}$ (Fig. 2.23).

The square roots of the principal values, $n_{(i)} := \varepsilon_{(i)}^{1/2}$, are denoted as refractive indices of the medium (Table 2.3). If all principal values are different, the medium is called, for reasons that will become clear below, biaxial; if two out of the three principal values are equal, and different from the third, the medium is called uniaxial; in this case, the axis corresponding to the deviating principal value is identified with the z-axis and $n_{(z)}$ is called extraordinary refractive index n_e, while $n_{(x)} = n_{(y)} = n_0$ is called ordinary refractive index.

2.3.1.3 Impermeability

In electro-optics and magneto-optics, the relation inverse to Eq. (1.8),

$$\mathbf{E} = \varepsilon_0^{-1} \eta \mathbf{D} \qquad (2.123)$$

Table 2.3 Refractive indices (principal values) of selected isotropic, uni- and biaxial materials

Material	$n_{(x)}$	$n_{(y)}$	$n_{(z)}$
Water	1.333	$n_{(x)}$	$n_{(x)}$
BK1 (Bor crown glass)	1.510	$n_{(x)}$	$n_{(x)}$
F3 (Flint glass)	1.613	$n_{(x)}$	$n_{(x)}$
NaCl	1.544	$n_{(x)}$	$n_{(x)}$
Quartz	1.544	$n_{(x)}$	1.553
Rutile	2.616	$n_{(x)}$	2.903
Calcite	1.658	$n_{(x)}$	1.486
Sapphire	1.768	$n_{(x)}$	1.660
Tourmaline	1.642	$n_{(x)}$	1.622
Mica	1.560	1.594	1.599
Kalium nitrate	1.335	1.505	1.506

Fig. 2.24 If **k** is parallel to a principal axis (i) of an anisotropic medium, the two polarization eigenstates are parallel to the (j, k)-axes, respectively, and propagate with the phase velocities $c_0/n_{(j,k)}$

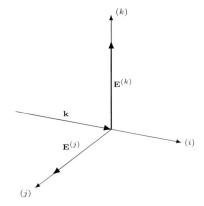

is used, where $\boldsymbol{\eta} = \boldsymbol{\varepsilon}^{-1}$ is known as the impermeability tensor; if $\boldsymbol{\varepsilon}$ is given in diagonal form, $\boldsymbol{\eta}$ is also diagonal and $\eta_{ii} = 1/\varepsilon_{ii}$

$$\boldsymbol{\eta} = \begin{bmatrix} \eta_{11} & 0 & 0 \\ 0 & \eta_{22} & 0 \\ 0 & 0 & \eta_{33} \end{bmatrix} = \begin{bmatrix} \varepsilon_{(x)}^{-1} & 0 & 0 \\ 0 & \varepsilon_{(y)}^{-1} & 0 \\ 0 & 0 & \varepsilon_{(z)}^{-1} \end{bmatrix} = \begin{bmatrix} n_{(x)}^{-2} & 0 & 0 \\ 0 & n_{(y)}^{-2} & 0 \\ 0 & 0 & n_{(z)}^{-2} \end{bmatrix}. \quad (2.124)$$

2.3.2   Propagation Along the Principal Axes

If the *wave vector* is parallel to a principal axis, then there are actually *two* polarization states that oscillate parallel to one of the remaining axes and represent polarization eigenstates (Fig. 2.24). If, for example, **k** is parallel to the y-axis, then the two eigenstates are polarized along the x and z-axis, respectively, having the wave numbers $k_0 n_{(x)}$ and $k_0 n_{(z)}$. A plate of thickness d, cut parallel to the x, z-plane, therefore serves as a linear retarder with

$$\Delta\phi_V = [n_{(z)} - n_{(x)}]k_0 d \tag{2.125}$$

as described in Sect. 1.5.2; to be consistent with Table 1.3, we identify the x-axis with the "fast" axis, i.e., the one with the smaller refractive index.

2.3.3   Propagation in Arbitrary Directions*

Consider a plane wave

$$\mathbf{E} = \mathbf{E}_0 e^{-j(\mathbf{k}\cdot\mathbf{x}-\omega t)} \tag{2.126}$$

with a wave vector

$$\mathbf{k} = nk_0\mathbf{e}, \tag{2.127}$$

where the unit vector $\mathbf{e}$ defines the direction of the wave vector. To find the corresponding propagation index n, we substitute this ansatz into the wave equation (1.15)

$$\nabla \times (\nabla \times \mathbf{E}) + \frac{\partial^2\mathbf{D}}{\varepsilon_0 c_0^2 \partial t^2} = \mathbf{0} \tag{2.128}$$

and use Eq. (1.33) to obtain

$$\mathbf{k} \times (\mathbf{k} \times \mathbf{E}) = -\frac{\omega^2}{\varepsilon_0 c_0^2}\mathbf{D}; \tag{2.129}$$

with Eq. (2.127) and $k_0 = \omega/c_0$, this can be written as

$$-\frac{\omega^2 n^2}{c_0^2}\mathbf{e} \times (\mathbf{e} \times \mathbf{E}) = \frac{\omega^2}{\varepsilon_0 c_0^2}\mathbf{D}. \tag{2.130}$$

Using the identity $\mathbf{a} \times (\mathbf{b} \times \mathbf{c}) = \mathbf{b}(\mathbf{a}\cdot\mathbf{c}) - \mathbf{c}(\mathbf{a}\cdot\mathbf{b})$ and the relation $\mathbf{D} = \varepsilon_0\boldsymbol{\varepsilon}\mathbf{E}$, we obtain

$$\varepsilon_0 n^2[\mathbf{E} - \mathbf{e}(\mathbf{e}\cdot\mathbf{E})] - \varepsilon_0\boldsymbol{\varepsilon}\mathbf{E} = \mathbf{0}; \tag{2.131}$$

note that $\mathbf{E} - \mathbf{e}(\mathbf{e}\cdot\mathbf{E}) = -\mathbf{e} \times (\mathbf{e} \times \mathbf{E})$ is simply the transverse component of $\mathbf{E}$. Equation (2.131) can be cast in matrix form

$$\boldsymbol{M}\mathbf{E} = \mathbf{0} \tag{2.132}$$

with

$$M = \begin{bmatrix} n^2(1-e_x^2) - \varepsilon_{(x)} & -n^2 e_x e_y & -n^2 e_x e_z \\ -n^2 e_x e_y & n^2(1-e_y^2) - \varepsilon_{(y)} & -n^2 e_y e_z \\ -n^2 e_x e_z & -n^2 e_y e_z & n^2(1-e_z^2) - \varepsilon_{(z)} \end{bmatrix}, \quad (2.133)$$

where we have assumed that ε is diagonal. A condition for a non-trivial solution $n \neq 0$ is

$$\det M = 0; \quad (2.134)$$

since the cubic terms cancel, this is a quadratic equation in the variable n^2. For a given direction $\mathbf{e}$, Eq. (2.134) provides two solutions $[n^{(1,2)}]^2$, corresponding to two wave vectors $\mathbf{k} = n^{(1,2)} k_0 \mathbf{e}$; the directions of the corresponding $\mathbf{E}$-vectors (which are the polarization eigenstates for propagation in the direction $\mathbf{e}$) result from Eq. (2.132) after substitution of $n^{(1,2)}$. As we shall show, the two eigenstates are mutually orthogonal because of the symmetry of ε.

An alternative way of finding n^2 is to write Eq. (2.131) in the form

$$(n^2 \mathbf{1} - \varepsilon)\mathbf{E} = n^2 \mathbf{e}(\mathbf{e} \cdot \mathbf{E}) \quad (2.135)$$

to obtain three equations

$$E_i = \frac{n^2 e_i}{n^2 - \varepsilon_{(i)}} \mathbf{e} \cdot \mathbf{E}. \quad (2.136)$$

Multiplying both sides with e_i and taking the sum of the resulting equations, we obtain

$$\mathbf{e} \cdot \mathbf{E} = \sum_i \frac{n^2 e_i^2}{n^2 - \varepsilon_{(i)}} \mathbf{e} \cdot \mathbf{E} \quad (2.137)$$

and finally the convenient (quadratic) equation

$$\frac{e_x^2 n^2}{n^2 - \varepsilon_{(x)}} + \frac{e_y^2 n^2}{n^2 - \varepsilon_{(y)}} + \frac{e_z^2 n^2}{n^2 - \varepsilon_{(z)}} = 1. \quad (2.138)$$

2.3.3.1 k-Surfaces

For a given frequency ω, we can represent all possible k-vectors in a cartesian coordinate system with coordinates k_x, k_y, k_z (the k-space): we select a direction $\mathbf{e}$, calculate the corresponding propagation indices $n^{(1,2)}$ and plot points at $n^{(1,2)}(\omega/c_0)\mathbf{e}$, respectively; varying $\mathbf{e}$, we obtain two surfaces that are called k-surfaces (Fig. 2.25) and are actually surfaces of constant ω in k-space. It is

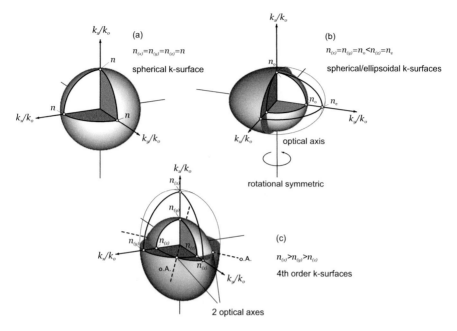

Fig. 2.25 Normalized k-surfaces for (**a**) isotropic, (**b**) uniaxial, and (**c**) biaxial media

instructive to calculate the intersection of these surfaces with one of the coordinate planes, say the k_y/k_z-plane (defined by $e_x = 0$). Equation (2.134) yields

$$\det \boldsymbol{M} = M_{11}(M_{22}M_{33} - M_{23}M_{32}) = 0, \qquad (2.139)$$

where M_{ij} are the components of $\boldsymbol{M}$. With $e_x^2 + e_y^2 + e_z^2 = 1$ and $\varepsilon_{(i)} = n_{(i)}^2$ we obtain two equations

$$n^2 e_y^2 + n^2 e_z^2 - n_{(x)}^2 = 0$$

$$\left(n^2 e_y^2 - n_{(z)}^2\right)\left(n^2 e_z^2 - n_{(y)}^2\right) - n^4 e_y^2 e_z^2 = 0 \qquad (2.140)$$

which, because of $k_j = n k_0 e_j$, are equivalent to

$$k_y^2 + k_z^2 = n_{(x)}^2 k_0^2 \qquad (2.141)$$

$$\frac{k_y^2}{n_{(z)}^2 k_0^2} + \frac{k_z^2}{n_{(y)}^2 k_0^2} = 1, \qquad (2.142)$$

that is a circle with radius $k_0 n_{(x)}$ and an ellipse with axes $k_0 n_{(y)}$ and $k_0 n_{(z)}$.

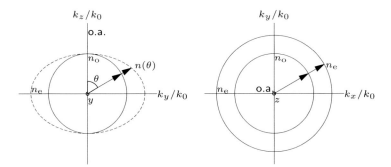

Fig. 2.26 Cross section through the (normalized) k-surfaces of a uniaxial medium; *solid circles* refer to ordinary waves, *dashed curves* are ellipses and refer to extraordinary waves

For a uniaxial medium ($n_{(x)} = n_{(y)} = n_o$ and $n_{(z)} = n_e$), the k-surface is rotationally symmetric around the k_z-axis, forming a sphere and an ellipsoid

$$k_x^2 + k_y^2 + k_z^2 = (n_o k_0)^2 \qquad (2.143)$$

$$\frac{k_x^2}{(n_e k_0)^2} + \frac{k_y^2}{(n_e k_0)^2} + \frac{k_z^2}{(n_o k_0)^2} = 1, \qquad (2.144)$$

respectively, that touch each other at the poles of the rotation axis (Fig. 2.25b). Light travelling along this axis has the propagation index n_o (the ordinary refractive index), independent of its polarization; such an axis is called optical axis (o.a.).

In any other propagation direction, there are two distinct polarization eigenstates with different propagation indices. Because of the rotational symmetry, the direction of the wave vector is fully characterized by the angle θ between $\mathbf{k}$ and the optical axis (Fig. 2.26a). According to Eq. (2.143), one propagation index is equal to the ordinary index n_o; evidently, the corresponding eigenvector is normal to the plane formed by $\mathbf{k}$ and the optical axis. The second value follows, with $k_y = n(\theta)k_0 \sin\theta$ and $k_z = n(\theta)k_0 \cos\theta$, from Eq. (2.144) to be

$$\frac{1}{n^2(\theta)} = \frac{\cos^2\theta}{n_o^2} + \frac{\sin^2\theta}{n_e^2}. \qquad (2.145)$$

Because of the orthogonality of the eigenstates (which will be proven below), the corresponding eigenvector is coplanar with $\mathbf{k}$ and the optical axis.

In the fully anisotropic case ($n_{(x)} \neq n_{(y)} \neq n_{(z)}$), the cross section of the normalized k-surface (Fig. 2.25c) with a plane normal to the principal axis (k) consists of a circle with radius $n_{(k)}$ and an ellipse with axes $n_{(i)}$ and $n_{(j)}$, where i, j, k are different from each other (Fig. 2.27). Obviously, circle and ellipse intersect

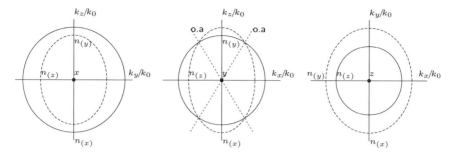

Fig. 2.27 Cross sections of the (normalized) k-surface of a biaxial medium with $n_{(x)} < n_{(y)} < n_{(z)}$ along planes containing two of the principal axes

each other only if $n_{(k)}$ is intermediate between $n_{(i)}$ and $n_{(j)}$. The four points of intersection define two directions for which the propagation index is independent of the polarization and are, by definition, optical axes. These axes lie in the plane normal to principal axis (k) and enclose the angle θ_a

$$\theta_a = \pm \arctan \frac{n_{(j)}}{n_{(i)}} \sqrt{\frac{n_{(i)}^2 - n_{(k)}^2}{n_{(k)}^2 - n_{(j)}^2}} \qquad (2.146)$$

with the axis (i); as the refractive indices depend on the wavelength of the light, so does the direction of the optical axes in biaxial media.

2.3.3.2 Polarization Eigenstates
For uniaxial media, the orientation of the eigenstates can be summarized as follows:

1. if $\mathbf{k} \parallel$ o.a., (Fig. 2.26a), any polarization is an eigenstate (the eigenstates are degenerate) with propagation index n_o
2. if $\mathbf{k} \perp$ o.a., (Fig. 2.26b), one eigenstate with propagation index $n_e = n_{(z)}$ is $\parallel$ o.a., the second, with index $n_o = n_{(x,y)}$, is $\perp$ o.a. and $\perp \mathbf{k}$
3. for any other direction of $\mathbf{k}$ (Fig. 2.26a), one eigenstate with the index n_o (the "ordinary" wave) is polarized perpendicular to the plane formed by o.a. and $\mathbf{k}$, the second one (the "extraordinary" wave) is polarized parallel to this plane, and its propagation index is given by Eq. (2.145).

In biaxial media, the polarization eigenstates in general must be found by substituting the eigenvalues n^2 into Eq. (2.132) and solving for the ratios E_i/E_j that determine the direction of the respective eigenvector. If the wave vector happens to be perpendicular to a principal axis (i), the situation is simplified (Fig. 2.27):

1. if $\mathbf{k} \parallel (j)$, one eigenstate, with propagation index $n_{(i)}$, is $\parallel (i)$, while the second, with index $n_{(k)}$ is $\parallel (k)$

Table 2.4 Geometric relations between various field vectors; each of the triples **H**, **D**, **k** and **H**, **E**, **S**, respectively, constitutes an orthogonal tripod (Fig. 2.28). $\mathbf{D}^{(1)}$ and $\mathbf{D}^{(2)}$ represent pairs of polarization eigenstates

$\mathbf{D} = -\frac{1}{\omega}\mathbf{k} \times \mathbf{H}$	$\rightarrow \mathbf{D} \perp \mathbf{H}$
	$\rightarrow \mathbf{D} \perp \mathbf{k}$
$\mathbf{H} = \frac{1}{\mu_0 \omega}\mathbf{k} \times \mathbf{E}$	$\rightarrow \mathbf{k} \perp \mathbf{H}$
	$\rightarrow \mathbf{E} \perp \mathbf{H}$
$\mathbf{S} = \mathbf{E} \times \mathbf{H}$	$\rightarrow \mathbf{S} \perp \mathbf{H}$
	$\rightarrow \mathbf{E} \perp \mathbf{S}$
$\varepsilon_{ij} = \varepsilon_{ji}$	$\rightarrow \mathbf{D}^{(1)} \perp \mathbf{D}^{(2)}$

2. if **k** ∥ o.a., any polarization is an eigenstate (the eigenstates are degenerate) with propagation index $n_{(i)}$
3. for any other **k** ⊥ (i), one eigenstate with the index $n_{(i)}$ is ∥ (i), while the second one (the "extraordinary" wave) lies in the plane ⊥ (i) and its propagation index is given by

$$\frac{1}{n^2(\theta)} = \frac{\cos^2 \theta}{n_{(j)}^2} + \frac{\sin^2 \theta}{n_{(k)}^2}, \tag{2.147}$$

where θ is the angle between **k** and (k).

2.3.3.3 Orthogonality of the Eigenstates

We now want to prove that the polarization eigenstates (1) and (2) of an anisotropic medium are mutually orthogonal. From Maxwell's equation follows that **E**, **D**, and $\mathbf{k} = k\mathbf{e}$ lie in a common plane normal to **H** (Table 2.4); this plane is also the plane of polarization (Fig. 2.28). Since **D** is normal to **k**, we have to show that $\mathbf{D}^{(1)} \perp \mathbf{D}^{(2)}$. The term $\mathbf{e}(\mathbf{e} \cdot \mathbf{E})$ in Eq. (2.131) is the longitudinal component $\mathbf{E}_L$ of **E**, so that $\mathbf{E}_T = [\mathbf{E} - \mathbf{e}(\mathbf{e} \cdot \mathbf{E})]$ is the transverse component, and Eq. (2.131) can be written as

$$\mathbf{D}^{(1,2)} = \varepsilon_0 \left[n^{(1,2)} \right]^2 \mathbf{E}_T^{(1,2)}. \tag{2.148}$$

From $\varepsilon_{ij} = \varepsilon_{ji}$ follows the reciprocity relation $E_i^{(1)} \varepsilon_{ij} E_j^{(2)} = E_j^{(2)} \varepsilon_{ji} E_i^{(1)}$ (in Einstein notation), i.e.,

$$\mathbf{E}^{(1)} \cdot \mathbf{D}^{(2)} = \mathbf{E}^{(2)} \cdot \mathbf{D}^{(1)}; \tag{2.149}$$

substituting Eq. (2.148) in Eq. (2.149) results in

$$\left[n^{(2)} \right]^2 \mathbf{E}^{(1)} \cdot \mathbf{E}_T^{(2)} = \left[n^{(1)} \right]^2 \mathbf{E}^{(2)} \cdot \mathbf{E}_T^{(1)}. \tag{2.150}$$

Fig. 2.28 Vectors **k**, **E**, **D**, **H**, and **S** in an anisotropic medium (compare Table 2.4): **D** and **H** are parallel to the phase front, which is normal to the wave vector **k**; **E**, **D**, **S**, and **k** are coplanar

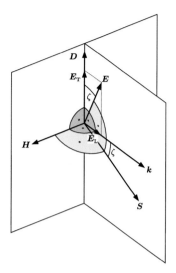

Since $\mathbf{E}^{(1)} \cdot \mathbf{E}^{(2)} = \left(\mathbf{E}_{\mathrm{L}}^{(1)} + \mathbf{E}_{\mathrm{T}}^{(1)}\right) \cdot \mathbf{E}_{\mathrm{T}}^{(2)} = \mathbf{E}_{\mathrm{T}}^{(1)} \cdot \mathbf{E}_{\mathrm{T}}^{(2)}$, we obtain

$$\left[\left[n^{(1)}\right]^2 - \left[n^{(2)}\right]^2\right] \mathbf{E}_{\mathrm{T}}^{(1)} \cdot \mathbf{E}_{\mathrm{T}}^{(2)} = 0. \tag{2.151}$$

For non-degenerate eigenstates $n^{(1)} \neq n^{(2)}$ follows $\mathbf{E}_{\mathrm{T}}^{(1)} \perp \mathbf{E}_{\mathrm{T}}^{(2)}$ and with Eq. (2.148) finally $\mathbf{D}^{(1)} \perp \mathbf{D}^{(2)}$.

2.3.3.4 Index Ellipsoid

The k-surface is just one out of a manifold of graphical descriptions of wave propagation in anisotropic materials. Another one is the so-called indicatrix or index ellipsoid (Fig. 2.29), a surface given by the equation

$$\frac{x^2}{n_{(x)}^2} + \frac{y^2}{n_{(y)}^2} + \frac{z^2}{n_{(z)}^2} = 1. \tag{2.152}$$

As can be shown (see, e.g., Nye 1985 or Born and Wolf 1999), this scheme allows us to determine, for a given direction **e** of the wave vector, the polarization eigenstates and their respective propagation index: the intersection of the ellipsoid with a plane $\perp$ **e** through the origin is an ellipse with half-axes that are parallel to the eigenstates and have a length equal to the corresponding n.

For uniaxial media, the indicatrix is rotationally symmetric, and the intersection with a plane normal to the **k**-vector is an ellipse with half-axes of length n_o and $(\cos^2 \theta / n_o^2 + \sin^2 \theta / n_e^2)^{-1/2}$, where θ is the angle between **e** and the optical axis; this result is in agreement with Eq. (2.145).

Fig. 2.29 Indicatrix (index
ellipsoid) of a biaxial
material; also shown is the
graphical construction of the
eigenstates and corresponding
propagation indices for a
given direction **e** of the wave
vector

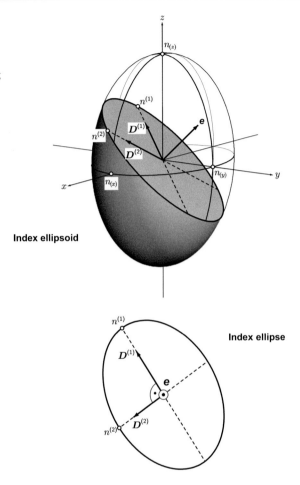

2.3.3.5 Anisotropic Media: Reflection and Refraction

The phase matching condition Eq. (2.5), $\mathbf{k}_\parallel^i = \mathbf{k}_\parallel^r = \mathbf{k}_\parallel^t$, at interfaces between
different media applies also to anisotropic media. The fact, however, that the
absolute value of the transmitted wave vector depends on the polarization, results
in the existence of two refracted waves (a phenomenon called birefringence).
Figure 2.30 demonstrates this effect for the simple example of an interface between
an isotropic medium and a uniaxial medium with the optical axis parallel to the
plane of incidence. The σ-component of the incident wave propagates as ordinary
transmitted wave and is refracted according to

$$\sin \theta^{t,o} = \frac{n_i}{n_o} \sin \theta^i, \tag{2.153}$$

Fig. 2.30 Refraction and reflection at an interface between an isotropic and a uniaxial medium: the refracted wave vectors terminate at the k-surfaces at $k_\parallel = k_\parallel^i$

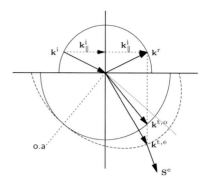

while the π-polarized component propagates as extraordinary wave with a wave vector direction given by

$$n(\theta^{t,e}) \sin \theta^{t,e} = n_1 \sin \theta^i. \tag{2.154}$$

If the optical axis is normal to the interface plane,

$$\sin \theta^{t,e} = \frac{n_e n_1}{\sqrt{n_o^2 n_e^2 + n_1^2 (n_e^2 - n_o^2) \sin^2 \theta^i}} \sin \theta^i. \tag{2.155}$$

As we shall see, the direction of the refracted extraordinary *beam* deviates from this direction; for extraordinary waves, Snell's law applies only to the phase front normal, not to the energy flow.

2.3.3.6 Anisotropic Media: Energy Transport

The energy transport of the electromagnetic field is given by the Poynting vector $\mathbf{S} = \mathbf{E} \times \mathbf{H}$ [Eq. (1.55)]. If $\mathbf{E}$ is not normal to $\mathbf{k}$, the direction of the energy flow deviates from that of the wave vector (Fig. 2.28). Since $\mathbf{D} \perp \mathbf{k}$ and $\mathbf{E} \perp \mathbf{S}$ (Table 2.4), the angle ζ between $\mathbf{S}$ and $\mathbf{k}$ is equal to that between $\mathbf{E} = \varepsilon^{-1}\mathbf{D}$ and $\mathbf{D}$

$$\zeta = \arccos \frac{\mathbf{D}\varepsilon^{-1}\mathbf{D}}{|\mathbf{D}||\varepsilon^{-1}\mathbf{D}|}. \tag{2.156}$$

Although the ray or beam velocity Eq. (1.46) $\mathbf{v}_{ray} = \nabla_k \omega(\mathbf{k})$, is defined without regard to the Poynting vector $\mathbf{S}$, the two must have the same direction because the energy flow is spatially confined to the beam. Since the k-surface is the surface of constant ω in k-space, the direction of the ray velocity is given by the k-surface normal (Fig. 2.31). The extraordinary k-surface Eq. (2.144) of a uniaxial material, for example, has a normal vector $[e_x/\varepsilon_e, e_y/\varepsilon_e, e_z/\varepsilon_o]$, where $\mathbf{e}$ is the unit vector in the direction of $\mathbf{k}$. Assuming $\mathbf{e}$ to lie in the x, z-plane, enclosing an angle of θ with

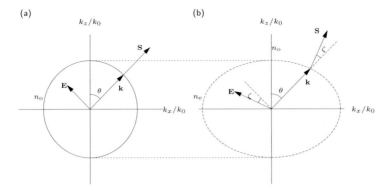

Fig. 2.31 Phase front normal ($\|\mathbf{k}$) and ray velocity ($\|\mathbf{S}$) in a uniaxial medium: (**a**) ordinary wave, (**b**) extraordinary wave

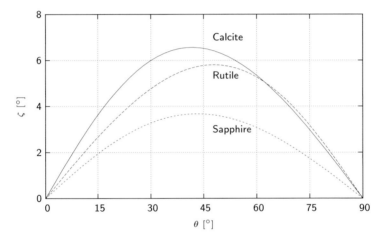

Fig. 2.32 Angle ζ between $\mathbf{k}$ and $\mathbf{S}$ for various uniaxial materials; θ is the angle between the $\mathbf{k}$-vector and the optical axis

the z-axis, we obtain

$$\zeta = \arccos \frac{\sin^2 \theta / \varepsilon_e + \cos^2 \theta / \varepsilon_o}{[(\sin \theta / \varepsilon_e)^2 + (\cos \theta / \varepsilon_o)^2]^{1/2}}, \qquad (2.157)$$

in agreement with Eq. (2.156); Fig. 2.32 shows the deviation for three different uniaxial materials as a function of θ.

The deviation between the direction of $\mathbf{k}$ and $\mathbf{S}$ is particularly striking if a beam is transmitted through an anisotropic medium under normal incidence: according to Snell's law, the *wave vector* of the transmitted light is not refracted, while the transmitted beam is split into an ordinary beam, normal to the boundary, and an

Fig. 2.33 Beam refraction in a uniaxial material under normal incidence

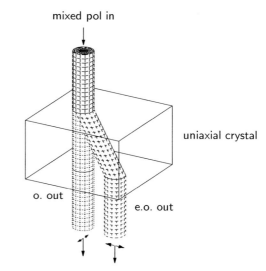

mixed pol in

uniaxial crystal

o. out

e.o. out

Fig. 2.34 A Glan–Thompson prism consisting of two trigonal prisms of calcite ($n_o > n_e$), connected by a thin layer of cement with a refractive index matching n_e; while the π-polarized component of the incoming wave is totally reflected, the σ-polarized component is almost completely transmitted

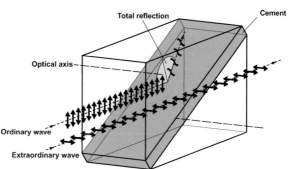

Total reflection   Cement

Optical axis

Ordinary wave

Extraordinary wave

extraordinary that propagates under the angle ζ given by Eq. (2.157) in respect to the interface normal (Fig. 2.33).

In Sect. 2.3.2, we have already described the realization of linear retarders using thin plates of anisotropic media, usually cut normal to one of the principal axes to avoid any beam walk-off between the two polarization eigenstates as described above.

The beam offset shown in Fig. 2.33 can be exploited to build a polarizer that separates the two polarization components of an incident beam into two parallel output beams, provided that the crystal is long enough. Another scheme to realize a polarization beam splitter is based on the polarization dependence of the critical angle of total reflection at an interface between a birefringent material and a medium of lower optical density. A high quality polarization beam splitter, the Glan–Thompson prism, relies on this effect; its design principle is shown in Fig. 2.34.

2.3.4   Electro-Optic Devices

The susceptibility and thus the permittivity tensor of a material can be altered by an external electrostatic field. This effect is called Pockels- or electro-optic Kerr effect, respectively, depending on whether the change is a linear or a quadratic function of he applied field. We will discuss these effects in more detail in Sect. 8.4.1; as an important application of the Pockels effect, we describe here an electrically controlled linear retarder, known as Pockels cell. Such a cell is typically a slab of KDP (KH_2PO_4), cut normal to its z-axis. In the absence of an electrostatic field, KDP is a uniaxial crystal; if an electric field E^{dc} is applied in z-direction it becomes biaxial with new principal axes x' and y' (Fig. 2.35) and corresponding principal

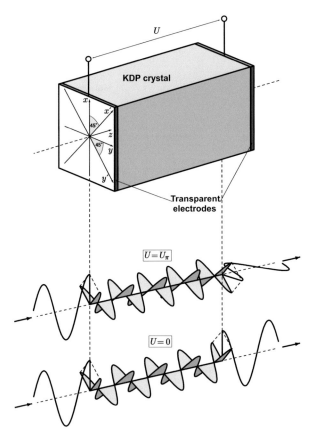

Fig. 2.35 Electro-optic wave plate (Pockels cell); the phase delay between the two polarization eigenstates is controlled by the applied voltage

values of the propagation index [Eq. (8.206)]

$$n_{(x',y')} = n_o \mp \frac{n_o^3}{2} r_{63} E_3^{dc}. \qquad (2.158)$$

For light propagating along the z-axis, the slab of thickness d acts as linear retarder inducing a phase difference of $\Delta\phi = n_o^3 r_{63} E_3^{dc} k_0 d = \pi(U/U_\pi)$, where U is the applied voltage so that $E_3^{dc} = U/d$, and $U_\pi = \frac{\lambda_0}{2n_o^3 r_{63}}$ is the voltage required to reach a phase difference of π. In the x'-y'-coordinate frame, the Jones matrix of the plate is then given by

$$T = \begin{bmatrix} 1 & 0 \\ 0 & e^{-j\pi(U/U_\pi)} \end{bmatrix}; \qquad (2.159)$$

note that the thickness of the slab does not influence the induced phase difference. With the values given in Table 8.4, we obtain $U_\pi = 14.6\,\text{kV}$ at a wavelength of $\lambda_0 = 1.064\,\mu\text{m}$. At $U = U_\pi$, the cell acts as a half-wave plate (Table 1.3) that converts an input polarization state into its mirror image.

Such a retarder, placed between two polarizers, can act as an electronically controlled optical shutter or modulator (Fig. 2.36). If the polarizers are oriented under $45°$ in respect to the x'-y'-system (see Table 1.3), the Jones matrix of this sequence in the x'/y' reference frame is

$$T = \frac{1}{4} \begin{bmatrix} 1 & 1 \\ 1 & 1 \end{bmatrix} \begin{bmatrix} 1 & 0 \\ 0 & e^{-j\pi(U/U_\pi)} \end{bmatrix} \begin{bmatrix} 1 & 1 \\ 1 & 1 \end{bmatrix}. \qquad (2.160)$$

With the input state $\mathbf{J}_{in} = \begin{bmatrix} 1 \\ 1 \end{bmatrix}$, the output state is

$$\mathbf{J}_{out} = \mathbf{J}_{in} \cos^2\left(\pi \frac{U}{2U_\pi}\right). \qquad (2.161)$$

The Pockels cell can be operated as a switch (with an extinction ratio of up to 10^{-4}) or a modulator; for the latter purpose the phase difference at $U = 0$ is shifted to $\pi/2$ by inserting a quarter wave plate so as to operate in the linear range of the transmission function.

There are other possible geometries to realize such a retarder in KDP, but the advantage of this longitudinal geometry ($\mathbf{E}^{dc} \| \mathbf{k}$) is that the aperture of the cell can be very large and does not influence the required voltage. The electrodes for the application of the electric field need, of course, to be transparent and are usually made out of transparent conductive oxides (TCOs).

Fig. 2.36 Pockels cell as
electro-optic switch or
modulator, respectively

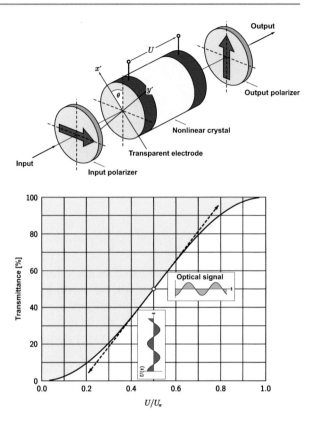

A transverse geometry can be realized by using the same orientation of the electric field but choosing the wave propagation along the y'-axis. The phase difference between the polarization components parallel to z and x'-axes, respectively, is then $2\pi \left[(n_{\mathrm{o}} - n_{\mathrm{e}}) - n_{\mathrm{o}}^3 r_{63} E_3^{\mathrm{dc}}/2 \right] l/\lambda_0$. In waveguides, with transverse dimensions in the μm-range, one can, with an applied voltage of a few Volt, obtain a phase modulation of π within an interaction length of a few mm. Electro-optic waveguide structures (Sect. 5.3) are usually based on lithium niobate, but the principle of operation is the same.

2.3.5 Liquid Crystal Devices

Liquid crystals (LC) are liquid phases of molecules that arrange themselves spontaneously in long-range order. In the nematic phase, for example, rod-like molecules align themselves along a common direction, which can be prescribed, e.g., by a glass substrate that is treated with some sort of brushing in the desired direction.

Fig. 2.37 Alignment of an
LC-molecule in an electric
field

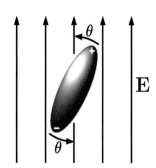

The principle of operation of LC-devices relies on the anisotropic molecular polarizability that gives rise to an anisotropic permeability of the nematic phase; a liquid crystal film can thus be understood as a thin uniaxial slab with the optical axis parallel to the long axis of the molecules. Light propagating normal to the film plane has two polarization eigenstates, one parallel to the optical axis, with refractive index n_o, the other one normal to it with refractive index n_e. An LC film of thickness d thus acts as a linear retarder with a phase difference of $2\pi(n_o - n_e)d/\lambda_0$.

A (quasi-static) electric field E normal to the glass substrate exerts a torque on the molecules that tends to align them parallel to the field (Fig. 2.37), even if the molecules do not have an intrinsic dipole moment: molecules that are (because of their thermal motion) oriented slightly out of plane experience a field component along their axis, which induces a dipole moment proportional to the field; as the torque is proportional to the field *and* to the dipole moment, it scales with the *square* of the field. Depending on the resulting average angle $\theta(E)$ of the molecular axis in respect to the surface normal, the propagation index of light polarized along the initial direction of the molecules is then given by Eq. (2.147)

$$\frac{1}{n^2(U)} = \frac{\cos^2\theta(U)}{n_o^2} + \frac{\sin^2\theta(U)}{n_e^2}.\tag{2.162}$$

The resulting refractive index anisotropy can therefore adjusted between the values $n_o - n_e$ and 0 (Fig. 2.38).

Arranged between two polarizers, or a polarizer and a mirror, such LC films can operate as optical shutters or modulators, similar to a Pockels cell; since the response is independent of the sign of the electric field, however, it rather resembles the (quadratic) electro-optic Kerr effect. As the variation of the refractive properties of LC films requires the rotation of molecules in a viscous environment, it is much slower than the Kerr effect, which is of purely electronic nature.

The most important application of such electro-optic LC-devices is in the field of display technology, where an array of small pixels of LCs is controlled by localized electrodes (see, e.g., Lueder 2010). In combination with polarizers, they act as spatially resolved transmittance modulators; in certain photonic applications,

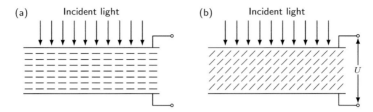

Fig. 2.38 Liquid crystal (LC) film between transparent electrodes: (**a**) without applied voltage, (**b**) with applied (AC) voltage

they are also employed as phase modulators that can modify the phase front of transmitted light.

2.4   Other Propagation Effects

2.4.1   Optical Activity

Optical activity is a manifestation of circular birefringence: for a given wave vector, there are two circularly polarized eigenstates $\sigma^{\pm}$ with different propagation indices $n^{\pm}$. An optically active medium of thickness d is therefore a circular retarder inducing a phase difference of

$$\Delta\phi_V = 2\pi(n^- - n^+)\frac{d}{\lambda_0} \tag{2.163}$$

between the two eigenstates. According to Eq. (1.124), such a retarder rotates any input state by

$$\varphi = \pi(n^+ - n^-)\frac{d}{\lambda_0}. \tag{2.164}$$

We restrict our discussion of optical activity to isotropic media; additional linear birefringence is possible in anisotropic media, and can be observed, for example, in crystalline quartz. The symmetry requirement for optical activity is, as we shall see, the lack of centrosymmetry; an example for an isotropic, optically active medium is a liquid solution of chiral molecules such as dextrose. The microscopic origin of optical activity is the rotating current induced in a molecule by the oscillating $\mathbf{B}$-field of the light. In chiral molecules (that structurally resemble a helix), this (helical) current produces an electric dipole moment proportional to $\partial\mathbf{B}/\partial t = -\nabla \times \mathbf{E}$; the total electric displacement density then has the form

$$\mathbf{D} = \varepsilon_0 \varepsilon \mathbf{E} + \varepsilon_0 \xi \nabla \times \mathbf{E}. \tag{2.165}$$

The appearance of the ∇-operator in this equation implies that the response of the medium in a point $\mathbf{x}$ does not depend only on the electric field in this point but also on the field in the neighborhood—just as the response at time t generally depends on the field in the past—in other words, the response of the medium is nonlocal. The general expression for the electric displacement density including nonlocal contributions is, in Einstein notation,

$$D_i = \varepsilon_0 \varepsilon_{ij} E_j + \varepsilon_0 \xi_{ijk} \nabla_j E_k \tag{2.166}$$

where $\boldsymbol{\xi}$ is a material property that relates three vectors—∇, $\mathbf{E}$, and $\mathbf{D}$, and is therefore a third rank tensor. In isotropic, centrosymmetric media, all components of $\boldsymbol{\xi}$ must satisfy $\xi_{ijk} = (-1)^3 \xi_{ijk}$ and are consequently zero; in non-centrosymmetric isotropic media, symmetry requires $\xi_{ijk} = \epsilon_{ijk} \xi$, where ϵ_{ijk} is the permutation symbol. Thus, $\xi_{123} = \xi_{231} = \xi_{312} = -\xi_{132} = -\xi_{213} = -\xi_{321} = \xi$, while all remaining components are zero. For a plane wave with wave vector $\mathbf{k}$, Eq. (1.33) allows us to replace ∇ with $-\mathbf{j}\mathbf{k}$; in isotropic media, we can, without loss of generality, choose the z- axis as propagation direction, so that $\mathbf{k} = [0, 0, k]$. Equation (2.166) then assumes the form

$$\mathbf{D} = \varepsilon_0 \begin{bmatrix} \varepsilon & -\mathbf{j}\xi k & 0 \\ \mathbf{j}\xi k & \varepsilon & 0 \\ 0 & 0 & \varepsilon \end{bmatrix} \mathbf{E}. \tag{2.167}$$

Substituting this result in wave equation Eq. (2.130), we obtain the eigenvalue equation $\boldsymbol{M}\mathbf{E} = \mathbf{0}$ with

$$\boldsymbol{M} = \begin{bmatrix} n^2 - \varepsilon & -\mathbf{j}\xi k & 0 \\ \mathbf{j}\xi k & n^2 - \varepsilon & 0 \\ 0 & 0 & \varepsilon \end{bmatrix}. \tag{2.168}$$

From $\det \boldsymbol{M} = 0$ we obtain the eigenvalues

$$n^{\pm} = \sqrt{\varepsilon \pm \xi k} \approx n_0 \pm \frac{\xi k}{2 n_0}, \tag{2.169}$$

with $n_0 = \sqrt{\varepsilon}$, corresponding to circularly polarized eigenstates $\sigma^{\pm}$. The resulting rotation angle is given by Eq. (2.164).

Analog to the case of anisotropic media, the two eigenstates are usually refracted in different directions at an interface of an optically active medium; circular polarizers can be realized by exploiting the different critical angles of total reflection at an interface.

2.4.2 Magneto-Optic Faraday Effect

Optical activity can be understood as resulting from the magnetic component of the
electromagnetic light wave; we now want to consider the effect of a static, external
magnetic field with flux density $\mathbf{B}^{\text{ext}}$ on wave propagation; similar to the treatment
of the electro-optic effect (Sect. 8.4.1) we start by expanding the permittivity as a
function of $\mathbf{B}^{\text{ext}}$:

$$\varepsilon_{ij}(\mathbf{B}^{\text{ext}}) = \varepsilon_{ij}^0 + \gamma_{ijk}B_k^{\text{ext}} + \dots, \tag{2.170}$$

where γ_{ijk} are the components of the magneto-optic tensor. Unlike the electro-
optic tensor or the nonlocal permittivity tensor ξ_{ijk}, the magneto-optic tensor does
not vanish in centrosymmetric media, because $\mathbf{B}^{\text{ext}}$, in contrast to $\mathbf{D}$ and $\mathbf{E}$, is a
pseudo-vector that does not change sign under inversion (this can by understood by
considering the fact that a circular loop current which produces a magnetic field is
also invariant under inversion). Isotropy requires $\gamma_{ijk} = -j\epsilon_{ijk}\gamma$, where ϵ_{ijk} is the
permutation symbol; in lossless media, ε must be Hermitian ($\varepsilon_{ij} = \varepsilon_{ji}^*$) for reasons
of energy conservation, so that γ_{ijk} must be imaginary.

Assuming $\mathbf{B}^{\text{ext}}$ to be parallel to the z-axis, the resulting electric displacement
density is

$$\mathbf{D} = \varepsilon_0 \begin{bmatrix} \varepsilon & -j\gamma B^{\text{ext}} & 0 \\ j\gamma B^{\text{ext}} & \varepsilon & 0 \\ 0 & 0 & \varepsilon \end{bmatrix} \mathbf{E}. \tag{2.171}$$

For propagation in the z-direction, this relation resembles Eq. (2.167)—with similar
consequences for the wave propagation: the eigenstates are circularly polarized $\sigma^\pm$,
with the propagation indices [compare Eq. (2.169)]

$$n^\pm = \sqrt{\varepsilon \pm \gamma B^{\text{ext}}} \approx n_0 \pm \frac{\gamma B^{\text{ext}}}{2n_0}. \tag{2.172}$$

According to Eq. (2.164), the Faraday effect results in a rotation of the input
polarization state by the angle

$$\varphi = \frac{\pi\gamma}{n_0\lambda_0}B^{\text{ext}}d =: VB^{\text{ext}}d, \tag{2.173}$$

where V is the material specific and frequency dependent Verdet constant. Typ-
ical values for glasses are on the order of $1\,\text{rad}\,\text{T}^{-1}\,\text{m}^{-1}$ at $1\,\mu\text{m}$ wavelength;
selected (paramagnetic) rare earth doped materials show values of more than
$-100\,\text{rad}\,\text{T}^{-1}\,\text{m}^{-1}$ at $1\,\mu\text{m}$ wavelength.

2.4.2.1 Faraday Isolator

While the polarization rotation (which is defined in respect to the propagation direction) in isotropic optically active materials is independent of the direction of **k**, the Faraday rotation depends on the orientation of **k** in respect to $\mathbf{B}^{\mathrm{ext}}$: if the propagation direction is reversed, the rotation also changes sign. The difference between the two effects becomes particularly obvious if we consider the combination of an optically active or magneto-optic medium, respectively, and a mirror, and look at the polarization state of the light reflected by this setup. The corresponding Jones matrix in the case of an optically active medium is

$$
\begin{bmatrix} \cos\varphi & -\sin\varphi \\ \sin\varphi & \cos\varphi \end{bmatrix} \begin{bmatrix} 1 & 0 \\ 0 & -1 \end{bmatrix} \begin{bmatrix} \cos\varphi & -\sin\varphi \\ \sin\varphi & \cos\varphi \end{bmatrix} = \begin{bmatrix} 1 & 0 \\ 0 & -1 \end{bmatrix}, \tag{2.174}
$$

which is equivalent to simple reflection; the rotation due to optical activity is compensated. Using a magneto-optic medium instead, the matrix is

$$
\begin{bmatrix} \cos(-\varphi) & -\sin(-\varphi) \\ \sin(-\varphi) & \cos(-\varphi) \end{bmatrix} \begin{bmatrix} 1 & 0 \\ 0 & -1 \end{bmatrix} \begin{bmatrix} \cos\varphi & -\sin\varphi \\ \sin\varphi & \cos\varphi \end{bmatrix} = \begin{bmatrix} 1 & 0 \\ 0 & -1 \end{bmatrix} \begin{bmatrix} \cos 2\varphi & \sin 2\varphi \\ \sin 2\varphi & \cos 2\varphi \end{bmatrix}, \tag{2.175}
$$

that is a rotation by 2φ, followed by a reflection. In combination with a linear input polarizer, a Faraday rotator with a rotation of $\varphi = 45°$ per pass blocks reflections completely, acting as so-called Faraday isolator (Fig. 2.39).

2.4.2.2 Drude–Lorentz Model of the Faraday Effect

We now want to understand the Faraday effect within the Drude–Lorentz model; for this purpose we supplement the force term in the equation of motion Eq. (2.51) with the Lorentz force $-e(\mathbf{v} \times \mathbf{B}^{\mathrm{ext}})$ that acts on an electron moving at velocity **v** in a magnetic field $\mathbf{B}^{\mathrm{ext}}$

$$
m_e \ddot{\mathbf{x}} + b\dot{\mathbf{x}} + a\mathbf{x} = -e\left[\mathbf{E}(t) + \dot{\mathbf{x}} \times \mathbf{B}^{\mathrm{ext}}\right]; \tag{2.176}
$$

as above, we assume that $\mathbf{B}^{\mathrm{ext}}$ and **k** are parallel to $\mathbf{e} = [0, 0, 1]$. We use complex amplitudes in the following and choose a circularly polarized base

$$
\tilde{\mathbf{E}}^{\pm} = \tilde{E}^{\pm}\sigma^{\pm} \tag{2.177}
$$

$$
\tilde{\mathbf{x}}^{\pm} = \tilde{x}^{\pm}\sigma^{\pm}, \tag{2.178}
$$

where the base vectors $\sigma^{\pm} = [1, \pm j, 0]\frac{1}{\sqrt{2}}$ have the property $\sigma^{\pm} \times \mathbf{e} = \pm j\sigma^{\pm}$. Equation (2.52) then becomes

$$
\tilde{\mathbf{x}}^{\pm}(\omega, B^{\mathrm{ext}}) = \frac{-e/m_e}{(\omega_0^2 - \omega^2) + j\omega\Gamma \mp \omega(e/m_e)B^{\mathrm{ext}}} \tilde{\mathbf{E}}^{\pm}, \tag{2.179}
$$

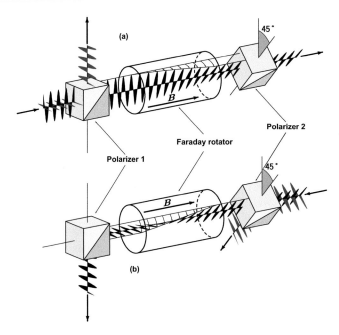

Fig. 2.39 Nonreciprocal transmittance of a Faraday isolator: (**a**) forward propagation, (**b**) backward propagation

and the susceptibility for circularly polarized light is given by the correspondingly modified Eq. (2.56)

$$\tilde{\chi}^{\pm}(B^{\text{ext}}) = \frac{n_e e^2}{\varepsilon_0 m_e} \frac{1}{[(\omega_0^2 - \omega^2) + j\omega\Gamma] \mp \omega(e/m_e)B^{\text{ext}}}. \tag{2.180}$$

This expression has the form $\frac{a}{b \mp x}$, which can be approximated by $\frac{a}{b} \pm \frac{a}{b}\frac{x}{b}$ for $x \ll b$. Since the Lorentz term is much smaller than the other terms in the denominator, we can write

$$\tilde{\chi}^{\pm}(B^{\text{ext}}) \approx \tilde{\chi}(0) \pm FB^{\text{ext}} = \tilde{\chi}(0) \pm \Delta\tilde{\chi}(B^{\text{ext}}), \tag{2.181}$$

where

$$F = \omega\frac{e}{m_e}\frac{1}{[\omega_0^2 - \omega^2 + j\omega\Gamma]}\tilde{\chi}(0) \approx \frac{\omega e}{\omega_0^2 m_e}\tilde{\chi}(0); \tag{2.182}$$

the approximation is valid well below the resonance, $\omega \ll \omega_0$. With $\Delta\tilde{\chi} = \Delta\tilde{\varepsilon}$, the permittivity tensor in the circular base is

$$\boldsymbol{\varepsilon} = \begin{bmatrix} \varepsilon + F(\omega)B^{\text{ext}} & 0 & 0 \\ 0 & \varepsilon - F(\omega)B^{\text{ext}} & 0 \\ 0 & 0 & \varepsilon \end{bmatrix}_c ; \qquad (2.183)$$

transformation into the linear base is obtained by Eq. (1.122), resulting in

$$\boldsymbol{\varepsilon} = \begin{bmatrix} \varepsilon & -jF(\omega)B^{\text{ext}} & 0 \\ jF(\omega)B^{\text{ext}} & \varepsilon & 0 \\ 0 & 0 & \varepsilon \end{bmatrix}_l ; \qquad (2.184)$$

comparison with Eq. (2.171) allows us to identify

$$\gamma = F(\omega) \qquad (2.185)$$

and to calculate the Verdet constant according to Eq. (2.173), using Eq. (2.182) and $\lambda_0 = 2\pi c_0/\omega$

$$V = \frac{\pi F(\omega)}{n_0 \lambda_0} = \left(\frac{\omega}{\omega_0}\right)^2 \frac{e}{m_e c_0} \frac{\tilde{\chi}(0)}{2 n_0}. \qquad (2.186)$$

The susceptibility of glass at a wavelength of 1 µm is about $\tilde{\chi} = n^2 - 1 = 1.25$; if we assume the resonance to be in the UV ($\omega/\omega_0 \approx 10^{-1}$), Eq. (2.186) yields a Verdet constant of about $1 \text{ rad T}^{-1} \text{ m}^{-1}$, in surprisingly good agreement with experimental values.

2.4.3   Wave Propagation in Moving Media*

A number of optical phenomena rely on effects that derive from the relative motion of source, medium, and/or detector. The treatment of these effects is provided by the theory of relativity and relies on the postulate that the vacuum velocity of light is identical in two reference systems that move relative to each other at a constant velocity. It is convenient to use a four-dimensional space time reference frame to describe relativistic phenomena, where a point is given by the four-vector

$$\mathbf{X} = \begin{bmatrix} x \\ y \\ z \\ jc_0 t \end{bmatrix}. \qquad (2.187)$$

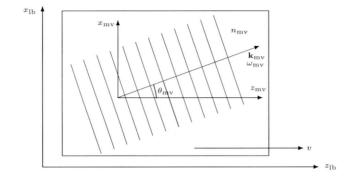

Fig. 2.40 A light wave propagating in a medium with optical density n_{mv} that moves at a velocity v relative to the observer; in the moving frame, the wave has the frequency ω_{mv} and the wave vector $\mathbf{k}_{mv}$; the observer in the lab reference frame measures ω_{lb} and $\mathbf{k}_{lb}$

Let us assume a medium that is, measured in our lab reference frame, moving along the z-axis at velocity v. A point with coordinates $\mathbf{x}_{mv}$ in the moving reference system has the coordinates $\mathbf{x}_{lb}$ in the lab system, with $\mathbf{x}_{mv}$ and $\mathbf{x}_{lb}$ related by the Lorentz transformation (see, e.g., Jackson 1999)

$$\mathbf{x}_{mv} = \mathbf{L}\mathbf{x}_{lb} \tag{2.188}$$

where

$$\mathbf{L} = \begin{bmatrix} 1 & 0 & 0 & 0 \\ 0 & 1 & 0 & 0 \\ 0 & 0 & \frac{1}{\sqrt{1-(v/c_0)^2}} & j\frac{(v/c_0)}{\sqrt{1-(v/c_0)^2}} \\ 0 & 0 & -j\frac{(v/c_0)}{\sqrt{1-(v/c_0)^2}} & \frac{1}{\sqrt{1-(v/c_0)^2}} \end{bmatrix}. \tag{2.189}$$

Consider a plane wave Eq. (1.26) that propagates with the phase velocity c_0/n_{mv} in the moving medium (Fig. 2.40); in four-vector notation, the wave function has the form

$$\tilde{\mathbf{E}}(\mathbf{x}_{mv}) = \tilde{\mathbf{E}}e^{-j\mathbf{k}_{mv}\mathbf{x}_{mv}}, \tag{2.190}$$

where

$$\mathbf{k}_{mv} = \frac{\omega_{mv}}{c_0}\left[n_{mv}\sin\theta_{mv}, 0, n_{mv}\cos\theta_{mv}, j \right] \tag{2.191}$$

is the wave four-vector and θ_{mv} is the angle between wave vector and z-axis.

For an observer in the lab reference frame, the wave has the frequency ω_{lb} and the wave vector $\mathbf{k}_{\text{lb}}$

$$\mathbf{k}_{\text{lb}} = \frac{\omega_{\text{lb}}}{c_0} \left[n_{\text{lb}} \sin \theta_{\text{lb}}, 0, n_{\text{lb}} \cos \theta_{\text{lb}}, j \right].$$

(2.192)

Since $\mathbf{x}_{\text{mv}}$ and $\mathbf{x}_{\text{lb}}$ describe identical points, the respective phase of the wave must be the same, $\mathbf{k}_{\text{mv}}\mathbf{x}_{\text{mv}} = \mathbf{k}_{\text{lb}}\mathbf{x}_{\text{lb}} = \mathbf{k}_{\text{mv}}\mathbf{L}\mathbf{x}_{\text{lb}}$, so that the wave four-vector in the lab system is

$$\mathbf{k}_{\text{lb}} = \mathbf{k}_{\text{mv}}\mathbf{L}.$$

(2.193)

Comparison of the components of the vectors on both sides of this equation allows us to extract the wave parameters in the lab system; for the frequency in the lab frame, we obtain

$$\omega_{\text{lb}} = \omega_{\text{mv}} \frac{1 + n_{\text{mv}}(v/c_0) \cos \theta_{\text{mv}}}{\sqrt{1 - (v/c_0)^2}},$$

(2.194)

the phase velocity is c_0/n_{lb} with

$$n_{\text{lb}} = \sqrt{1 + \frac{(n_{\text{mv}}^2 - 1)[1 - (v/c_0)^2]}{[1 + n_{\text{mv}}(v/c_0) \cos \theta_{\text{mv}}]^2}},$$

(2.195)

and the direction of the wave vector in respect to the z-axis is

$$\tan \theta_{\text{lb}} = \frac{n_{\text{mv}} \sqrt{1 - (v/c_0)^2} \sin \theta_{\text{mv}}}{n_{\text{mv}} \cos \theta_{\text{mv}} + (v/c_0)}.$$

(2.196)

The dependence of the frequency on the relative velocity is known as Doppler effect; an important case is $\theta_{\text{mv}} = 0$ (longitudinal Doppler effect); if $n_{\text{mv}} = 1$, we obtain

$$\omega_{\text{lb}} = \omega_{\text{mv}} \sqrt{\frac{1 + v/c_0}{1 - v/c_0}} \approx \omega_{\text{mv}}(1 + v/c_0),$$

(2.197)

where the approximation is valid for $|v/c_0| \ll 1$. A frequency shift is also observed if the propagation direction in the moving frame is orthogonal to the motion of the system ($\theta_{\text{mv}} = 90°$): this transverse Doppler-shift is given by

$$\omega_{\text{lb}} = \omega_{\text{mv}} \frac{1}{\sqrt{1 - (v/c_0)^2}}.$$

(2.198)

The propagation index of a medium moving parallel to the wave vector, measured in the lab frame, finally is

$$n_{lb} = \frac{n_{mv} + v/c_0}{n_{mv}v/c_0 + 1} \tag{2.199}$$

and the phase velocity in the lab frame amounts to

$$v_{lb} = c_0/n_{lb} = \frac{c_0/n_{mv} + v}{1 + v/n_{mv}c_0} \approx c_0/n_{mv} + v(1 - 1/n_{mv}^2); \tag{2.200}$$

the term $1 - 1/n_{mv}^2$ is known as Fresnel's drag coefficient.

2.5   Summary

The electric field of a light wave induces oscillations of the electrons in matter, and the oscillating electrons generate an electromagnetic wave: this is the core of the classical theory of light–matter interaction. Reflection and refraction at boundaries, absorption, dispersion, birefringence, optical activity, or the magneto-optic effect are examples for the enormous variety of optical phenomena that result from this interplay. The fundamental response of matter to a light wave is described by the polarization density; the relation between electric field and polarization density is provided by the susceptibility.

The optical polarization density is a wave, and coupling of incident and outgoing waves, for example, at an interface, requires phase matching between polarization wave and electromagnetic wave. The laws of reflection and refraction at a smooth interface, and that of diffraction at a periodic grating follow immediately from this condition.

Mathematically, the representation of harmonically oscillating real quantities by complex amplitudes turns out to be extremely advantageous; provided that the response of a medium to the electric field is linear, the theory of linear systems can be utilized to describe light–matter interactions. The susceptibility, for example, can be understood as a complex transfer function, with the imaginary part being responsible for the energy transfer between light field and matter, and the real part essentially determining the propagation velocity of the light wave. The Kramers–Kronig relations are a special case of the Hilbert transformation that relates the real and imaginary part of the transfer function of a causal system and constitute an important tool for the analysis of photonic elements.

The response of the electrons can be modeled, with astonishing success, by a simple harmonic oscillator (Drude–Lorentz model). The restoring force and the damping constant determine the magnitude, resonance frequency, and phase behavior of the susceptibility. A possible anisotropy of the restoring force results in optical birefringence. In centrosymmetric lossless media, the susceptibility tensor can always be represented by a diagonal matrix and the eigenstates of propagation are linearly polarized; in non-centrosymmetric media, additional nonlocal terms can lead to imaginary off-diagonal elements, giving rise to optical activity. The presence of a static magnetic field has a similar effect, exhibited by the Faraday rotation of the transmitted polarization state.

The basic optical response of metals can also be understood in the framework of the Drude-Lorentz model as that of media without restoring force; in particular, the very high reflectivity of metals up to the plasma frequency, and a phase velocity exceeding c_0 are correctly predicted by the model.

Finally, some consequences of special relativity on the optical properties of moving media, as observed from a system at rest, are derived. The results are of particular interest for sensing applications, but also for Doppler effects in laser media.

2.6   Problems

1. Assume a plane wave incident on a surface at an oblique angle of incidence; what is the velocity of the phase fronts, measured in the surface plane?
2. Assume a stack of plane parallel plates of index n_2 to n_{j-1} between two media with n_1 and n_j. Calculate the angle of transmission into medium j as a function of the angle of incidence in medium 1.
3. Calculate Brewster's angle for $n_i > n_t$ by finding the angle of incidence where $r_\pi = 0$. If it exists, compare it to Brewster's angle for reverse propagation and check whether the two angles are related by Snell's law.
4. A line grating can be used to "retroreflect" light incident under an oblique angle into itself (in a laser resonator, for example). For a given wavelength and grating period, calculate the angular condition for retroreflection. Also calculate the derivative of the retroreflected wavelength with respect to the angle of incidence; derive the spectral resolution of the "grating mirror" in terms of the spread of the angle of incidence. In Chap. 3, we will see that the angular spread of a collimated light field is related to its transverse dimensions; use Eq. (3.19) to express the spectral resolution as a function of the incident beam diameter.
5. For σ-polarized light, the total electric field $E(z)$ and its derivative dE/dz must be continuous at the interface between two media (z is normal to the interface): (a) derive this statement from Maxwell's equations; (b) based upon these boundary conditions, calculate the phase shift (the complex reflection coefficient) of totally reflected σ-polarized light; plot the field amplitude in a manner similar to Fig. 2.16.
6. Assume a material having a volumetric heat capacity of $2 \times 10^6 \, \mathrm{J\,m^{-3}\,K^{-1}}$; further assume a 1 ns long laser pulse of 100 mJ energy, focussed to a circular area of 1 mm diameter, being completely absorbed by the medium within an absorption length of $20 \, \mu\mathrm{m}$. Neglecting heat diffusion, what is the temperature increase of the irradiated volume of the medium? The radiation pressure

associated with the electromagnetic field is I/c_0; calculate the peak pressure on the irradiated medium.

7. Looking into a fish tank with vertical windows, what is the maximum angle in respect to the window surface normal under which you can observe total reflection at the horizontal water surface? Take the influence of the window glass (refractive index 1.6) into account. If you lie on your back on the floor of a pool, with your diving goggle glasses horizontal, can you observe total reflection at the water surface? Can you see people sitting at the pool side?

8. The complex refractive index of silicon (silver) at a wavelength of 500 nm is $4.298-0.073j$ $(0.050-3.13j)$. Calculate the reflectance at 45° angle of incidence for π and σ polarized light. Calculate the absorption length under this angle of incidence (vertical penetration depth) using the imaginary part of the normal component of the transmitted wave vector.

9. Design a Glan–Thompson prism from calcite (orientation and cutting angles).

10. Design a Fresnel rhomb from a glass with refractive index of 1.6.

References and Suggested Reading

Born, M., & Wolf, E. (1999). *Principles of optics*. New York: Cambridge University Press.

Haus, H. A. (1984). *Waves and fields in optoelectronics*. Englewood Cliffs, NJ: Prentice Hall.

Haussuehl, S. (2008). *Physical properties of crystals*. New York: John Wiley.

Hodgson, J. N. (1970). *Optical absorption and dispersion in solids*. New York: Springer.

Jackson, J. D. (1999). *Classical electrodynamics*. New York: Wiley.

Klein, M. V., & Furtak, T. E. (1986). *Optics*. New York: John Wiley.

Kojima, N., & Sugano, S. (2000). *Magneto-optics* (Vol. 128). Berlin: Springer.

Lipson, S. G., & Lipson, H. (1969). *Optical physics*. London: Cambridge University Press.

Lucarini, V., Peiponen, K. E., Saarinen, J. J., & Vartiainen, E. M. (2005). *Kramers-Kronig relations in optical materials research*. New York: Springer.

Lueder, E. (2010). *Liquid crystal displays: Addressing schemes and electro-optical effects*. New York: John Wiley & Sons.

Nye, J. F. (1985). *Physical properties of crystals*. New York: Oxford University Press.

Palik, E. D. (1997). *Handbook of optical constants of solids*. New York: Academic Press.

Petit, R., (Ed.). (1980). *Electromagnetic theory of gratings*. New York: Springer.

Saleh, B. E., & Teich, M. C. (2007). *Fundamentals of photonics*. New York: Wiley.

Toll, J. S. (1956). Causality and the dispersion relation: logical foundations. *Physical Review, 104*(6), 1760.

Wooten, F. (1972). *Optical properties of solids*. Amsterdam: Elsevier.

Yariv, A., & Yeh, P. (1983). *Optical waves in crystals*. New York: John Wiley.

Optical Beams and Pulses

3

The treatment of optical wave propagation given so far was restricted to monochromatic plane waves. These simple solutions allow us to study effects such as reflection and refraction or propagation in birefringent media. Plane waves are, however, somewhat unrealistic because they extend over the entire space, with a constant amplitude, and carry infinite total energy; monochromatic waves, moreover, extend over infinite times.

Optical beams and pulses are electromagnetic waves concentrated in space and/or time; they have finite energy content and can be produced by optical sources such as lasers. The following discussion relates to the propagation of coherent beams and pulses, that are characterized by completely controlled spatial and temporal phase.

3.1 Beam Propagation

3.1.1 Paraxial Wave Equation

As we have seen earlier, the time- and space dependence of an optical wave function can be treated separately. In the following description of optical *beams*, we will assume a harmonic (monochromatic) time dependence. One way to construct a beam-like wave function is to multiply a plane carrier wave with a transverse profile function $A(\mathbf{x})$ that is concentrated along an axis (parallel to the wave vector of the carrier wave) and falls off rapidly with increasing distance from the beam axis (which we identify with the z-axis)

$$\mathbf{E}(\mathbf{x}, t) = A(\mathbf{x})\mathbf{n}e^{-j(kz-\omega t)}; \qquad (3.1)$$

© Springer International Publishing Switzerland 2016
G.A. Reider, *Photonics*, DOI 10.1007/978-3-319-26076-1_3

n is a unit vector defining the polarization state of the wave.[1] Substituting this ansatz into the scalar Helmholtz equation Eq. (1.23), we obtain the differential equation

$$\nabla^2 A - 2jk\frac{\partial A}{\partial z} = 0 \tag{3.2}$$

for $A(\mathbf{x})$. We assume that $A(x, y, z)$ changes only slowly along z on the scale of a wavelength, so that $|(\partial A/\partial z)| \ll 2\pi|A|/\lambda$ and

$$\left|\frac{\partial\left(\frac{\partial A}{\partial z}\right)}{\partial z}\right| \ll \frac{2\pi}{\lambda}\left|\frac{\partial A}{\partial z}\right|. \tag{3.3}$$

Under this so-called slowly varying envelope approximation, we can neglect $(\partial^2 A/\partial z^2)$ in comparison to $k(\partial A/\partial z) = (2\pi/\lambda)(\partial A/\partial z)$ and obtain the paraxial Helmholtz equation

$$\nabla_T^2 A - 2jk\frac{\partial A}{\partial z} = 0 \tag{3.4}$$

where $\nabla_T^2 := \partial^2/\partial x^2 + \partial^2/\partial y^2$. This equation shows immediately that in a homogeneous medium, where k is constant, the transverse variation of the amplitude entails an axial variation and vice versa. In general, the transverse profile changes its shape during propagation, and the profiles at two distant points of propagation may show hardly any similarity. Some selected profiles, however, are conserved during propagation and only change their spatial extension. One of these profiles is the Gaussian profile $|A| \propto \exp[(x^2 + y^2)/w^2(z)]$. As we shall see in Sect. 3.1.6, propagation over a large distance converts an initial profile into its Fourier transform; the Gaussian profile is one out of the set of functions that are similar to their Fourier transform.

3.1.2 Gaussian Beams

The wave function of a Gaussian beam—i.e., a beam with a Gaussian profile—can be obtained from a paraxial approximation of a spherical wave $(A_0/|\mathbf{x}|)\,e^{-jk|\mathbf{x}|}e^{j\omega t}$, combined with a complex coordinate transformation of the z-component (an alternative derivation of the wave function will be given in Sect. 3.1.6). The phase

---

[1] In general, **n** must also be a function of **x** for Eq. (3.1) to be a solution of the Helmholtz equation; here, we neglect this fact and assume that $\mathbf{n} \perp \mathbf{k}$, so that Eq. (3.1) describes actually the transverse component of the field (which we assume to dominate). If one interprets Eq. (3.1) as vector potential, one obtains exact solutions for the fields (Haus 1984).

Fig. 3.1 Spherical wave: surfaces of constant phase

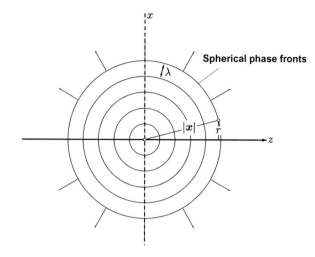

Spherical phase fronts

fronts ($k|\mathbf{x}| = $ const.) of a spherical wave are concentric spheres (Fig. 3.1) that expand (or collapse) with the phase velocity $v_{\mathrm{ph}} = \omega/k$. The prefactor $1/|\mathbf{x}|$ reduces the amplitude so that the total power flowing through a given sphere is conserved. Looking for a beam-like solution, we approximate the term $|\mathbf{x}| = \sqrt{x^2 + y^2 + z^2}$ in the vicinity of the z-axis ($x^2 + y^2 \ll z^2$) using $\sqrt{1+u} \approx 1 + u/2$

$$|\mathbf{x}| = z\sqrt{1 + \frac{r^2}{z^2}} \approx z + \frac{r^2}{2z}, \qquad (3.5)$$

where $r := \sqrt{x^2 + y^2}$ is the distance from the z-axis. The spatial part of the wave function is then

$$\frac{A_0}{|\mathbf{x}|} e^{-jk|\mathbf{x}|} \approx \frac{A_0}{z} e^{-jkr^2/2z} e^{-jkz}, \qquad (3.6)$$

where the second order expansion was used only in the highly sensitive phase term. This coincides with the carrier wave ansatz Eq. (3.1) with the amplitude function

$$A(\mathbf{x}) = \frac{A_0}{z} e^{-jkr^2/2z}; \qquad (3.7)$$

it is easy to verify that this amplitude function satisfies the paraxial Helmholtz equation Eq. (3.4).

The singularity at $z = 0$ can be removed by the complex transformation (Kogelnik and Li 1966)

$$z \to q := z + \mathrm{j}z_0; \tag{3.8}$$

the meaning of z_0 will become obvious immediately. To analyze the resulting amplitude function

$$A(\mathbf{x}) = \frac{A_0}{q} \mathrm{e}^{-\mathrm{j}kr^2/2q}$$

$$= \frac{A_0}{z + \mathrm{j}z_0} \exp\left[-\mathrm{j}\frac{kr^2}{2(z + \mathrm{j}z_0)}\right]; \tag{3.9}$$

we split $1/q$ into its real and imaginary part

$$\frac{1}{q} = \frac{1}{z + \mathrm{j}z_0} = \frac{z - \mathrm{j}z_0}{z^2 + z_0^2} =: \frac{1}{R} - \mathrm{j}\frac{2}{kw^2(z)}, \tag{3.10}$$

where $w(z)$ and $R(z)$ are given by

$$w^2(z) = w_0^2\left[1 + \left(\frac{z}{z_0}\right)^2\right] \tag{3.11}$$

with

$$w_0^2 = 2z_0/k, \tag{3.12}$$

and

$$R(z) = z\left[1 + \left(\frac{z_0}{z}\right)^2\right]. \tag{3.13}$$

Thus, the exponential term

$$\mathrm{e}^{-\mathrm{j}kr^2/2q} = \exp\left[-\frac{r^2}{w^2(z)}\right]\exp\left[-\mathrm{j}k\frac{r^2}{2R(z)}\right] \tag{3.14}$$

in Eq. (3.9) is the product of a Gaussian amplitude profile of width w and a parabolic phase term that represents an approximately spherical with curvature $1/R$. The prefactor A_0/q in Eq. (3.9) can be written as

$$\frac{A_0}{z + jz_0} = A_0' \frac{w_0}{w} e^{j\xi(z)}, \tag{3.15}$$

where $A_0' := A_0/jz_0$ and

$$\xi(z) = \arctan \frac{z}{z_0} \tag{3.16}$$

is a slowly varying phase term called Gouy phase that is, as we shall see, responsible for a z-dependent phase velocity in the focal region. The entire wave function (including the carrier) is then

$$a(\mathbf{x}) = A_0' \underbrace{\frac{w_0}{w(z)}}_{\text{amplitude}} \exp\underbrace{\left[-\frac{r^2}{w^2(z)}\right]}_{\text{profile}} \exp\underbrace{\left[-jk\frac{r^2}{2R(z)}\right]}_{\text{phase curvature}} \underbrace{e^{-j[kz-\xi(z)]}}_{\text{carrier}}. \tag{3.17}$$

3.1.2.1 Axial and Radial Field Distribution

The first factor in Eq. (3.17) denotes the amplitude on the axis ($r = 0$); it peaks at $z = 0$ and falls off with $1/z$ for $|z| \gg z_0$. Radially, the amplitude follows a Gauss function (second factor); the third and fourth factors are transverse and axial phase terms, respectively. The intensity distribution $I(\mathbf{x}) \propto aa^*$ is given by

$$I(\mathbf{x}) = I_0 \frac{w_0^2}{w^2(z)} e^{-2r^2/w^2(z)}, \tag{3.18}$$

where $I_0 := I(0)$ is the intensity maximum; Fig. 3.2 shows axial and transverse beam cross sections.

Because of the quadratic intensity–amplitude relation, the width of the transverse intensity profile is smaller than the amplitude profile by a factor of $\sqrt{2}$. The radial distance $w(z)$ denotes the $1/e$ ($1/e^2$) point of the amplitude (intensity) profile and is called beam radius; it is a hyperbolic function of z, with the minimum beam radius w_0 at $z = 0$ defining the so-called beam waist.

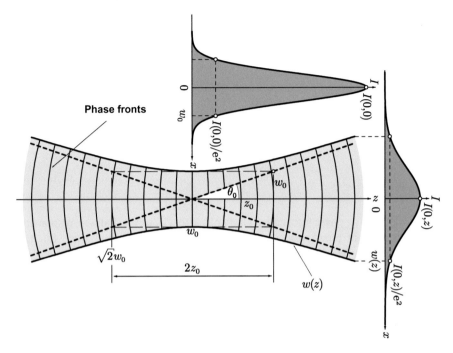

Fig. 3.2 Transverse and axial intensity profiles of a Gaussian beam; the transverse profiles are taken at $z = 0$ and $z \approx 2z_0$, respectively. The $1/e^2$-radius $w(z)$ of the profile follows a hyperbola with the axes z_0 and w_0

The axial range $|z| \leq z_0$ is called confocal range, and z_0 is called confocal parameter (frequently in the literature, z_0 is called Rayleigh range and the confocal parameter is defined as $2z_0$); compared to the peak intensity in the waist, the axial intensity drops to one half at $z = \pm z_0$. Note that z_0 and w_0 are not independent, but related by Eq. (3.12), so that only one of the two parameters can be chosen at a given wavelength. In the far-field $|z| \gg z_0$, the beam radius grows approximately linearly, $w(z) \approx z\theta_0$, where

$$2\theta_0 = 2\arctan\frac{w_0}{z_0} \approx 2\frac{w_0}{z_0} = \frac{2\lambda}{\pi w_0} = 2\sqrt{\frac{\lambda}{\pi z_0}} \qquad (3.19)$$

is the angle of divergence in [rad]; the axial intensity decreases with $1/z^2$ like that of any other light source in the far field. At a given wavelength, $w_0\theta_0 =$ const, implying a trade-off between the waist diameter and the beam divergence: well collimated beams necessarily have a waist much bigger than the wavelength, while a small beam waist implies a large divergence.

Strictly speaking, the field amplitude of a Gaussian beam does not vanish at any distance from the axis. The definition of the beam radius and diameter (and thus the beam divergence) is therefore somewhat arbitrary and a matter of convention; an important practical measure is the full width of the intensity profile at half maximum (FWHM), that can be readily related to $w(z)$

$$d_{\text{FWHM}} = w(z)\sqrt{2\ln 2}. \tag{3.20}$$

There is a variety of other measures for the beam width, for example the diameter of the circle that contains a certain percentage of the total beam power. For a Gaussian beam, all these measures are related to the beam radius $w(z)$ by some characteristic number ($\sqrt{2\ln 2}$ in the case of FWHM). As there are different definitions of the beam diameter, there are also different definitions of the beam divergence, which generally is the ratio of beam diameter to confocal parameter. The confocal parameter, however, is uniquely defined as the distance of the point of maximum phase front curvature (see below) from the beam waist.

Beams emerging from real light sources can be compared to the "ideal" beam by comparing the product of beam radius and beam divergence with that of a Gaussian beam; this ratio is called M^2-parameter and is a spatial quality measure of the light source (ISO-Standard 2005).

In a lossless medium, the power transported by the beam is independent of z and is obtained by integration over the cross section

$$P = \int_0^\infty I(r,0)\, 2r\pi\, dr = \tfrac{1}{2} I_0 \pi w_0^2; \tag{3.21}$$

this allows us to relate the peak intensity to beam power

$$I_0 = \frac{2P}{\pi w_0^2}. \tag{3.22}$$

Equation (3.18) can therefore be written as

$$I(\mathbf{x}) = \frac{2P}{\pi w^2(z)} e^{-2r^2/w^2(z)}. \tag{3.23}$$

3.1.2.2 Phase Front Curvature

The phase distribution of an optical wave function does not show up in the local intensity; nonetheless, it is a very important property that has a critical influence on the spatial evolution. The phase fronts of a Gaussian beam are determined by the factor $e^{-jkr^2/2R(z)}$ in Eq. (3.17) and are paraboloid, or approximately spherical with

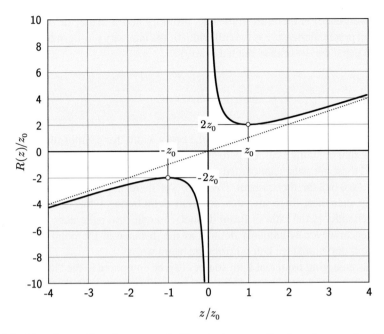

Fig. 3.3 Phase front radius of curvature of a Gaussian beam as a function of propagation distance

the curvature $1/R(z)$, where the radius $R(z)$ is given by Eq. (3.13): $|R|$ is infinite in the beam waist, drops to its minimum value at $z = z_0$, and approaches $R \approx z$ for $z \gg z_0$ (Fig. 3.3).

If the phase front radii R_1 and R_2 are given at two different axial positions, separated by a distance d, Eq. (3.13) yields the equations

$$R_1 = \frac{z^2 + z_0^2}{z} \qquad (3.24)$$

$$R_2 = \frac{(z + d)^2 + z_0^2}{z + d}, \qquad (3.25)$$

where z is the (unknown) distance of the first mirror from the beam waist; after subtraction of the two equations, we obtain

$$z = \frac{d(d - R_2)}{R_2 - R_1 - 2d} \qquad (3.26)$$

Table 3.1 Gaussian beam parameters; for expressions in terms of q, see Table 3.2

$w^2(z) = w_0^2 \left[1 + \left(\frac{z}{z_0} \right)^2 \right]$	$R(z) = z \left[1 + \left(\frac{z_0}{z} \right)^2 \right]$	$I(\mathbf{x}) = \frac{2P}{\pi w^2} e^{-2r^2/w^2}$
$w_0^2 = 2z_0/k$	$z_0 = kw_0^2/2$	$\theta_0 = \frac{\lambda}{\pi w_0} = \sqrt{\frac{\lambda}{\pi z_0}}$
$q = z + jz_0$	$\mathrm{Re}\,[1/q] = 1/R(z)$	$\mathrm{Im}\,[1/q] = -2/kw^2(z)$

and, with Eq. (3.24), the confocal parameter

$$z_0^2 = \frac{d(d - R_2)(d + R_1)(R_2 - R_1 - d)}{(R_2 - R_1 - 2d)^2}. \tag{3.27}$$

This result allows us, for example, to calculate the mode parameters of a laser resonator, where the curvature of the phase front of a mode must match that of the (spherical) mirrors.

3.1.2.3 Characteristic Parameters

Table 3.1 lists various parameters characterizing a Gaussian beam. Once the wavelength and the location of the beam waist (defining $z = 0$) are given, there is only one free parameter left; this can be w_0, z_0, or one of the parameters R, w at a particular distance z from the beam waist. The q-parameter Eq. (3.8), as a complex number, actually contains two parameters: if $q = z + jz_0$ is given at some point on the axis, the distance from the beam waist is given by the real part of q, while the confocal parameter is equal to the imaginary part

$$z = \mathrm{Re}\,[q]$$
$$z_0 = \mathrm{Im}\,[q]. \tag{3.28}$$

The local values of R and w then follow from Eq. (3.10)

$$R = \frac{1}{\mathrm{Re}\,[q^{-1}]}$$
$$w^2 = \frac{2}{k|\mathrm{Im}\,[q^{-1}]|}. \tag{3.29}$$

If, on the other hand, R and w are given at a point on the axis, Eq. (3.10) yields

$$q = \left[\frac{1}{R} - j \frac{\lambda}{\pi} \frac{1}{w^2(z)} \right]^{-1} = z + jz_0; \tag{3.30}$$

comparing the respective real and imaginary parts, we find

$$z = \frac{R}{1 + (\lambda R/\pi w^2)^2} \tag{3.31}$$

$$z_0 = \frac{\lambda R^2/\pi w^2}{1 + (\lambda R/\pi w^2)^2} \tag{3.32}$$

and, with Eq. (3.12)

$$w_0 = \frac{w}{[1 + (\pi w^2/\lambda R)^2]^{1/2}}. \tag{3.33}$$

As we will see in Sect. 3.1.4, q is a particularly valuable beam parameter that helps to simplify the treatment of beam propagation problems to a great extent.

3.1.3   Optical Components and Gaussian Beams

A Gaussian beam can be modified by optical elements and transformed into another one, with different waist location and waist radius; in this way, the output beam of a given laser can be matched to the requirements of a particular application.

3.1.3.1 Amplitude Modification

If a Gaussian beam Eq. (3.17) is transmitted through a "soft" aperture with the transmission coefficient

$$t(r) = e^{-r^2/w_a^2}, \tag{3.34}$$

the beam profile changes from e^{-r^2/w^2} in front of the aperture to e^{-r^2/w'^2} immediately behind it, where

$$\frac{1}{w'^2} = \frac{1}{w^2} + \frac{1}{w_a^2}, \tag{3.35}$$

while the phase curvature remains unchanged, $R' = R$. With Eqs. (3.31) and (3.33), the new waist location and radius can be readily calculated.

3.1.3.2 Phase Front Modification

Beam transformation by shaping the amplitude profile implies beam power losses. It is therefore more efficient to locally change the phase front curvature while leaving the local beam profile unchanged. To see how this works, we start with a (thin) plane parallel, dielectric plate of thickness d and propagation index n. A paraxial plane wave $e^{-j\mathbf{k}\cdot\mathbf{x}}$ incident on such a plate experiences a phase retardation that can be expressed by a transmission factor

$$t \approx e^{-jnk_0 d}. \tag{3.36}$$

An optical lens (Fig. 3.4) is a dielectric plate with varying thickness $d(r)$, where r is the distance from the axis. Usually, the propagation index n within the plate is constant and the surfaces are spherical (or planar). We consider a plano-convex lens (Fig. 3.5) with an axial thickness d_0 and a front surface radius of curvature R_1: in addition to the phase shift $-k_0 d_0$ that an empty slice of thickness d_0 imposes on a transmitted wave, the dielectric medium contributes the phase shift $-k_0(n-1)d(r)$. As can be seen from Fig. 3.5, the local thickness $d(r)$ is given by

$$d(r) = \sqrt{R_1^2 - r^2} - (R_1 - d_0) \approx d_0 - \frac{r^2}{2R_1}, \tag{3.37}$$

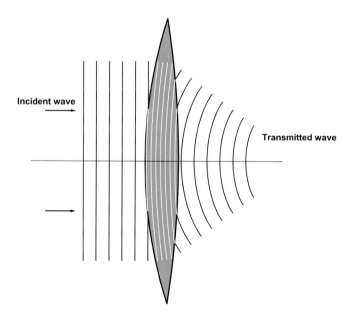

Fig. 3.4   The phase retardation by a dielectric lens induces a spherical deformation of the incident phase front

Fig. 3.5 Thin spherical
plano convex lens

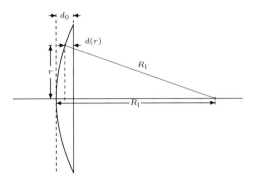

so that the phase change is approximately

$$-\left[k_0 d_0 + (n-1)k_0\left(d_0 - \frac{r^2}{2R_1}\right)\right] = -\left[d_0 n k_0 - k_0\frac{(n-1)r^2}{2R_1}\right] \qquad (3.38)$$

and can be expressed by the transmission factor

$$e^{j(n-1)k_0 r^2/2R_1} = e^{jk_0 r^2/2f}, \qquad (3.39)$$

where

$$f := \frac{R_1}{n-1} \qquad (3.40)$$

is called focal length and its reciprocal value

$$\frac{1}{f} = \frac{n-1}{R_1} \qquad (3.41)$$

is known as focusing power (measured in diopters). The constant phase factor $e^{-jd_0 n k_0}$ has been dropped in Eq. (3.39) as it has no impact on the shape of the phase front.

The assumption of a thin lens implies that the incident beam profile (with radius w) will emanate from the lens unchanged,

$$w'(z) = w(z). \qquad (3.42)$$

The phase front, however, is transformed from $e^{-jk_0 r^2/2R}$ to

$$e^{-jk_0 r^2/2R} e^{jk_0 r^2/2f} = e^{-jk_0 r^2/2R'}, \qquad (3.43)$$

constituting a new beam with the local phase front curvature

$$\frac{1}{R'} = \frac{1}{R} - \frac{1}{f};$$

(3.44)

the new beam waist follows from Eq. (3.33) to be

$$w_0' = \frac{w}{[1 + (\pi w^2/\lambda R')^2]^{1/2}}$$

(3.45)

and the distance of the new waist from the lens is, according to Eq. (3.31),

$$z' = -\frac{R'}{1 + (\lambda R'/\pi w^2)^2}$$

(3.46)

(note that z is measured from the waist, which accounts for the negative sign in this expression).

3.1.3.3 Gradient Index Lens

The radial phase change introduced by a thin phase object is given, in the paraxial approximation, by the radial variation of $k_0(n - 1)d$; in a conventional lens, n is constant and d is a function of r. The same effect on the phase front can be achieved by a phase object with constant d but varying propagation index,

$$n(r) = n_0 \left(1 - \tfrac{1}{2}\alpha_g^2 r^2\right)$$

(3.47)

(Fig. 3.6). If the thickness of such a so-called gradient index lens (GRIN-lens) is so small that the beam radius does not significantly change during the propagation, the action of a GRIN-lens on a paraxial wave function can be expressed by the transmission factor $e^{j k_0 n_0 \alpha_g^2 r^2 d/2}$ (where a constant phase factor has been dropped); comparison with Eq. (3.39) shows that the GRIN-lens has the effect of a conventional lens with the focusing power

$$\frac{1}{f_{\text{grin}}} = n_0 \alpha_g^2 d.$$

(3.48)

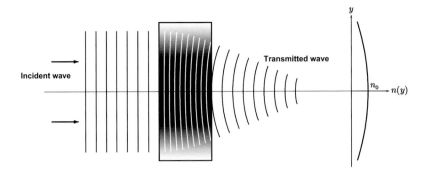

Fig. 3.6 Effect of a thin gradient index lens on an incident plane wave

3.1.3.4 Thin Kerr Lens

An interesting variant of a GRIN-lens is the so-called Kerr lens, where the radial index variation is not built in, but dynamically induced by the transmitted beam. As we shall see in Sect. 8.3.2, the propagation index depends slightly on the intensity,

$$n(I) = n_0 + n_2 I, \qquad (3.49)$$

where the coefficient n_2 is positive in most materials and on the order of several $10^{-20}\,\mathrm{m^2\,W^{-1}}$ in glasses; obviously, it takes high values of intensity to get noticeable index changes, but such intensities are available, for example, inside laser resonators or from pulsed lasers.

Using $\exp(-x^2) \approx 1 - x^2$, the radial intensity distribution Eq. (3.23) of a Gaussian beam can be approximated by

$$I(r) \approx \frac{2P}{\pi w^2}\left(1 - \frac{2r^2}{w^2}\right); \qquad (3.50)$$

the propagation index in the range $r < w$ is therefore

$$n(r) \approx n_0 + \frac{2n_2 P}{\pi w^2}\left(1 - \frac{2r^2}{w^2}\right) \approx n_0\left(1 - \tfrac{1}{2}\alpha_{\mathrm{g\,kerr}}^2 r^2\right), \qquad (3.51)$$

where the (small) term $2n_2 P/\pi w^2$ has been dropped as it does not depend on r. For a given beam power and radius, this corresponds to a GRIN-lens [Eq. (3.47)] with

$$\alpha_{\mathrm{g\,kerr}} = \frac{2}{w^2}\sqrt{\frac{2n_2 P}{\pi n_0}} \qquad (3.52)$$

and a focusing power of

$$\frac{1}{f_{\text{kerr}}} = \frac{8n_2 P d}{w^4 \pi}. \tag{3.53}$$

At sufficiently high power, this effect is strong enough to focus the beam by (over)compensating its natural divergence, a phenomenon called self-focusing (Sect. 8.3.2). Inside the medium, the beam then induces a channel of increased propagation index that acts as a gradient index waveguide (Sect. 5.2.3). For an estimate of the required power, we treat the beam as a superposition of plane waves (see Sect. 3.1.6) whose wave vectors enclose an angle θ with the axis, ranging from 0 to $a\theta_0 = a\lambda/\pi w_0$, where θ_0 is the beam divergence (Fig. 3.2) and a is a beam profile dependent factor of order 1; to ensure guiding, the index increase in the channel must be large enough to provide total internal reflection for all wave vectors; according to Eq. (2.10), this is the case if the axial component $\cos\theta k_0 [n_0 + 2n_2 P/\pi w_0^2]$ is larger than the wave number $k_0 n_0$ in the surrounding medium. Using $\cos x \approx 1 - x^2/2$, we obtain the inequality $(1 - a^2\theta_0^2/2)k_0[n_0 + 2n_2 P/\pi w_0^2] > k_0 n_0$, from which follows the so-called critical power for the onset of self-focusing,

$$P_{\text{crit}} = a^2 \frac{\lambda_0^2}{4\pi n_2 n_0}. \tag{3.54}$$

Typical values for P_{crit} are several MW in glass materials at $\lambda_0 = 1\,\mu\text{m}$; in air, with an n_2 of $4 \times 10^{-23}\,\text{m}^2\,\text{W}^{-1}$, the critical power is in the GW-range. Note that the self-focusing condition is independent of the beam diameter and refers to the power and not to the beam intensity, as Eq. (3.49) might suggest.

3.1.3.5 Spherical Mirror

Another optical component that modifies the phase front is a curved mirror; Fig. 3.7 shows a plane wave impinging on a spherical concave mirror. The incident phase fronts are first reflected by the rim of the mirror; the axial sections of the wavefront have to travel a further distance of $d(r)$ before they arrive at the mirror apex, while the outer sections have already travelled the same distance in backward direction, so that the total phase difference amounts to $2k_0 d(r)$. In the approximation $\sqrt{1 - x^2} \approx 1 - x^2/2$, we can set $d(r) = R_s - \sqrt{R_s^2 - r^2} \approx -r^2/2R_s$, where R_s is the radius of curvature of the mirror (the curvature of a concave surface is negative by convention). Thus, the action of a spherical mirror on a phase front is represented by the phase factor $e^{-jk_0 r^2/R_s}$. This is equivalent to the action of a lens with focusing power

$$\frac{1}{f} = -\frac{2}{R_s}, \tag{3.55}$$

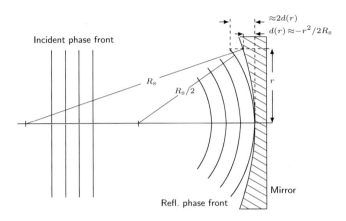

Fig. 3.7 Phase front modification by a spherical mirror

and the formulas obtained for a thin lens apply accordingly.

3.1.3.6 Dielectric Half Space

If a Gaussian beam propagates from a dielectric half space of index n into a half space of index n', separated by a plane boundary normal to the beam axis, the radial wave function [Eq. (3.17)] at the interface does not change, so that

$$e^{-r^2/w'^2} e^{-jk'r^2/2R'} = e^{-r^2/w^2} e^{-jkr^2/2R}. \qquad (3.56)$$

Nonetheless, the beam is modified, as a comparison of real and imaginary parts of the exponents shows:

$$w' = w, \qquad R' = \frac{k'}{k}R = \frac{n'}{n}R. \qquad (3.57)$$

Equation (3.31) yields the distance of the new waist from the interface

$$z' = \frac{R'}{1 + (\lambda'R'/\pi w^2)^2} = \frac{n'}{n}z, \qquad (3.58)$$

where $\lambda'/\lambda = n/n'$ was used and z is the position of the interface relative to the original beam waist; note that the original and the new beam waists are always on the same side of the interface; only one of them is real, the other one is "virtual" (Fig. 3.8). From Eq. (3.33) follows that the waist radius remains unaltered

$$w_0' = \frac{w}{[1 + (\pi w^2/\lambda'R')^2]^{1/2}} = w_0 \qquad (3.59)$$

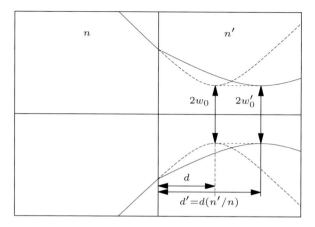

Fig. 3.8 Beam transition between different media: the beam is stretched axially by the factor n'/n, while the waist radius remains unchanged

since $\lambda'R' = \lambda R$, while the new confocal parameter is

$$z_0' = (n'/n)z_0. \tag{3.60}$$

In other words, the effect of the transition is an axial stretching of the beam by the factor n'/n (Fig. 3.8).

3.1.4 ABCD-Transformation of Gaussian Beams

The effect of the optical components discussed above can also be expressed in terms of a transformation $q \to q'$ of the q-parameter measured in the in- and output plane of the component, respectively (Fig. 3.9). As we shall see, this so-called ABCD-transformation greatly simplifies the calculation of beam propagation in optical systems.

From the definition of $q = z + jz_0$ follows immediately that free propagation over the distance d is, independent of the refractive index of the medium, equivalent to

$$q' = q + d, \tag{3.61}$$

and the transition between two dielectric media [Eq. (3.57)] can be described, using Eq. (3.10) by

$$q' = \frac{n'}{n}q, \tag{3.62}$$

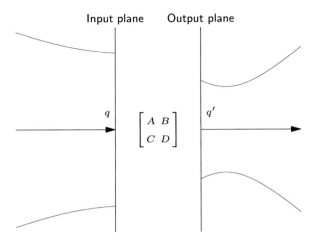

Fig. 3.9 A paraxial optical system consisting of lenses, spherical mirrors, and free space, arranged between an input and an output plane, can be represented by a single ABCD matrix that relates the q-parameter of the incident beam, measured in the input plane, to the q'-parameter of the transmitted beam, measured in the output plane

if the input- and output planes are located immediately at the interface. With the same choice of reference planes, Eqs. (3.10), (3.42), and (3.44) allow us to describe the action of a thin lens (or a spherical mirror) by

$$\frac{1}{q'} = \frac{1}{q} - \frac{1}{f}, \tag{3.63}$$

while a "soft" aperture [Eq. (3.35)] results in the transformation

$$\frac{1}{q'} = \frac{1}{R} - j\frac{2}{kw'^2(z)} = \frac{1}{q} - j\frac{2}{kw_{\mathrm{a}}^2(z)}. \tag{3.64}$$

These transformations can be cast in the generalized form

$$q' = \frac{Aq + B}{Cq + D}, \tag{3.65}$$

where the four coefficients constitute the so-called ABCD matrix

$$\boldsymbol{M} = \begin{bmatrix} A & B \\ C & D \end{bmatrix} \tag{3.66}$$

Table 3.2 ABCD-matrices of selected optical elements, and relations between q and other characteristic beam parameters; the spherical mirror resonator is treated in Sect. 4.3.1

Element	ABCD matrix
Free propagation	$\begin{bmatrix} 1 & d \\ 0 & 1 \end{bmatrix}$
Transition $n \rightarrow n'$	$\begin{bmatrix} 1 & 0 \\ 0 & n/n' \end{bmatrix}$
Dielectric plate	$\begin{bmatrix} 1 & (n/n')d \\ 0 & 1 \end{bmatrix}$
Gaussian aperture	$\begin{bmatrix} 1 & 0 \\ -j2/kw_a^2 & 1 \end{bmatrix}$
Thin lens	$\begin{bmatrix} 1 & 0 \\ -1/f & 1 \end{bmatrix}$
Spherical mirror	$\begin{bmatrix} 1 & 0 \\ 2/R & 1 \end{bmatrix}$
GRIN-lens	$\begin{bmatrix} \cos\alpha_g d & (1/n_0\alpha_g)\sin\alpha_g d \\ -n_0\alpha_g \sin\alpha_g d & \cos\alpha_g d \end{bmatrix}$
Single lens system	$\begin{bmatrix} 1 - d'/f & d + d' - dd'/f \\ -1/f & 1 - d/f \end{bmatrix}$
Sph. mirror resonator	$\begin{bmatrix} (1 + 2d/R_{s1})(1 + 2d/R_{s2}) + 2d/R_{s2} & d(2 + 2d/R_{s1}) \\ 2(1 + 2d/R_{s2})/R_{s1} + 2/R_{s2} & 2d/R_{s1} + 1 \end{bmatrix}$

$q' = \frac{Aq+B}{Cq+D}$	$z = \mathrm{Re}\,[q]$	$R = \frac{1}{\mathrm{Re}[q^{-1}]}$	$\theta_0^2 = \frac{2}{k\mathrm{Im}[q]}$
$q'^{-1} = \frac{C+Dq^{-1}}{A+Bq^{-1}}$	$z_0 = \mathrm{Im}\,[q]$	$w^2 = \frac{2}{k\|\mathrm{Im}[q^{-1}]\|}$	$w_0^2 = \frac{2\mathrm{Im}[q]}{k}$

given in Table 3.2; to show the validity for transformations that modify the phase curvature (lens and spherical mirror), it is convenient to write Eq. (3.65) in the form

$$\frac{1}{q'} = \frac{C + D/q}{A + B/q}. \tag{3.67}$$

The power of this formalism lies in the fact that a sequence of optical elements can be represented by the product of the elementary matrices: if a beam is propagated through a series of optical elements with matrices $M_1, M_2, \ldots, M_n$ (in this sequence), then the system matrix is given by

$$M_t = \begin{bmatrix} A_t & B_t \\ C_t & D_t \end{bmatrix} =: M_n \ldots M_2 M_1 \qquad (3.68)$$

and the relation between input and output parameter is

$$q' = \frac{A_t q + B_t}{C_t q + D_t}; \qquad (3.69)$$

the proof of this statement is left to the reader as an exercise.

3.1.4.1 Transformed Parameters

We now want to derive a few useful formulas for the transformation of Gaussian beams by systems that consist of lenses (or spherical mirrors) and sections of free propagation. Since we are free in the choice of the input plane, we put it in the waist of the incident beam, so that the input q-parameter is $q = jz_0$; the output parameter is then

$$q' = \frac{jz_0 A + B}{jz_0 C + D} = \frac{(BD + ACz_0^2) + jz_0(AD - BC)}{D^2 + z_0^2 C^2}. \qquad (3.70)$$

The distance d' of the new beam waist, measured from the output plane, is the negative value of the real part of q',

$$d' = -\mathrm{Re}\left[q'\right]. \qquad (3.71)$$

The imaginary part of q' is the output confocal parameter z_0',

$$z_0' = \frac{z_0}{D^2 + z_0^2 C^2} =: \mathcal{M}^2 z_0, \qquad (3.72)$$

where we have used the fact that $\det M = AD - BC = 1$ (provided that the input and output medium have the same propagation index). The factor

$$\mathcal{M} := \frac{1}{\sqrt{D^2 + z_0^2 C^2}} \qquad (3.73)$$

has the meaning of a magnification

$$M = \sqrt{\frac{z_0'}{z_0}} = \frac{w_0'}{w_0} = \frac{\theta_0}{\theta_0'} \tag{3.74}$$

as follows from Eq. (3.12) in the form $w_0'/w_0 = \sqrt{z_0'/z_0}$.

As an example, we consider a single lens of focusing power $1/f$, positioned at a distance d from the input beam waist. The matrix of the system, stretching from the input beam waist to the backside of the lens, is

$$\boldsymbol{M}_{df} = \begin{bmatrix} 1 & 0 \\ -1/f & 1 \end{bmatrix} \begin{bmatrix} 1 & d \\ 0 & 1 \end{bmatrix} = \begin{bmatrix} 1 & d \\ -1/f & 1 - d/f \end{bmatrix}, \tag{3.75}$$

from which we obtain the magnification

$$M = \sqrt{\frac{f^2}{(d-f)^2 + z_0^2}}. \tag{3.76}$$

The distance of the new beam waist from the lens, $d' = -\text{Re}\,[q']$ is related to d by the equation

$$\frac{d'-f}{d-f} = M^2. \tag{3.77}$$

If $d = 0$ (Fig. 3.10), the input wave at the lens has a planar phase front, and the new beam waist with the radius [Eq. (3.74)]

$$w_0' = \frac{w_0}{[1 + (z_0/f)^2]^{1/2}} = \frac{w_0}{\left[1 + (w_0^2\pi/\lambda f)^2\right]^{1/2}} \tag{3.78}$$

is formed at the distance

$$d' = \frac{f}{1 + (f/z_0)^2} < f, \tag{3.79}$$

which is, somewhat surprisingly, shorter than the focal length f of the lens unless $z_0 \gg f$.

As an inspection of Eq. (3.78) shows, the strategy to minimize w_0' is to use an input beam with large waist radius w_0 and a lens with large focusing power $1/f$; w_0' is then approximately $\lambda f/\pi w_0$. Since the input beam radius w_0 is limited by the lens radius $D/2$,

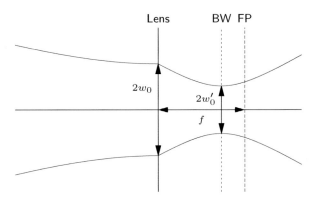

Fig. 3.10 Beam transformation by a single lens; note that the location of the beam waist (BW) does not coincide with the focal plane (FP) but is shifted towards the lens; this effect is negligible only if $f \ll z_0$

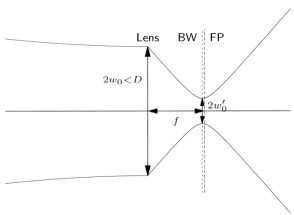

Fig. 3.11 Beam focusing with a lens: the beam waist is minimized if the input beam fills the entire lens aperture D

$$w_0' > \frac{2\lambda f}{\pi D};\tag{3.80}$$

this limit is proportional to the f-number (f/D) of the lens, which should be as small as possible for that purpose (Fig. 3.11). The FWHM [Eq. (3.20)] of the focus is given by

$$d_{\text{FWHM}} > \frac{2\sqrt{2\ln 2}}{\pi}\left(\frac{f}{D}\right)\lambda \approx 0.75\left(\frac{f}{D}\right)\lambda.\tag{3.81}$$

3.1.4.2 Thick Dielectric Plate

We now turn to the derivation of the ABCD matrix of several important optical elements. First we consider a dielectric plate of thickness d and propagation index n'. The propagation through such a plate can be split into three steps: transition $n \to n'$, free propagation over d, and finally transition $n' \to n$; the system matrix is

$$M = \begin{bmatrix} 1 & 0 \\ 0 & \frac{n'}{n} \end{bmatrix} \begin{bmatrix} 1 & d \\ 0 & 1 \end{bmatrix} \begin{bmatrix} 1 & 0 \\ 0 & \frac{n}{n'} \end{bmatrix} = \begin{bmatrix} 1 & d\frac{n}{n'} \\ 0 & 1 \end{bmatrix} \tag{3.82}$$

which is equivalent to a free space propagation in the original medium over a distance $d(n/n')$.

3.1.4.3 Thick Lens

A thick lens can be understood as a set of two thin lenses separated by a thick dielectric plate of refractive index n', and its matrix is obtained by multiplication of the respective matrices shown in Table 3.2. Since the expression Eq. (3.41) for the focusing power of a thin lens implies empty space in front and behind the lens, we have to introduce an infinitely thin layer of empty space between the lenses and the plate, and set $n = 1$ in matrix Eq. (3.82). The system matrix is therefore, with $C_{1,2} = -\frac{1}{f_{1,2}}$,

$$M_t = \begin{bmatrix} 1 & 0 \\ C_2 & 1 \end{bmatrix} \begin{bmatrix} 1 & d/n' \\ 0 & 1 \end{bmatrix} \begin{bmatrix} 1 & 0 \\ C_1 & 1 \end{bmatrix} = \begin{bmatrix} A_t & B_t \\ C_t & D_t \end{bmatrix}, \tag{3.83}$$

where $A_t, D_t = 1 + \frac{d}{n'}C_{1,2}$, $B_t = \frac{d}{n'}$, and

$$C_t = C_1 + C_2 + \frac{d}{n'}C_1 C_2. \tag{3.84}$$

The input and output planes of this "system" are the front and rear faces of the thick lens. We can normalize this matrix by choosing alternate reference planes $H_{1,2}$, where the distance of H_1 to the front face is h_1 and the distance from the rear face to H_2 is h_2. The new system matrix is then

$$M^* = \begin{bmatrix} A^* & B^* \\ C^* & D^* \end{bmatrix} = \begin{bmatrix} A_t + h_2 C_t & h_1 A_t + B_t + h_1 h_2 C_t + h_2 D_t \\ C_t & h_1 C_t + D_t \end{bmatrix}. \tag{3.85}$$

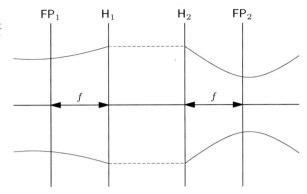

Fig. 3.12 Any system with a non-vanishing C_t coefficient can be represented by a set of two principal planes $H_{1,2}$ in respect to which the system performs like a thin lens

We choose $h_{1,2}$ so that $A^* = D^* = 1$; then it turns out that $B^* = 0$, because the determinant of M_t (and all other involved matrices) is equal to 1. With

$$h_1 = \frac{1 - D^*}{C^*}$$

$$h_2 = \frac{1 - A^*}{C^*}, \tag{3.86}$$

the ABCD matrix of the thick lens,

$$M^* = \begin{bmatrix} 1 & 0 \\ -\frac{1}{f_t} & 1 \end{bmatrix}, \tag{3.87}$$

is equal to that of a thin lens with focal length $f_t = -1/C_t$, with the only difference that the reference planes are the so-called principal planes $H_{1,2}$ (Fig. 3.12). Note that this normalization can be performed for *any* system with a non-vanishing C^*, i.e., any such system is equivalent to a thin lens.

3.1.4.4 Beam Expander

Setting $n' = 1$ in Eq. (3.83), we obtain the matrix of a system of two lenses, separated by a distance d. The case $d = f_1 + f_2$ is of particular interest, because the resulting $C_t = 0$; such a system is a telescope with the matrix

$$M_t = \begin{bmatrix} -\frac{f_2}{f_1} & d \\ 0 & -\frac{f_1}{f_2} \end{bmatrix} \tag{3.88}$$

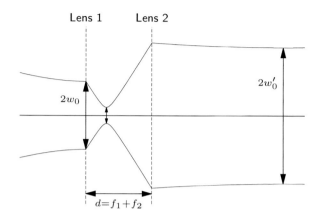

Fig. 3.13 A confocal two-lens system acting as beam expander with magnification $\mathcal{M} = \frac{f_2}{f_1}$

and cannot be replaced by a thin lens. If we position the input beam waist (radius w_0) at the location of the first lens (Fig. 3.13), the magnification according to Eq. (3.73) is $\mathcal{M} = \frac{f_2}{f_1}$ and the output beam waist follows from (3.74) to be

$$w_0' = \mathcal{M}w_0 = \frac{f_2}{f_1}w_0. \tag{3.89}$$

Such systems, with $|f_2| \gg |f_1|$, are frequently used to expand a beam waist and reduce the divergence

$$\theta_0' = \theta_0/\mathcal{M} = \frac{f_1}{f_2}\theta_0. \tag{3.90}$$

3.1.4.5 Thick GRIN-Lens

Another interesting system is the GRIN-lens of arbitrary thickness d, where the change of the beam radius due to the propagation cannot be neglected. We slice the GRIN-lens into m thin sections of thickness $\delta z = d/m$. Focusing and propagation effects of a single slice are treated separately by embedding a thin lens with focusing power $n_0\alpha_g^2\delta z$ [Eq. (3.48)] between two homogeneous slices of thickness $\delta z/2$ and index n_0; the matrix of such a slice is then

$$M_\delta = \begin{bmatrix} 1 & \delta z/2n_0 \\ 0 & 1 \end{bmatrix} \begin{bmatrix} 1 & 0 \\ -n_0\alpha_g^2\delta z & 1 \end{bmatrix} \begin{bmatrix} 1 & \delta z/2n_0 \\ 0 & 1 \end{bmatrix}$$

$$\approx \begin{bmatrix} 1 - \delta z^2\alpha_g^2/2 & \delta z/n_0 \\ -n_0\alpha_g^2\delta z & 1 - \delta z^2\alpha_g^2/2 \end{bmatrix}. \tag{3.91}$$

The GRIN-lens is composed of m such slices and is accordingly represented by

$$M_{\mathrm{grin}} = M_\delta^m. \tag{3.92}$$

For the evaluation of this expression, we use the relation

$$\begin{bmatrix} \cos\theta & (1/K)\sin\theta \\ -K\sin\theta & \cos\theta \end{bmatrix}^m = \begin{bmatrix} \cos m\theta & (1/K)\sin m\theta \\ -K\sin m\theta & \cos m\theta \end{bmatrix}, \tag{3.93}$$

whose validity can be shown by induction, and substitute $\theta = \alpha_g\,\delta z$ and $K = \alpha_g n_0$. Since δz is assumed to be small, $\sin\theta \approx \theta$ and $\cos\theta \approx 1 - \theta^2/2$, so that the resulting matrix is approximately equal to M_δ^m. With $m\,\delta z = d$ we obtain

$$M_{\mathrm{GRIN}} = \begin{bmatrix} \cos\alpha_g d & (1/n_0\alpha_g)\sin\alpha_g d \\ -n_0\alpha_g\sin\alpha_g d & \cos\alpha_g d \end{bmatrix}. \tag{3.94}$$

3.1.5   Hermite–Gaussian Beams

The Gaussian beam belongs to an infinite set of solutions of the paraxial wave equation with spherical phase fronts. Related solutions can be generated, for example, by multiplying the Gaussian wave function Eq. (3.17) with transverse functions $X(u)$ and $Y(v)$

$$a(\mathbf{x}) = A_0' X(u) Y(v) e^{\mathrm{j}Z(z)} \frac{w_0}{w(z)} \exp\left[-\frac{r^2}{w^2(z)}\right] \exp\left[-\mathrm{j}k\frac{r^2}{2R(z)}\right] e^{-\mathrm{j}kz + \mathrm{j}\xi(z)}, \tag{3.95}$$

where $u = \sqrt{2}x/w(z)$ and $v = \sqrt{2}y/w(z)$ are transverse coordinates normalized by the local radius $w(z)$ of the Gaussian profile. Substituting this ansatz in Eq. (3.4) yields differential equations for X and Y whose solutions are the Hermite polynomials of order l

$$H_0(u) = 1$$

$$H_1(u) = 2u$$

$$H_2(u) = 4u^2 - 2$$

$$\cdots$$

$$H_{l+1}(u) = 2uH_l(u) - 2lH_{l-1}(u); \tag{3.96}$$

l is also the number of real valued roots of the polynomial. For a given set of polynomials $X = H_l$ and $Y = H_m$, the additional phase term Z is given by

$$Z(z) = (l+m)\xi(z), \tag{3.97}$$

where $\xi = \arctan(z/z_0)$ [Eq. (3.16)].

The intensity distribution of a Hermite–Gaussian beam of order (l, m) is

$$I = I_0 \left[\frac{w_0}{w(z)}\right]^2 H_l^2\left(\frac{\sqrt{2}x}{w(z)}\right) H_m^2\left(\frac{\sqrt{2}y}{w(z)}\right) e^{-2[x^2+y^2]/w^2(z)} \tag{3.98}$$

and is characterized by l and m nodal lines parallel to the y- and x-axis, respectively, as shown in Fig. 3.14. Since the polynomials diverge for large values of x, y, the

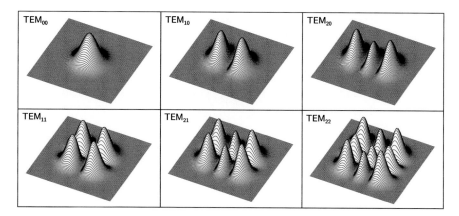

Fig. 3.14 Transverse intensity profiles of Hermite–Gaussian beams of order (0,0) to (2,2); the order is equal to the number of nodes

beam power is redistributed to larger radial distances in comparison to a Gaussian beam.

The phase front curvature of a Hermite–Gaussian beam is determined by the confocal parameter z_0 of the underlying Gaussian beam. Hermite–Gaussian beams are therefore fully characterized by the order l, m of the polynomials and the q-parameter of the Gaussian "carrier." Because of their small longitudinal electric field components, these waves are also called TEM_{lm}-modes. Since any linear combination of Hermite–Gaussian beams is also a solution of the paraxial Helmholtz equation, any superposition of Hermite–Gaussian wave functions with identical q-parameter forms a beam with spherical phase fronts of curvature $\text{Re}\,[1/q]$. Note, however, that the radial profile of such a superposition is generally not conserved during propagation, because Hermite–Gaussians of different order have different axial phases Eq. (3.97). Since laser resonators (Sect. 4.3) control primarily the phase front curvature of the generated laser beam, lasers tend to produce such superpositions.

3.1.6 Fourier Optical Treatment of Beam Propagation

Because of the linearity of Maxwell's equations, any propagating electromagnetic field, and beams in particular, can be synthesized by a superposition of plane waves. Neglecting polarization, each of these waves is characterized by its wave vector $\mathbf{k}$ and a complex amplitude $A(\mathbf{k})$. At a given frequency ω, the wave vector is constrained by the dispersion relation $|\mathbf{k}| = \sqrt{k_x^2 + k_y^2 + k_z^2} = k = \omega/c$, so that only two of its three components, say k_x and k_y are free variables.

3.1.6.1 Spatial Fourier Transform

Consider a monochromatic beam with the (scalar) wave function $a(x, y; z)$ propagating in the z-direction. The semicolon separating the z-variable indicates that we study the transverse field distribution in the plane $z = $ const. In such a plane, the function can be written as two-dimensional Fourier integral

$$a(x, y; z) = \mathcal{F}^{-1}\left\{A_{k_x,k_y}(z)\right\} = \frac{1}{(2\pi)^2} \iint\limits_{-\infty}^{\infty} A_{k_x,k_y}(z) e^{-jk_x x} e^{-jk_y y} \, \mathrm{d}k_x \, \mathrm{d}k_y,$$

$$(3.99)$$

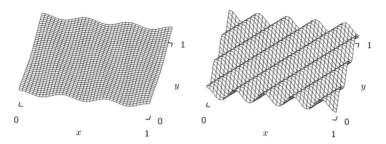

Fig. 3.15 Two-dimensional Fourier components: *left panel* $[k_x, k_y] = [6\pi, 2\pi]$, *right panel* $[k_x, k_y] = [-9\pi, 6\pi]$ and larger amplitude

where k_x and k_y have the meaning of spatial frequency components (Fig. 3.15) and

$$A_{k_x,k_y}(z) = \mathcal{F}\{a(x, y; z)\} = \iint\limits_{-\infty}^{\infty} a(x, y; z) e^{jk_x x} e^{jk_y y} \, dx \, dy \qquad (3.100)$$

is the two-dimensional Fourier transform of $a(x, y; z)$.

Equation (3.99) (and its integrand) has to satisfy the Helmholtz equation Eq. (1.22) in the form

$$\left[\frac{\partial^2}{\partial x^2} + \frac{\partial^2}{\partial y^2} + \frac{\partial^2}{\partial z^2} + k^2 \right] a(\mathbf{x}) = 0. \qquad (3.101)$$

Since $A_{k_x,k_y}(z)$ depends only on z, this can be simplified to

$$\left[k^2 - k_x^2 - k_y^2 + \frac{\partial^2}{\partial z^2} \right] A_{k_x,k_y}(z) = 0 \qquad (3.102)$$

with the solution

$$A_{k_x,k_y}(z) = A_{k_x,k_y}(0) e^{-j\sqrt{k^2 - k_x^2 - k_y^2} \, z}. \qquad (3.103)$$

The integrand in Eq. (3.99) can thus be written as

$$A_{k_x,k_y}(0) e^{-j(k_x x + k_y y + k_z z)} \qquad (3.104)$$

with $k_z = \sqrt{k^2 - k_x^2 - k_y^2}$. This is exactly the plane wave mentioned above, with the wave vector $[k_x, k_y, k_z]$ and the complex amplitude $A_{k_x,k_y}(0)$, and (3.99) is the aforementioned superposition of plane waves. Note that a (monochromatic) beam is therefore completely determined if its complex amplitude in an arbitrary plane is known.

3.1.6.2 Transfer Function

In the following we assume that $A_{k_x,k_y}(0)$ is negligible outside the paraxial range $k_x^2 + k_y^2 \ll k^2$; this restriction to low transverse spatial frequencies implies that $a(x, y; 0)$ varies only slowly on the scale of $1/k = \lambda/2\pi$ and that, in particular, the beam radius is significantly larger than the wavelength. The angle θ between the corresponding wave vectors and the beam axis is then also small and can be approximated by

$$\theta \approx \sin\theta = \frac{\sqrt{k_x^2 + k_y^2}}{k} ; \tag{3.105}$$

the axial component of the wave vector is approximately

$$k_z = \sqrt{k^2 - k_x^2 - k_y^2} \approx k - \frac{k_x^2 + k_y^2}{2k}. \tag{3.106}$$

With these approximations, Eq. (3.103) can be written as

$$A_{k_x,k_y}(z) = A_{k_x,k_y}(0)e^{j\frac{k_x^2+k_y^2}{2k}z}e^{-jkz}. \tag{3.107}$$

The factor

$$H(k_x, k_y; z) = e^{j\frac{k_x^2+k_y^2}{2k}z}e^{-jkz} \tag{3.108}$$

relating the output-Fourier transform to the input is called the transfer function of a system, in our case of the free space propagation over the distance z.

We can now cast (3.99) in the form

$$a(x, y; z) = \frac{1}{(2\pi)^2} \iint\limits_{-\infty}^{\infty} A_{k_x,k_y}(0)e^{j\frac{k_x^2+k_y^2}{2k}z}e^{-j(k_x x + k_y y)} \, dk_x \, dk_y e^{-jkz} \tag{3.109}$$

or

$$a(x, y; z) = \mathcal{F}^{-1}\left\{H(k_x, k_y; z)A_{k_x,k_y}(0)\right\}. \tag{3.110}$$

Knowing the complex amplitude profile in an arbitrary plane (and the propagation direction), Eq. (3.110) allows calculating the complete wave function.

3.1.6.3 Example: Gaussian Beam

Let the spatial amplitude profile at $z = 0$ be a two-dimensional Gaussian function

$$a(x, y; 0) = a'_0 e^{-(x^2+y^2)/w_0^2}; \tag{3.111}$$

the Fourier transform

$$A_{k_x,k_y}(0) = \pi w_0^2 A'_0 e^{-(k_x^2+k_y^2)w_0^2/4} \tag{3.112}$$

is a Gaussian in the k_x, k_y spatial frequency plane, with a $1/e$ half width of $4/w_0^2$. According to Eq. (3.105), the angle between the corresponding wave vectors and the beam axis is distributed between 0 and

$$\theta_0 = \frac{2}{w_0 k} = \frac{\lambda}{\pi w_0}; \tag{3.113}$$

note that this angle coincides exactly with the beam divergence Eq. (3.19).

With this spectrum, Eq. (3.110) yields

$$a(x, y; z) = \frac{A'_0}{(2\pi)^2} \int\!\!\!\int_{-\infty}^{\infty} e^{-(k_x^2+k_y^2)(w_0^2/4-\mathrm{j}z/2k)} e^{-\mathrm{j}(k_x x+k_y y)} \, dk_x \, dk_y$$

$$= \frac{A'_0 w_0^2}{C^2} e^{-(x^2+y^2)/C^2}, \tag{3.114}$$

where $C^2 = w_0^2 - 2\mathrm{j}z/k$. To perform the integration, the integrand was multiplied with $e^{(x^2+y^2)/C^2} e^{-(x^2+y^2)/C^2}$ to obtain a quadratic exponent and the identity $\int_{-\infty}^{\infty} e^{-x^2/a^2} \, dx = \sqrt{a\pi}$ was used. As can be easily shown, $\mathrm{Re}\left[1/C^2\right] = 1/w^2$ as given by Eq. (3.11) and $\mathrm{Im}\left[1/C^2\right] = -k/2R$ [Eq. (3.13)]; moreover, $w_0^2/C^2 = w_0/w(z)e^{-\mathrm{j}\xi(z)}$ with $\xi(z)$ given by Eq. (3.16). With these substitutions, Eq. (3.114) agrees completely with the wave function Eq. (3.17).

The power of Eq. (3.110) lies in the fact that it allows calculating the beam resulting from *any* paraxial amplitude distribution.

3.1.6.4 Point Spread Function

An optical propagation system such as free space or a lens system transforms an input amplitude distribution $a_{\mathrm{in}}(x, y; z_{\mathrm{in}})$ into an output distribution $a_{\mathrm{out}}(x, y; z_{\mathrm{out}})$, where $z_{\mathrm{in}}, z_{\mathrm{out}}$ are the positions of input and output planes, respectively. This can be formally written as

$$a_{\mathrm{out}} = \mathcal{S}\{a_{\mathrm{in}}\}, \tag{3.115}$$

where the operator S represents the system. If S is linear, one can apply concepts of the theory of linear systems to relate this real-space description to the frequency-space description above.[2]

Exploiting the properties of the Dirac δ-distribution, we can write $a_{\text{in}}(x, y)$ in the form

$$a_{\text{in}}(x, y) = \iint\limits_{-\infty}^{\infty} a_{\text{in}}(x', y')\delta(x - x', y - y') \, dx' \, dy', \tag{3.116}$$

which is a superposition of weighted δ-distributions located at all possible points x', y' of the input plane. Assuming that the system is linear and the response [that is the output function at (x, y)] to a δ-distribution in (x', y') is

$$h(x, y, x', y') = S\left\{\delta(x - x', y - y')\right\}, \tag{3.117}$$

the output is

$$a_{\text{out}}(x, y) = \iint\limits_{-\infty}^{\infty} a_{\text{in}}(x', y')h(x, y, x', y') \, dx' \, dy'; \tag{3.118}$$

$h(x, y, x', y')$ is called point spread function of the system.

If we further assume that the system is invariant under a transverse shift, the point spread function is not an explicit function of the coordinates, but only of their respective differences, $h(x, y, x', y') = h(x - x', y - y')$, and we obtain

$$a_{\text{out}}(x, y) = \iint\limits_{-\infty}^{\infty} h(x - x', y - y')a_{\text{in}}(x', y') \, dx' \, dy'. \tag{3.119}$$

Thus, the output amplitude of a linear, shift invariant optical system is the convolution of the input function with the point spread function h.

Since the Fourier transform of a convolution of two functions is equal to the product of their respective Fourier transforms, we can write

$$\mathcal{F}\{a_{\text{out}}\} = \mathcal{F}\{h\}\,\mathcal{F}\{a_{\text{in}}\}. \tag{3.120}$$

Comparison with Eq. (3.110) shows that $\mathcal{F}\{h\} = H$, and the point spread function is consequently the inverse Fourier transform $h = \mathcal{F}^{-1}\{H\}$ of the transfer function.

---

[2]See, e.g., Goodman (1996).

For free space propagation within the paraxial approximation, we obtain, using Eq. (3.108), $h(x - 0, y - 0) = \mathcal{F}^{-1} \left\{ \exp(\mathrm{j}kz) \exp\left[\mathrm{j}\pi\lambda z(k_x^2 + k_y^2) \right] \right\}$, so that

$$h(x - 0, y - 0) = h_0 \exp\left[-\mathrm{j}k \frac{x^2 + y^2}{2z} \right], \qquad (3.121)$$

where $h_0 = (\mathrm{j}/\lambda z) \exp(-\mathrm{j}kz)$. Thus, the point spread function of free space is a spherical wave Eq. (3.6), centered at the point (x', y'). The response of free space to an arbitrary input distribution is therefore the convolution

$$a_{\mathrm{out}}(x, y; z) = h_0 \iint\limits_{-\infty}^{\infty} a_{\mathrm{in}}(x', y') \exp\left[-\mathrm{j}k \frac{(x - x')^2 + (y - y')^2}{2z} \right] \mathrm{d}x' \, \mathrm{d}y'.$$

$$(3.122)$$

3.1.6.5 Fourier Transformation by Far-Field Propagation

Equation (3.122) can be cast in the form

$$a_{\mathrm{out}}(x, y; z) = \mathrm{e}^{-\mathrm{j}k \frac{x^2 + y^2}{2z}} \iint\limits_{-\infty}^{\infty} \mathrm{e}^{-\mathrm{j}k \frac{x'^2 + y'^2}{2z}} a_{\mathrm{in}}(x', y') \mathrm{e}^{\mathrm{j}2\pi \frac{xx' + yy'}{\lambda z}} \mathrm{d}x' \, \mathrm{d}y', \qquad (3.123)$$

where the constant factor h_0 has been dropped as irrelevant. In the far field, i.e., for propagation distances d satisfying $\frac{x'^2 + y'^2}{\lambda d} \ll 1$,

$$a_{\mathrm{out}}(x, y; d) \sim \exp\left[-\mathrm{j}\pi \frac{x^2 + y^2}{\lambda d} \right] \iint\limits_{-\infty}^{\infty} a_{\mathrm{in}}(x', y') \exp\left[\mathrm{j}2\pi \frac{xx' + yy'}{\lambda d} \right] \mathrm{d}x' \, \mathrm{d}y'.$$

$$(3.124)$$

Associating the output coordinates x, y with the spatial frequencies

$$k_x \leftrightarrow 2\pi \frac{x}{\lambda d}, \quad k_y \leftrightarrow 2\pi \frac{y}{\lambda d}, \qquad (3.125)$$

a_{out} can be interpreted as the Fourier transform of the input function, multiplied with a parabolic phase factor,

$$a_{out}(x, y) \sim e^{-jk(x^2+y^2)/2d} A_{k_x,k_y};$$ (3.126)

the output *intensity* distribution does not include the phase factor and represents the undistorted two-dimensional power spectrum of the input function. It should be noted that the far-field condition $\frac{x'^2+y'^2}{\lambda d} \ll 1$ is quite restrictive: for an input-diameter of 1 mm, $d \gg 1$ m for visible light.

3.1.6.6 Fourier Transformation by a Lens

As we have seen in Sect. 3.1.3.2, a thin lens adds a phase term Eq. (3.39) to an input function a_1, so that the output function a_1' immediately behind the lens is

$$a_1'(x', y') = a_1 e^{jk(x'^2+y'^2)/2f}.$$ (3.127)

The further propagation behind the lens can be described by the convolution Eq. (3.122). In the focal plane $z = f$, in particular,

$$a_{out}(x, y; f) = e^{-jk(x^2+y^2)/2f} \iint\limits_{-\infty}^{\infty} a_1(x', y') \exp\left[j2\pi \frac{xx' + yy'}{\lambda f} \right] dx'\, dy',$$ (3.128)

since the quadratic terms under the integral cancel. If we again associate the coordinates x, y of the focal plane with the spatial frequencies

$$k_x \leftrightarrow \frac{2\pi x}{\lambda f} = k\frac{x}{f}, \quad k_y \leftrightarrow \frac{2\pi y}{\lambda f} = k\frac{y}{f},$$ (3.129)

we obtain a result similar to Eq. (3.126): the amplitude in the focal plane is equal to the Fourier transform of the wave function at the entrance of the lens, multiplied with a parabolic phase factor

$$a_{out}(x, y; f) = e^{-jk(x^2+y^2)/2f} \mathcal{F}\left\{a_1(x', y')\right\}.$$ (3.130)

The phase distortion can be omitted if we move the input plane to the front focal plane of the lens (creating a so-called 2f-system): let a_{in} be the input amplitude distribution in the front focal plane; then we know from Eq. (3.108) that the Fourier transform of the field at the entrance of the lens is

$$\mathcal{F}\{a_1\} = H\mathcal{F}\{a_{\text{in}}\}, \tag{3.131}$$

where H is

$$H = e^{jk(x^2+y^2)/2f}. \tag{3.132}$$

Substituting Eq. (3.131) in Eq. (3.130), we obtain

$$a_{\text{out}} = \mathcal{F}\{a_{\text{in}}\}. \tag{3.133}$$

Thus, a $2f$-system produces an undistorted Fourier transform of the front focal plane in the rear focal plane.

3.1.6.7 4f-System

Since the inverse Fourier transform differs from the direct transform only by the sign of the coordinates, a second $2f$-system can be used for the inverse transformation. Such a $4f$-system images the front focal plane of the first lens onto the rear focal plane of the second plane with a magnification of -1. In the joint central focal plane, the Fourier transform is accessible and can be manipulated with phase- and amplitude objects, in particular with electronically controlled spatial light modulators (SLMs) made of liquid crystal panels (Sect. 2.3.5). In this way, an SLM can realize an arbitrary transfer function with an appropriate transmission function $H(k_x, k_y) = t(x, y) \exp j\phi(x, y)$.

3.1.6.8 Grating Spectrometer

The spatial Fourier transform capabilities of a lens (or a concave mirror) can also be employed to perform spectrum analysis of light. The concept of the spectrum of polychromatic light will be discussed in detail in Sect. 4.4.1. Roughly speaking, a polychromatic light field can be decomposed into monochromatic waves with complex amplitudes a_ω. The function $|a_\omega|^2$ (or $|a_{\lambda_0}|^2 = |a_\omega|^2 \frac{d\omega}{d\lambda_0}$) is called the power spectrum of the signal and is measured with a spectrometer. A grating spectrometer consists of a grating that disperses the individual frequency components of the incoming light into plane waves with wave vectors pointing in different directions (Fig. 3.16). If a collimated optical signal is incident orthogonally ($\mathbf{k}_\parallel^i = 0$) on a grating with period Λ, the diffracted wave vectors (with length $|\mathbf{k}| = \omega/c_0 = 2\pi/\lambda_0$) have a common transverse component of $k_x^d = m(2\pi/\Lambda)$, where the integer m is the order of diffraction (Fig. 2.5). According to Eq. (3.129), a

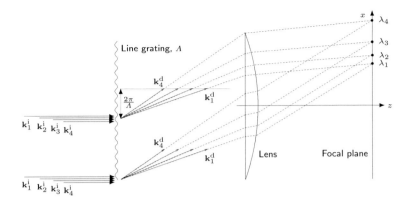

Fig. 3.16 A spectrometer consists of a dispersive element (grating) and a lens acting as Fourier transformer. The polychromatic signal is incident from the *left*, the *arrows* indicate wave vectors with length $2\pi/\lambda_0$

lens images these waves onto the position

$$x = \frac{k_x^{\mathrm{d}}}{k_0}f = m\frac{\lambda_0}{\Lambda}f \qquad (3.134)$$

in its rear focal plane. Because of the linear relation between x and λ_0, $\mathrm{d}\lambda_0 \propto \mathrm{d}x$, and the power density $|a|^2(x)$ in the focal plane (measured with a linear detector array) can be identified with $|a_{\lambda_0}|^2$.

3.2 Pulse Propagation

Similar to a beam that is a concentration of electromagnetic radiation in space, a light pulse is a concentration of radiation in time. The mathematical treatment of both phenomena is similar; in particular, both can be understood as superpositions of monochromatic waves that propagate through a medium. Since the phase velocity in dielectrics is frequency dependent, the relative phase of the Fourier components changes along the propagation, and the pulse shape (and duration) generally changes during propagation; the underlying mechanism is the frequency dependent susceptibility of dielectrics (dispersion). In contrast, the transverse profile of a beam changes because its spatial Fourier components have different axial phase velocities for geometric reasons.

Pulse broadening due to dispersion is of particular relevance for the propagation of pulses in (very long) glass fibers in optical communications since the resulting temporal overlap of consecutive pulses limits the data rate. For extremely short pulses, even the transmission through a glass plate or lens can significantly increase the pulse duration and thus reduce the peak power, which can deteriorate, for example, ultrafast laser ablation of materials.

3.2.1 Dispersive Propagation Effects

Consider an electromagnetic wave $\mathbf{E}(z, t) = \mathrm{Re}\,[a(z, t)]$, where $a(z, t)$ is the product of a carrier wave $e^{-j(k^0 z - \omega_0 t)}$ and a slowly varying (complex) envelope $A(z, t)$

$$a(z, t) = A(z, t)e^{-j(k^0 z - \omega_0 t)}; \qquad (3.135)$$

ω_0 and k^0 are the frequency and wave number of the carrier, respectively. The pulse intensity according to Eq. (1.71) is $I = n|A|^2/2Z_0$, where n is the propagation index and $|A(z, t)|^2$ is the z-dependent pulse shape.

3.2.1.1 Temporal Fourier Transform

The Fourier transform of the envelope function $A(z, t)$,

$$A_{\Delta\omega}(z) = \int_{-\infty}^{\infty} A(z, t)e^{-j\Delta\omega t}\, dt, \qquad (3.136)$$

is concentrated around $\Delta\omega = 0$ within a bandwidth that we assume to be much smaller than ω_0. The inverse transform is

$$A(z, t) = \frac{1}{2\pi} \int_{-\infty}^{\infty} A_{\Delta\omega}(z)e^{j\Delta\omega t}\, d\Delta\omega. \qquad (3.137)$$

We can also express the complete wave function Eq. (3.135) as a Fourier integral

$$a(z, t) = \frac{1}{2\pi} \int_{-\infty}^{\infty} a_\omega(z)e^{j\omega t}\, d\omega, \qquad (3.138)$$

where

$$a_\omega(z) = \int_{-\infty}^{\infty} A(z, t)e^{-j(k^0 z - \omega_0 t)}e^{-j\omega t}\, dt. \qquad (3.139)$$

With the substitution $\omega = \omega_0 + \Delta\omega$, this can be written as

$$a_{\omega_0 + \Delta\omega}(z) = e^{-jk^0 z} \int_{-\infty}^{\infty} A(z, t) e^{-j\Delta\omega t}\, dt, \tag{3.140}$$

so that

$$a_\omega(z) = A_{\Delta\omega}(z) e^{-jk^0 z}; \tag{3.141}$$

apart from a common phase factor, the Fourier component of the complex wave function at $\omega_0 + \Delta\omega$ is equal to the Fourier component of the envelope at $\Delta\omega$.

3.2.1.2 Spectral Characterization

According to Parseval's theorem, the pair of functions $A(t), A_{\Delta\omega}$ is related by

$$\int_{-\infty}^{\infty} |A(t)|^2\, dt = \int_{-\infty}^{\infty} |A_{\Delta\omega}|^2\, d\Delta\omega. \tag{3.142}$$

Since the left-hand side of this equation is the pulse energy, the integrand on the right-hand side, $|A_{\Delta\omega}|^2\, d\omega$, can be interpreted as the differential energy in the frequency interval $[\omega, \omega + d\omega]$, and

$$S(\Delta\omega) := |A_{\Delta\omega}|^2 \tag{3.143}$$

represents the spectral energy distribution, or energy spectrum of the pulse.

As an example, we choose a (real) Gaussian envelope

$$A(0, t) = A_0 e^{-t^2/\tau_0^2}, \tag{3.144}$$

where $2\tau_0$ denotes the time between the $1/e$ $(1/e^2)$ points of the amplitude (intensity) envelope. In practice, the pulse duration is frequently given as the FWHM-width of the intensity $I(t) = |A(0, t)|^2 \propto e^{-2t^2/\tau_0^2}$,

$$\tau_{\mathrm{FWHM}} = \sqrt{2 \ln 2}\, \tau_0 = 1.1774\, \tau_0. \tag{3.145}$$

The Fourier transform of the envelope is also Gaussian

$$A_{\Delta\omega} = \sqrt{\pi}\, \tau_0 A_0 e^{-(\Delta\omega)^2 \tau_0^2/4}, \tag{3.146}$$

and the energy spectrum

$$|A_{\Delta\omega}|^2 \propto \tau_0^2 e^{-(\Delta\omega)^2 \tau_0^2/2} \tag{3.147}$$

has an FWHM-width of

$$\Delta\omega_{\mathrm{FWHM}} = 2\sqrt{2\ln 2}\,\frac{1}{\tau_0}, \tag{3.148}$$

which scales with $1/\tau_0$.

It is important to note that a given power spectrum $|A_{\Delta\omega}|^2$ allows for an infinite number of amplitude spectra $A_{\Delta\omega}$, differing by the relative phases of the individual frequency components. Since the Fourier integral (3.137) is very phase sensitive, the corresponding temporal pulse profiles $|A(t)|^2$ may vary greatly in shape and duration. It is quite obvious, however, that if all Fourier components happen to have the same phase (0, for example), the pulse shape $|A(t)|^2$ reaches the highest possible peak value. Since the pulse energy (given by the area under $|A(t)|^2$) is independent of the phase, this pulse is also the shortest possible, and is called Fourier limited for that reason. The product of the pulse duration and the width of the power spectrum has therefore a lower limit; for a Gaussian spectrum, it can be expressed as

$$\tau_{\mathrm{FWHM}} \Delta\omega_{\mathrm{FWHM}} \geq 4\ln 2 \approx 0.44 \times 2\pi. \tag{3.149}$$

3.2.1.3 Propagation Effects in the Frequency Domain

The Fourier component a_ω of a propagating pulse corresponds to the plane wave $a_\omega(0)e^{-j(kz-\omega t)} = \left[a_\omega(0)e^{-jkz}\right]e^{j\omega t}$, so that

$$a_\omega(z) = a_\omega(0)e^{-jkz}. \tag{3.150}$$

Substituting Eq. (3.141) and introducing $\Delta k := k - k^0$, we obtain the analog relation for the envelope

$$A_{\Delta\omega}(z) = A_{\Delta\omega}(0)e^{-j\Delta kz}. \tag{3.151}$$

This equation has the structure

$$A_{\Delta\omega}(z) = A_{\Delta\omega}(0)H(\Delta\omega), \tag{3.152}$$

where $H(\Delta\omega) = \mathrm{e}^{-\mathrm{j}\Delta k z}$ is the transfer function of the propagation process. To account for the dispersion $k = k(\omega)$, we use the expansion

$$\Delta k = \frac{\mathrm{d}k}{\mathrm{d}\omega}\Delta\omega + \frac{\mathrm{d}^2 k}{2\,\mathrm{d}\omega^2}(\Delta\omega)^2 + \dots \tag{3.153}$$

$$= \frac{1}{v_g}\Delta\omega + \frac{D_\omega}{2}(\Delta\omega)^2 + \dots \tag{3.154}$$

to obtain

$$H(\Delta\omega) = \mathrm{e}^{-\mathrm{j}\left(\Delta\omega/v_g + D_\omega(\Delta\omega)^2/2 + \dots\right)z}. \tag{3.155}$$

The first term of the exponent is responsible for a propagation delay of the pulse envelope by the time z/v_g and can be taken into account by a coordinate transformation to a system co-propagating with the pulse at v_g [see Eq. (3.165)]. The second term is equivalent to the spatial transfer function Eq. (3.108) for beam propagation and has the effect of changing the envelope during propagation. The envelope $A(z, t)$ at an arbitrary distance is the inverse Fourier transformation of Eq. (3.152),

$$A(z, t) = \frac{1}{2\pi}\int_{-\infty}^{\infty} A_{\Delta\omega}(z)\mathrm{e}^{\mathrm{j}\Delta\omega t}\,\mathrm{d}\Delta\omega. \tag{3.156}$$

Note that the power spectrum

$$|A_{\Delta\omega}(z)|^2 = |A_{\Delta\omega}(0)|^2 \tag{3.157}$$

is conserved during propagation, provided that the expansion Eq. (3.154) is real valued.

The term

$$\frac{1}{v_g} = \frac{\mathrm{d}k}{\mathrm{d}\omega} \tag{3.158}$$

is the inverse of the group velocity v_g as introduced in Eq. (1.41) and denotes the group delay l/v_g per unit length. The so-called dispersion coefficient

$$D_\omega := \frac{\mathrm{d}^2 k}{\mathrm{d}\omega^2} = \frac{\mathrm{d}}{\mathrm{d}\omega}\left(\frac{1}{v_g}\right) \tag{3.159}$$

represents the frequency dependence of the group delay and is a measure of the group velocity dispersion (GVD); D_ω is given in $[(\text{ps})^2 \, \text{km}^{-1}]$ (1 ps $= 10^{-12}$ s). An alternative definition is $D_\lambda := (\text{d}/\text{d}\lambda_0)(1/v_\text{g})$ which represents the *wavelength* dependence of the group delay. Since $D_\lambda \, \text{d}\lambda_0 = D_\omega \, \text{d}\omega$ and $\omega = 2\pi c_0/\lambda_0$, the two quantities are related by

$$D_\lambda = -2\pi \frac{c_0}{\lambda_0^2} D_\omega. \tag{3.160}$$

If the propagation index is given as a function of λ_0, it follows from Eq. (1.43) that

$$\frac{1}{v_\text{g}} = \frac{1}{c_0}\left(n - \lambda_0 \frac{\text{d}n}{\text{d}\lambda_0}\right) \tag{3.161}$$

and

$$D_\lambda = \frac{\text{d}(1/v_\text{g})}{\text{d}\lambda_0} = -\frac{\lambda_0}{c_0}\frac{\text{d}^2 n}{\text{d}\lambda_0^2}, \tag{3.162}$$

given in units of $[\text{ps}\,\text{nm}^{-1}\,\text{km}^{-1}]$. Depending on the sign of the dispersion coefficient D_ω, one distinguishes normal (or positive) GVD ($D_\omega > 0$) and anomalous (negative) GVD ($D_\omega < 0$); note that D_ω and D_λ have opposite sign.

The dispersion coefficient is frequency dependent; according to Eq. (3.162), D_λ is a measure of the curvature of the function $n(\lambda_0)$, which can change from positive to negative (Fig. 3.17). At the inflection points, the GVD is zero, which has important consequences for pulse propagation.

3.2.1.4 Propagation Effects in the Time Domain

Differentiation of Eq. (3.152) yields

$$\frac{\partial}{\partial z} A_{\Delta\omega} = -\text{j}\left[\frac{1}{v_\text{g}}\Delta\omega + \frac{D_\omega}{2}(\Delta\omega)^2\right] A_{\Delta\omega}, \tag{3.163}$$

where cubic and higher terms in the transfer function Eq. (3.155) have been neglected. According to Eq. (3.156), the differentiation $\partial^n/\partial t^n$ of the envelope $A(z,t)$ is equivalent to a multiplication of its Fourier components with $(\text{j}\Delta\omega)^n$ and vice versa. We therefore can convert Eq. (3.163) into a differential equation for $A(z,t)$

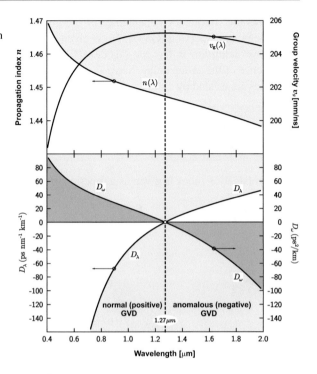

Fig. 3.17 Refractive index, group velocity, and dispersion coefficient of quartz glass; $n(\lambda)$ exhibits an inflection point at $\lambda_0 = 1.27\,\mu\text{m}$, so that v_g assumes an extremal value and D_ω (D_λ) is zero at this wavelength

using the substitution $j\Delta\omega \rightarrow \partial/\partial t$ and $(\Delta\omega)^2 \rightarrow -\partial^2/\partial t^2$

$$\left[\frac{\partial}{\partial z} + \frac{1}{v_g}\frac{\partial}{\partial t} - \frac{jD_\omega}{2}\frac{\partial^2}{\partial t^2}\right]A(z,t) = 0. \tag{3.164}$$

With the aforementioned transformation to a moving frame

$$\tau := t - \frac{z}{v_g}, \qquad \zeta := z, \tag{3.165}$$

and using $\frac{\partial}{\partial z} = \frac{\partial}{\partial \zeta}\frac{\partial \zeta}{\partial z} + \frac{\partial}{\partial \tau}\frac{\partial \tau}{\partial z} = \frac{\partial}{\partial \zeta} - \frac{1}{v_g}\frac{\partial}{\partial \tau}, \frac{\partial}{\partial t} = \frac{\partial}{\partial \tau}$, Eq. (3.164) assumes the form

$$\left[\frac{\partial}{\partial \zeta} - \frac{jD_\omega}{2}\frac{\partial^2}{\partial \tau^2}\right]A(\zeta,\tau) = 0. \tag{3.166}$$

In the absence of GVD, Eq. (3.166) reduces to $\partial A(\zeta,\tau)/\partial \zeta = 0$ and the envelope propagates without change[3] at the velocity v_g (Fig. 3.18),

---

[3]This statement is, of course, only valid if the higher terms in Eq. (3.154) are negligible.

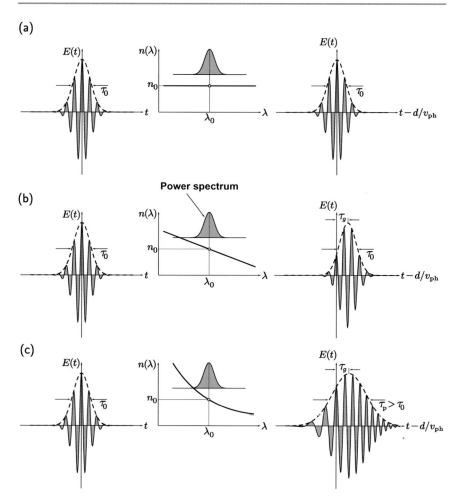

Fig. 3.18 Development of a light pulse during propagation over a distance d: (**a**) if n is constant over the entire pulse spectrum, the pulse remains completely unchanged and is delayed by d/v_{ph}; (**b**) in the absence of GVD ($D_\lambda = 0$), the temporal pulse profile is conserved, but experiences a propagation delay d/v_{g}, different from the phase delay d/v_{ph} of the carrier; (**c**) in the general case ($D_\lambda \neq 0$), the pulse profile changes and the momentary frequency becomes time dependent

$$A(\zeta, \tau) = A(0, \tau). \tag{3.167}$$

3.2.1.5 Gaussian Pulses

For $D_\omega \neq 0$, Eq. (3.166) can be written in the form

$$\left[\frac{\partial^2}{\partial \tau^2} + 2\mathrm{j}\frac{1}{D_\omega}\frac{\partial}{\partial \zeta}\right] A(\zeta, \tau) = 0. \tag{3.168}$$

With the substitutions $x \to \tau, z \to \zeta, k \to -1/D_\omega$, this corresponds to the paraxial Helmholtz equation Eq. (3.4), reduced, however, to two dimensions

$$\left[\frac{\partial^2}{\partial x^2} - 2jk\frac{\partial}{\partial z} \right] A(z,x) = 0. \tag{3.169}$$

A solution of the three-dimensional Helmholtz equation was discussed as Gaussian beam Eq. (3.9) in Sect. 3.1.2. As can be easily shown, Eq. (3.168) has a very similar solution

$$A(\zeta,\tau) = A_0 \frac{\sqrt{j\zeta_0}}{\sqrt{\zeta + j\zeta_0}} \exp\left[j\frac{\tau^2}{2D_\omega(\zeta + j\zeta_0)} \right]$$

$$= A_0 \sqrt{\frac{j\zeta_0}{\zeta + j\zeta_0}} \exp\left[\frac{\tau^2}{2D_\omega} \frac{\zeta_0}{\zeta^2 + \zeta_0^2} \right] \exp\left[j\frac{\tau^2}{2D_\omega} \frac{\zeta}{\zeta^2 + \zeta_0^2} \right]. \tag{3.170}$$

Apart from the square root in the leading factor, there is a one-to-one correspondence with Eq. (3.9). In particular, the real valued exponential factor can be interpreted as Gaussian envelope $\exp(-\tau^2/\tau_p^2)$ with the $1/e$ half width τ_p given by

$$\tau_p^2(\zeta) := \tau_0^2\left[1 + \frac{\zeta^2}{\zeta_0^2} \right], \tag{3.171}$$

where $\tau_0 = \tau_p(0)$ is related to ζ_0 by

$$\zeta_0 := -\frac{\tau_0^2}{2D_\omega}; \tag{3.172}$$

note that while ζ_0 can be positive or negative, depending on the sign of D_ω, ζ_0/D_ω in Eq. (3.170) is always negative.

Before discussing the phase terms, we calculate the pulse intensity $n|A|^2/2Z_0$

$$I(\zeta,\tau) = I_0 \frac{\tau_0}{\tau} \exp\left[-\frac{2\tau^2}{\tau_p^2} \right], \tag{3.173}$$

where $I_0 = I(0,0)$ [note the different prefactor in Eq. (3.18)]; this is a Gaussian pulse with an FWHM-width of $\tau_p\sqrt{2\ln 2}$. At a distance $|\zeta_0|$ from the point of minimum pulse width, the pulse duration has increased by a factor $\sqrt{2}$ (Fig. 3.19).

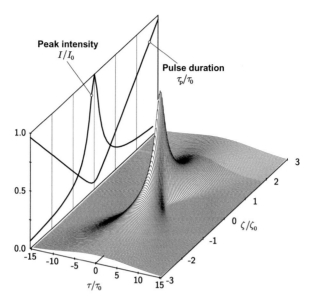

Fig. 3.19 Pulse duration and peak intensity of a Gaussian pulse as a function of propagation distance in a dispersive medium

This characteristic distance

$$|\zeta_0| = \frac{\tau_0^2}{2|D_\omega|}, \tag{3.174}$$

plays the same role as the confocal parameter z_0 for beam propagation and is called dispersion length. For $|\zeta| \gg |\zeta_0|$, the pulse duration grows almost linearly

$$\tau_p(\zeta) \approx \tau_0 \left| \frac{\zeta}{\zeta_0} \right| = \left| \frac{2D_\omega}{\tau_0} \zeta \right| = \frac{\Delta\omega_{\text{FWHM}}}{\sqrt{2\ln 2}} |D_\omega| \zeta; \tag{3.175}$$

the increase is proportional to D_ω and $1/\tau_0$, which is essentially the spectral width of the pulse [Eq. (3.148)].

The dispersion length depends on the propagation medium, the wavelength, and the (Fourier limited) pulse duration; in silica at a wavelength of $1\,\mu$m, for example, the dispersion coefficient is $D_\omega \approx 20\,\text{ps}^2\,\text{km}^{-1}$ (Fig. 3.17); the dispersion length for a 1 ns pulse is 5×10^4 km, and pulse broadening is negligible in practice. For a 1 ps pulse, it amounts to 50 m, implying that long distance optical communications in silica fibers is not feasible in this operating regime, since at 1 km the pulse broadening is already 20-fold. For ultrashort (10 fs) pulses, the dispersion length is reduced to a few mm, so that even the transmission through a lens results in significant pulse broadening.

3.2.1.6 Frequency Chirp
With Eqs. (3.171) and (3.172), Eq. (3.170) assumes the form

$$A(\zeta, \tau) = A_0 \sqrt{\frac{j\zeta_0}{\zeta + j\zeta_0}} \, \exp\left[-\frac{\tau^2}{\tau_p^2}\right] \exp\left[-j\frac{\tau^2}{\tau_0^2} \frac{(\zeta/\zeta_0)}{(\zeta/\zeta_0)^2 + 1}\right]. \tag{3.176}$$

To understand the implications of the phase term, it is useful to introduce the concept of momentary frequency,

$$\omega(\tau) = \partial\phi/\partial\tau; \tag{3.177}$$

for a monochromatic plane wave with the phase $-kz + \omega_0 t$, the momentary frequency is, of course, constant $\omega(\tau) = \omega_0$. For a pulse Eq. (3.135) with a Gaussian envelope Eq. (3.176), we find

$$\omega(\tau) = \omega_0 + \frac{\partial}{\partial\tau}\left[-\frac{\tau^2}{\tau_0^2} \frac{(\zeta/\zeta_0)}{(\zeta/\zeta_0)^2 + 1}\right] = \omega_0 - \frac{(\zeta/\zeta_0)}{(\zeta/\zeta_0)^2 + 1} \frac{2\tau}{\tau_0^2}; \tag{3.178}$$

during the pulse duration τ_p, the momentary frequency varies between the values $\omega(-\frac{\tau_p}{2})$ and $\omega(\frac{\tau_p}{2})$; for $\frac{\zeta}{|\zeta_0|} \gg 1$, the sweep covers the range

$$\omega_0 \pm \frac{|\zeta_0|}{\zeta} \frac{\tau_p}{\tau_0} \frac{1}{\tau_0} = \omega_0 \pm \frac{1}{\tau_0}, \tag{3.179}$$

which according to (3.148) is approximately equal to the entire pulse spectrum. Long distance pulse propagation thus provides spectral pulse analysis, just as far-field propagation of beams provides spatial Fourier transformation.

Depending on the sign of the slope, one refers to the sweep as positive or negative chirp; for normal GVD and $\zeta > 0$, the chirp is positive (from low to high frequencies), as shown in Fig. 3.20. This is what one might expect, because for normal GVD, higher frequencies travel slower than lower ones; note, however, that it is the dispersion of the *group* velocity, not of the phase velocity that matters.

3.2.1.7 Chirp Compensation and Pulse Compression
As can be seen from Eq. (3.171) and Fig. 3.19, a pulse can actually also get shorter during propagation. In a medium with normal dispersion, for example, a pulse that starts with a negative chirp contracts for some distance until it reaches its minimum duration and begins to broaden, acquiring a positive chirp. If such a positively

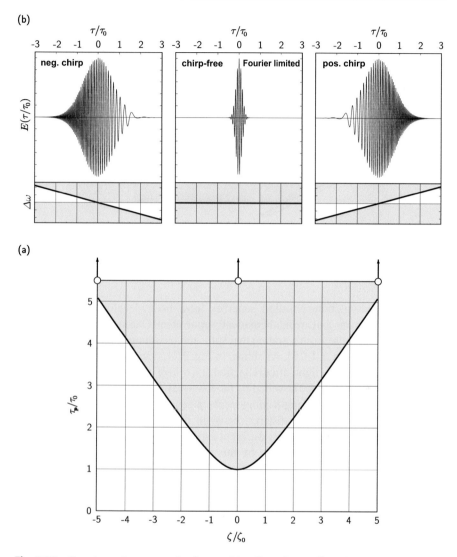

Fig. 3.20 Gaussian pulse propagating in a positive dispersive medium: (**a**) pulse duration, (**b**) normalized wave function at three selected points; for $\zeta < 0$ ($\zeta > 0$), the pulse exhibits a negative (positive) "chirp"; at $\zeta = 0$, the pulse is chirp-free and the pulse duration is minimal (Fourier limit)

chirped pulse is launched into a medium of anomalous dispersion, the broadening is reversed and the pulse, after a certain distance, reaches its minimum duration again. Thus, the chirp (and the broadening) introduced by one medium can, in principle, always be compensated by another medium. In practice, however, media with sufficient anomalous dispersion are not always available.

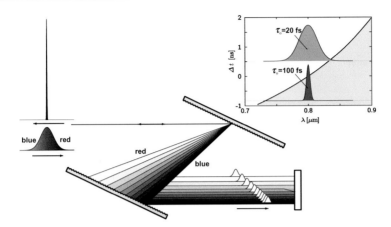

Fig. 3.21 Compression of a positively chirped pulse with the aid of a pair of gratings; the *inset* shows the group delay as a function of wavelength, and the (normalized) power spectrum of two short pulses

For that reason, compensation schemes have been developed that employ different mechanisms, but are mathematically equivalent. One such scheme relies on diffraction of the pulse from a grating that splits the incoming signal into its Fourier components [see Sect. 3.1.6.8]. As shown in Fig. 3.21, a second diffraction grating collimates the diverging light; after reflection at a mirror, the signal path is reversed. Since, however, the distance travelled by the long-wavelength components is longer than that of the short-wavelength components, a positive chirp can be compensated by proper adjustment of the distance between the gratings. As frequency chirp and envelope broadening are just different aspects of the same phenomenon, the incoming pulse is also compressed by this scheme, ideally to its Fourier limited duration. Schemes like that are indispensable for the generation and application of ultrashort pulses.

Pulse compression is only possible if the phases of the spectral components of a pulse are not randomly distributed; incoherent pulses therefore exhibit a much larger time–bandwidth product than coherent, Fourier limited pulses [Eq. (3.149)].

3.2.2 Nonlinear Propagation Effects

The optical Kerr effect Eq. (3.49),

$$n(I) = n_0 + n_2 I, \tag{3.180}$$

that we have encountered in the context of beam propagation is also responsible for interesting pulse propagation effects. The intensity envelope of the pulse produces a time dependent variation δn of the refractive index and thus of the wave number; this effect is called self-phase modulation (SPM) and will be discussed in more detail in Sect. 8.3.2. We can include it in the propagation equation by appending the term $\delta k(I) = k_0 \delta n = \frac{\omega}{c_0} \delta n$ in the expansion Eq. (3.154). With $I = n|A|^2/2Z_0$ we obtain

$$\delta k = n_2 \frac{\omega}{c_0} \frac{n}{2Z_0} |A(\zeta, \tau)|^2 = \kappa_k |A(\zeta, \tau)|^2, \tag{3.181}$$

where

$$\kappa_k := n_2 \frac{\omega}{c_0} \frac{n}{2Z_0}. \tag{3.182}$$

Equation (3.154) then assumes the form

$$\Delta k = \frac{1}{v_g} \Delta \omega + \frac{D_\omega}{2}(\Delta \omega)^2 + \kappa_k |A|^2 + \dots. \tag{3.183}$$

If we incorporate the Kerr term in Eq. (3.166), we obtain the nonlinear propagation equation

$$\left[\frac{\partial}{\partial \zeta} - \frac{jD_\omega}{2} \frac{\partial^2}{\partial \tau^2} + j\kappa_k |A(\zeta, \tau)|^2 \right] A(\zeta, \tau) = 0. \tag{3.184}$$

3.2.2.1 Spectral Broadening

Let us first discuss the case of vanishing GVD: Eq. (3.184) then is reduced to

$$\frac{\partial}{\partial \zeta} A(\zeta, \tau) = -j\kappa_k |A(\zeta, \tau)|^2 A(\zeta, \tau); \tag{3.185}$$

with the trial assumption that the *intensity* profile is not affected by the nonlinear propagation, we can integrate the equation to obtain

$$A(\zeta, \tau) = A(0, \tau) e^{-j\kappa_k |A(0,\tau)|^2 \zeta}, \tag{3.186}$$

consistent with our assumption of ζ-independent pulse shape,

$$|A(\zeta, \tau)|^2 = |A(0, \tau)|^2. \tag{3.187}$$

The pulse *amplitude*, however, acquires a time and space dependent phase (Fig. 3.22) that results in the production of new frequency components.

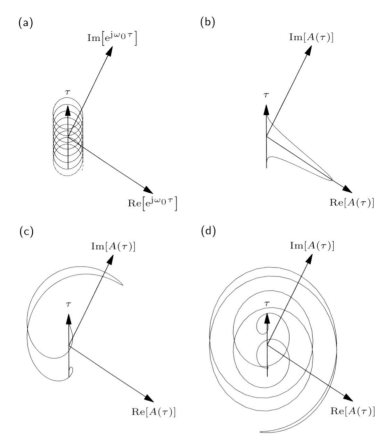

Fig. 3.22 Self-phase modulation of a short intense light pulse, shown in the complex amplitude plane as a function of time: (**a**) carrier; (**b**)–(**d**) envelope at $\zeta = 0$, $\zeta = 5\,\zeta_{NL}$, $\zeta = 20\,\zeta_{NL}$. During propagation, the intensity dependent phase $-\kappa_k|A(0,\tau)|^2\zeta$ is added to the complex amplitude, "wrapping" it up and resulting in a rotating amplitude phasor (note the rotation reversal at $\tau = 0$); the time dependent phase is added to the carrier, shifting the momentary frequency up ($\tau > 0$) or down ($\tau < 0$)

For a Gaussian pulse $A(0,\tau) = A_0 e^{-\tau^2/\tau_0^2}$, Eq. (3.177) yields

$$\omega(\zeta,\tau) = \omega_0 + \frac{\partial}{\partial\tau}\left(-\kappa_k|A|^2\zeta\right) = \omega_0 + \frac{4\tau}{\tau_0^2}\frac{\zeta}{\zeta_{NL}}e^{-2\tau^2/\tau_0^2}, \tag{3.188}$$

where

$$\zeta_{NL} := \frac{1}{\kappa_k|A_0|^2} = \frac{c_0}{n_2\omega n I_0} = \frac{\lambda_0}{2\pi n_2 n I_0} \tag{3.189}$$

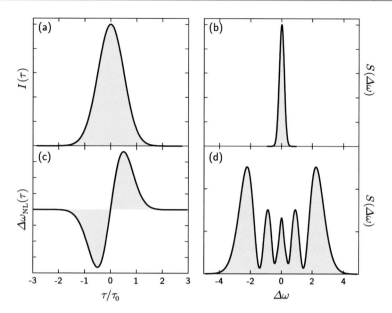

Fig. 3.23 Propagation of a Gaussian pulse in a nonlinear, dispersion free medium: (**a**) temporal intensity profile, (**b**) initial power spectrum, (**c**) instantaneous frequency due to the Kerr effect, (**d**) power spectrum after propagation over $15\zeta_{NL}$ (note that spectra (**b**) and (**d**) are scaled differently)

is called nonlinear length and $I_0 = n|A_0|^2/2Z_0$ is the peak intensity of the pulse (Fig. 3.23). Since n_2 is usually positive, the momentary frequency in the rising (falling) section of the pulse is red (blue)-shifted (Fig. 3.23c). Approximating the central section of the pulse by $\exp(-x^2) \approx 1 - x^2$, we obtain a linear, positive chirp

$$\omega(\tau) \approx \omega_0 + \frac{4\tau}{\tau_0^2}\frac{\zeta}{\zeta_{NL}}. \tag{3.190}$$

The manifestations of dispersion and SPM are somewhat complementary: dispersion conserves the power spectrum but modifies the pulse shape, while SPM conserves the pulse shape and modifies the power spectrum. The resulting spectrum is not only broader than the original, but may also show oscillatory features (Fig. 3.23d) that result from spectral interference, since the same frequency may be generated at different times. As the total pulse energy is conserved, the new frequencies are created at the expense of others.

To estimate the spectral broadening, we determine the maximum frequency excursion, that appears, according to Eq. (3.188), at the steepest points $\tau = \pm\tau_0/2$ of the intensity envelope $|A|^2$. The frequency excursion grows linearly with propagation distance and amounts to $\pm(2/\sqrt{e})(\zeta/\zeta_{NL})\tau_0^{-1}$. Propagation of an originally chirp-free pulse over $\zeta = \zeta_{NL}$ approximately doubles the spectral width of the original pulse, τ_0^{-1}. Under proper conditions, the emerging spectral width can span an entire octave, an effect that is called white light generation. Since SPM is a coherent process, the temporal and spatial phase of the input pulse is transferred to the output pulse so that, for example, two white light pulses, derived from the same initial pulse, can interfere with each other.

For propagation distances much smaller than ζ_{NL}, SPM can be neglected; in this sense, ζ_{NL} plays a similar role for nonlinear propagation effects as ζ_0 for the onset of significant dispersion.

As a numerical example, let us calculate the nonlinear length of pulse propagation in a silica fiber ($n_2 = 3.2 \times 10^{-20}$ m^2 W^{-1}) of $100\,\mu$m^2 core area: a pulse with a wavelength of $1\,\mu$m and a peak power of 1 W (peak intensity $I_0 = 10^{10}$ W m^{-2}) results in a nonlinear length of 330 m. To ensure that nonlinear effects are negligible over a fiber distance of several 10 km, the peak power must be kept below some 10 mW.

3.2.2.2 Combined Dispersive and Nonlinear Effects, Solitons

The frequency chirp resulting from SPM is usually positive; in a normally dispersive medium, the dispersion induced chirp is also positive, leading to pulse broadening and a reduction of peak intensity. Accordingly, SPM becomes less and less important during the propagation in such a medium. Nonetheless, the spectrum is broadened, providing spectral width for a pulse that is potentially shorter than the original pulse. If the accumulated chirp is compensated after the passage through the nonlinear medium (ideally to the Fourier limit), a pulse that is up to 100 times shorter than the input pulse can be obtained; this is a very powerful technique to produce, for example, femtosecond pulses from picosecond lasers.

If the nonlinearity is combined with negative dispersion, the two chirp contributions can cancel each other so that not only the power spectrum, but also the envelope remains unchanged during propagation. Such pulses are eigenfunctions of the nonlinear propagation equation and are called solitons.[4] With $D_\omega < 0$, Eq. (3.184) in the form

$$\left[j\frac{\partial}{\partial\zeta} + \frac{D_\omega}{2}\frac{\partial^2}{\partial\tau^2} - \kappa_k|A|^2 \right] A(\zeta, \tau) = 0 \qquad (3.191)$$

---

[4]For a more precise definition see, e.g., Hasegawa (2003) and Agrawal (2012).

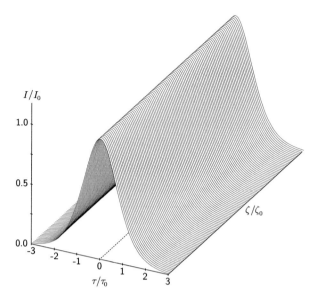

Fig. 3.24 Soliton propagation in a nonlinear, dispersive medium: the intensity profile *and* the power spectrum remain unchanged (compare Fig. 3.19)

has the same structure as the Schrödinger equation Eq. (6.1)

$$\left[-j\frac{\partial}{\partial t} + \frac{\hbar^2}{2m}\frac{\partial^2}{\partial x^2} - V(x) \right] \Phi(x, t) = 0, \tag{3.192}$$

that we will encounter in Sect. 6.1 as the wave equation of an electron in a potential V. In Eq. (3.191), the Kerr term $\propto -\kappa_k|A|^2$ plays the role of an attractive potential that prevents the wave function from dispersing; since the potential in turn depends on the wave function, Eq. (3.184) is called nonlinear Schrödinger equation. The simplest solution of Eq. (3.184) has a hyperbolic secans envelope

$$A(\zeta, \tau) = A_0 \mathrm{sech}\left[\frac{\sqrt{2}\tau}{\tau_0} \right] e^{j\zeta/2\zeta_0}, \tag{3.193}$$

where $\mathrm{sech}(x) := 2/(e^x + e^{-x})$ and the pulse duration[5] τ_0 is related to the dispersion length ζ_0 by Eq. (3.172). Cancellation of dispersive and nonlinear effects requires a subtle balance between spectral width and peak amplitude; substitution

---

[5]The FWHM duration of a $\mathrm{sech}^2\left(\sqrt{2}\tau/\tau_0\right)$ pulse is given by $\sqrt{2}\ln\left(1 + \sqrt{2}\right)\tau_0 = 1.247\tau_0$ for comparison, the FWHM duration of a Gaussian pulse is $1.177\tau_0$.

of Eq. (3.193) in Eq. (3.184) yields the soliton condition

$$A_0 = \frac{1}{\tau_0} \sqrt{\frac{2|D_\omega|}{\kappa_k}}. \tag{3.194}$$

With these pulse parameters, dispersion length and nonlinear length match exactly, $\zeta_0/\zeta_{NL} = 1$. Figure 3.24 shows the propagation of such a soliton. The pulse energy is proportional to $|A_0|^2 \tau_0$ and, because of Eq. (3.194) proportional to $1/\tau_0$. Shorter solitons therefore require higher pulse energy; even more interestingly, if a soliton looses energy during propagation, it adjusts itself by loosing bandwidth (by broadening) to conserve its soliton status. On the other hand, if a pulse with the right shape but too much energy is launched, it gets rid of the excess energy by splitting into an "ordinary," dispersive pulse (which fades away by broadening) and a soliton.

Silica glass fibers offer a spectral range with negative dispersion and very low transmission losses: at a wavelength of $1.5\,\mu m$, the dispersion coefficient is $D_\omega \approx -20\,ps^2\,km^{-1}$ so that a 10 ps pulse has a dispersion length of about 5 km. Assuming a core cross section of $100\,\mu m^2$, a peak power of just $\approx 100\,mW$ is necessary to meet the soliton condition.

3.3   Summary

Coherent light sources allow for the controlled generation of light pulses and beams, i.e., the concentration of electromagnetic energy in space and time. The electrodynamic wave equation requires the envelope of the pulse or the transverse profile of the beam, respectively, to change during propagation. While temporal broadening of pulses is due to the dispersion of the propagation medium, the divergence of optical beams is a geometric effect possible only in space.

A deeper understanding of pulse and beam evolution is provided by (Fourier) transformation of the wave function into the (temporal or spatial) frequency domain where propagation can be described by a multiplicative transfer function. Neglecting dissipation, the transfer function acts exclusively on the phases of the Fourier components; in real space and time this is equivalent to a change, usually a broadening, of the pulse or beam envelope during propagation. The power spectrum is conserved during propagation; for a given power spectrum, it is possible to find, in a unique way, the pulse shape that reaches the highest possible peak power and thus represents the shortest possible pulse; such a pulse is called Fourier limited.

Laser beams are technologically very important and their transformation during propagation in free space, by lenses or curved mirrors is a frequently encountered task. The replacement of the propagation coordinate by a complex q-parameter, and a bilinear transformation acting on this parameter (ABCD formalism) greatly facilitates the treatment of such problems; any sequence of lenses or curved mirrors and sections of free space is represented by a specific ABCD matrix; the question, for example, whether a Gaussian mode exists that "fits" between the two curved mirrors of a laser resonator, is reduced to the condition that the absolute value of trace of the ABCD matrix is less than 2, as we shall show in Sect. 4.3.1. Initially introduced for Gaussian beams, the ABCD formalism can also be applied to the family of Hermite–Gaussian beams and others.

The high intensities provided by laser sources introduce nonlinear propagation effects in addition to dispersion; while the entire Chap. 8 is dedicated to nonlinear optics, nonlinear propagation effects relying on the intensity dependence of the propagation index are described in the present Chapter; spectral broadening of pulses, soliton propagation, and self-focusing are important manifestations of this class of effects.

3.4   Problems

1. Prove Eq. (3.68) by induction.
2. Assume a Gaussian laser beam ($\lambda = 1064\,nm$) having an FWHM-diameter (intensity) of 5 mm. By transmission through a thin nonlinear crystal, a new wave is generated that is proportional to the square of the incoming field (and therefore radiates at twice the frequency, or 532 nm wavelength). This second harmonic (SH) beam co-propagates with the fundamental laser beam. What is the FWHM-diameter of the SH beam? What are the confocal parameter and the divergence of the two beams? What is the respective FWHM-diameter after a distance of 1 km?
3. Using the ABCD formalism and appropriate graphical software (gnuplot), reproduce Fig. 3.11 and vary the location of the input beam waist and the diameter of the input beam at the lens.
4. With the "beam tracing" software developed in Problem 3, reproduce Fig. 3.13 and vary the distance between the lenses, the location of the input beam waist, and other parameters and observe what happens.
5. What is the duration of a Fourier limited Gaussian pulse with FWHM duration 10 ns (10 fs) and $\lambda_0 = 800\,nm$ after propagation through 1 km of air ($D_\omega = 40\,fs^2\,m^{-1}$)?
6. Calculate the electric field amplitude of a focused, Fourier limited ultrashort light pulse (spatial and temporal Gaussian) with pulse energy of 10 nJ, center wavelength 800 nm, spectral width (FWHM) 150 nm, focal length of focusing lens 50 mm, beam diameter at the lens entrance 20 mm; neglect the dispersion of the lens material.
7. Assume that the lens in Problem 6 is made of BK7 glass with a refractive index of 1.5 and a dispersion coefficient of $-130\,ps/nm\,km$ at 800 nm wavelength; the lens diameter is 25 mm. Calculate the minimum thickness of the lens and the pulse duration behind the lens; compare the electric field amplitude in the focus with the result of problem 6.

8. A Fourier limited Gaussian pulse of 1 ps duration is propagated through 50 m of dispersion free fiber and experiences spectral broadening by self-phase modulation. Assuming perfect chirp compensation of the resulting output pulse, calculate the shape and duration of the output pulse after compression. The pulse energy is 1 nJ, the effective area of the core is $50 \, \mu m^2$, $n_{eff} = 1.5$, $n_2 = 3 \times 10^{-20} \, m^2 \, W^{-1}$, $\lambda = 1.3 \, \mu m$.

9. 1 ps pulses ($\lambda = 1.3 \, \mu m$) are transmitted through a dispersion free fiber with $20 \, \mu m^2$ effective core area; what is the maximum number of photons/pulse, so that the nonlinear length is more than 100 km?

References and Suggested Reading

Agrawal, G. P. (2012). *Nonlinear fiber optics*. New York: Academic Press.

Arnaud, J. A. (1976). *Beam and fiber optics*. New York: Academic Press.

Belanger, P. A. (1991). Beam propagation and the ABCD ray matrices. *Optics Letters, 16*(4), 196–198.

Brillouin, L. (1960). *Wave propagation and group velocity*. New York: Academic Press.

Cerullo, G., Longhi, S., Nisoli, M., Stagira, S., Svelto, O. (2001). *Problems in laser physics*. New York: Springer.

Gerrard, A., & Burch, J. M. (1994). *Introduction to matrix methods in optics*. New York: Dover.

Goodman, J. W. (1996). *Introduction to Fourier-optics*. New York: McGraw-Hill.

Hasegawa, A. (2003). *Optical solitons in fibers*. New York: Springer.

Haus, H. A. (1984). *Waves and fields in optoelectronics*. Englewood Cliffs, NJ: Prentice Hall.

ISO-Standard (2005). Lasers and laser-related equipment—test methods for laser beam widths, divergence angles and beam propagation ratios. *ISO-Standard 11146*.

Kogelnik, H., & Li, T. (1966). Laser beams and resonators. *Applied Optics, 5*(10), 1550–1567.

Saleh, B. E., & Teich, M. C. (2007). *Fundamentals of photonics*. New York: Wiley.

Svelto, O. (2010). *Principles of lasers*. New York: Plenum Press.

Wartak, M. S. (2012). *Computational photonics*. New York: Cambridge University Press.

Optical Interference

4

As a consequence of the linearity of Maxwell's equations, the total electromagnetic field that results from a superposition of fields is the vector sum of the fields; practically all optical detectors, however, respond to the light *intensity*, i.e., to the absolute square of the field. The linear superposition principle generally applies only to the fields, but not to the intensity of a superposition of fields. Deviations from the linear superposition of intensities are called interference; in the following we will discuss important manifestations thereof.

4.1 Two Field Interference

For convenience, we introduce a complex vector amplitude $\tilde{\mathbf{U}}$ normalized such that the intensity is

$$I(\mathbf{x}) = \tilde{\mathbf{U}}(\mathbf{x}) \cdot \tilde{\mathbf{U}}^*(\mathbf{x}). \tag{4.1}$$

A superposition of two fields $\tilde{\mathbf{U}}_1$, $\tilde{\mathbf{U}}_2$ results in the intensity

$$I(\mathbf{x}) = (\tilde{\mathbf{U}}_1 + \tilde{\mathbf{U}}_2)(\tilde{\mathbf{U}}_1 + \tilde{\mathbf{U}}_2)^*$$

$$= I_1 + I_2 + 2\mathrm{Re}\left[\tilde{\mathbf{U}}_1(\mathbf{x}) \cdot \tilde{\mathbf{U}}_2^*(\mathbf{x})\right]. \tag{4.2}$$

The first two terms are the intensities of the isolated individual fields, while the third one is the so-called interference term, which can be positive (constructive interference) or negative (destructive interference), depending on the phase difference between the two fields.

A given detector can follow temporal changes only up to a certain frequency—above this frequency it measures only the time average of the signal. If the phase between the two fields—and thus the sign of the interference term—changes too

© Springer International Publishing Switzerland 2016
G.A. Reider, *Photonics*, DOI 10.1007/978-3-319-26076-1_4

quickly, the average over the third term may vanish and the total intensity tends towards the sum of the individual intensities; if the two fields are completely uncorrelated, the interference term vanishes and the linear superposition principle applies to the intensity.

The interference term also vanishes if the two amplitude vectors are orthogonal, i.e., if the two fields represent orthogonal polarization states (Sect. 1.5.1.1). It is very important to note, however, that what matters for interference phenomena is the polarization at the *detector*: a polarization filter in front of the detector (that projects the two fields onto a common polarization state) can render interferences visible that are not detected in the absence of the filter.

In the following, we restrict ourselves to monochromatic, coherent fields of equal polarization, so that we can use a scalar description $\tilde{U}_{1,2}$

$$\tilde{U}_{1,2} = A_{1,2}e^{j\phi_{1,2}}, \tag{4.3}$$

where $A_{1,2}$ is real and positive. The intensity of the superposition is then

$$I(\mathbf{x}) = I_1(\mathbf{x}) + I_2(\mathbf{x}) + 2\sqrt{I_1(\mathbf{x})I_2(\mathbf{x})}\cos(\phi_2 - \phi_1). \tag{4.4}$$

If we further assume fields of equal intensity, $I_{1,2} := I_0$, we obtain

$$I(\mathbf{x}) = 2I_0(\mathbf{x})\left[1 + \cos(\Delta\phi)\right] = 4I_0\cos^2(\Delta\phi/2), \tag{4.5}$$

where $\Delta\phi = \phi_2 - \phi_1$; the total intensity then can assume any value between 0 and $4I_0$. The maximum intensity $I_{max} = 4I_0$ is obtained if $\Delta\phi = 2m\pi$, where m is an integer. For the sake of simplicity, we use in the following plane waves Eq. (1.26) with the wave function $e^{-j(\mathbf{k}\cdot\mathbf{x}-\omega t)}$. Any difference in the path length, wave vector, frequency, or propagation time between the two partial waves results in a phase difference

$$\Delta\phi = -\Delta(\mathbf{k}\cdot\mathbf{x} - \omega t) \tag{4.6}$$

and may give rise to interference.

Phase correlated fields can be generated by splitting a (coherent) field with the help of a semi-transparent mirror (beam splitter). The two emerging fields can then propagate different paths before they are recombined on a detector. Important implementations of this scheme are the Michelson, Mach–Zehnder, and Sagnac interferometers (Fig. 4.1).

4.1.1 Michelson Interferometer

The beam splitter used to produce two phase correlated fields is usually a 50 % splitter (also called 3 dB splitter) that converts the incoming field of intensity I_0

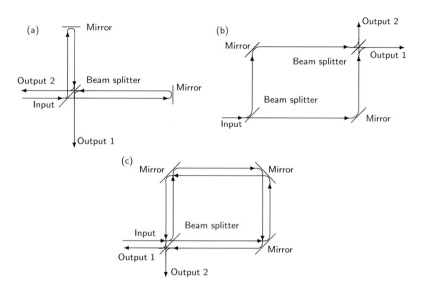

Fig. 4.1 Optical path in a (**a**) Michelson interferometer, (**b**) Mach–Zehnder interferometer, and (**c**) Sagnac interferometer

into two partial fields of intensity $I_0/2$ (note that the amplitude of the partial fields is not $U_0/2$, but $U_0\sqrt{2}/2$). In the Michelson scheme (Fig. 4.1a), the two waves travel different paths, are retroreflected by mirrors, and reach the beam splitter again, where they are partially transmitted and reflected; with proper geometric alignment, the fields can be overlapped to produce two output fields; the beam splitter thus also serves to recombine the partial waves. Neglecting, for the moment, possible phase shifts by the beam splitter, the phase difference between the superimposed fields is due to different path lengths

$$\Delta\phi = 2k\Delta s, \tag{4.7}$$

where Δs is the geometric length difference of the two interferometer branches. The output intensity is given by

$$I = I_0 \cos^2 \frac{2\pi\Delta s}{\lambda_0} = \frac{I_0}{2}\left(1 + \cos\frac{4\pi\Delta s}{\lambda_0}\right), \tag{4.8}$$

where $k = 2\pi/\lambda_0$ was used (Fig. 4.2). A variation of the length difference by only $\lambda_0/4$ is required to change the output intensity from a maximum to a minimum. Michelson interferometers are therefore employed for position measurements with nm-resolution. Usually, one arm is used as reference branch with constant length,

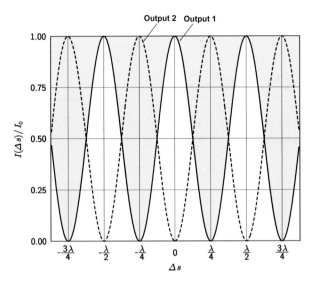

Fig. 4.2 Output power at the two output ports of a Michelson interferometer as a function of the length difference of the two branches

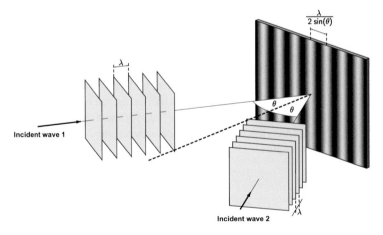

Fig. 4.3 Interference of two non-collinear plane waves

while the mirror in the second arm is attached to the object whose position is to be measured.

4.1.1.1 Tilted Wavefronts

If the wave vectors of the two waves incident on the detector are not parallel, there is a second source for a phase difference; let us assume a symmetric situation where the two wave vectors include an angle of $\pm\theta$ with the surface normal of the detector (Fig. 4.3). The two wave vectors can then be written as $\mathbf{k}_{1,2} = (\pm k \sin\theta, 0, k \cos\theta)$

and the phase difference $\Delta\phi(x) = 2kx \sin\theta$ is a function of the lateral coordinate of the detector plane. The resulting intensity pattern is

$$I(x) = 4I_0 \cos^2(kx \sin\theta), \tag{4.9}$$

a periodic pattern of bright and dark stripes with the period $\lambda_0/2\sin\theta$, which can be recorded by a spatially resolving detector such as a film or a camera chip.

4.1.1.2 Doppler Effect

Equation (4.7) represents the phase difference between two stationary positions of the interferometer mirrors. In practice, however, the sensing mirror moves between two positions at some finite velocity v, giving rise to a Doppler shift of the reflected light frequency. To calculate this effect, we assume a coordinate axis z along the sensing arm of the interferometer. Let the mirror, moving towards the beam splitter, be at position z_1 when a selected phase front strikes it and is reflected; the consecutive phase front, initially a distance of λ_0 behind the first one, strikes the mirror after a delay time τ, during which the mirror has moved to the position $z_2 = z_1 - v\tau$; τ follows from the equation $c_0\tau = \lambda_0 - v\tau$ to be $\tau = \lambda_0/(c_0 + v)$. Since the previous phase front has propagated to position $z_3 = z_1 - c_0\tau$ in the meantime, the distance between the two reflected phase fronts (which is, by definition, the wavelength of the reflected light) is $\lambda_0' = z_2 - z_3 = c_0\tau - v\tau = \lambda_0(c_0 - v)/(c_0 + v)$, and the Doppler shifted frequency of the reflected light is, accordingly,

$$\omega' = \omega \frac{1 + v/c_0}{1 - v/c_0}. \tag{4.10}$$

An identical result is obtained using the relativistic factor Eq. (2.197), which has to be applied twice since the moving mirror acts both as receiver and transmitter; note, however, that a relativistic treatment is not necessary in this case, since the final "observer" and the light source do not move relative to each other.

On the detector, the Doppler shifted wave from the sensing mirror and the wave from the reference arm (with frequency ω) are superimposed, resulting in a signal $\propto |e^{j\omega t} + e^{j\omega' t}|^2$ that varies with the beat frequency $\Delta\omega = |\omega' - \omega|$. Moving the mirror over the distance Δs at velocity v takes the time $t = \Delta s/v$, during which the phase difference between reference and Doppler shifted light adds up to

$$\Delta\phi = \Delta\omega t = \omega \left[\frac{1 + v/c_0}{1 - v/c_0} - 1 \right] \frac{\Delta s}{v} \approx 2\frac{\omega}{c_0}\Delta s, \tag{4.11}$$

where the approximation is valid for $v \ll c_0$ and agrees with Eq. (4.7). Since the beat frequency is a direct measure of the velocity of the mirror, the Michelson interferometer can also be used as a velocimeter.

4.1.2   Mach–Zehnder and Sagnac Interferometers

In a Mach–Zehnder interferometer, a separate beam splitter is used to recombine the partial waves (Fig. 4.1b). The geometric path length is equal in both branches and a possible phase difference can only result from a different *optical* path length, i.e., from different propagation indices in the two branches. This structure is frequently used for sensor application, where the propagation index in the reference branch is kept constant, while the sensing branch is exposed to some external influence that changes the propagation index.

Still another interferometric structure is the Sagnac interferometer (Fig. 4.1c), where both partial fields travel the same path (a loop), but in different directions. Only effects that depend on the propagation direction can give rise to a phase difference. In a more general way one can say that the Sagnac interferometer is sensitive to effects that are not invariant under time reversal, such as a rotation of the interferometer or the magnetooptic Faraday effect (Section 2.4.2).

Interferometric sensors are frequently implemented in integrated optics, i.e., as waveguide structures. We return to this important issue in Sect. 5.3.4.

4.1.3   S-Matrix

Equation (4.8) regarding the output intensity of the Michelson interferometer was obtained neglecting possible phase shifts by the beam splitter. All the interferometers discussed here have two output ports (corresponding to the two "ports" of the recombination beam splitter); if both ports would deliver an output according to Eq. (4.8), the total power would not be conserved. We have therefore to conclude that the beam splitter necessarily introduces a phase shift between the two emerging partial waves, so that the two output powers of the interferometer add up to the input power.

To understand the properties of a beam splitter, we describe it as a linear system with two complex input amplitudes $a_{1,2}$ and two output amplitudes $b_{1,2}$ (Fig. 4.4); in the case of a partially transmitting mirror, the outputs are the reflected and

Fig. 4.4 Complex in- and output signals at a beam splitter, measured in reference planes 1,2

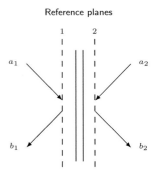

transmitted wave, respectively, and the inputs are waves incident on the mirror either from the front or the rear side. The in- and outputs are electromagnetic fields and we assume the relations between them to be linear

$$b_1 = S_{11}a_1 + S_{12}a_2$$
$$b_2 = S_{21}a_1 + S_{22}a_2. \tag{4.12}$$

Since the signals are waves in space, we have to define two reference planes (front and rear), where the signals (in particular their phase) are measured; these planes can be the physical surface of the mirror or any other plane parallel to the mirror surface. In matrix notation, Eq. (4.12) has the form

$$b = Sa, \tag{4.13}$$

where

$$a := \begin{bmatrix} a_1 \\ a_2 \end{bmatrix}, \qquad b := \begin{bmatrix} b_1 \\ b_2 \end{bmatrix}, \tag{4.14}$$

and

$$S := \begin{bmatrix} S_{11} & S_{12} \\ S_{21} & S_{22} \end{bmatrix} \tag{4.15}$$

is the so-called scattering matrix.

Let the signals be normalized such that their absolute square is equal to the energy flux density. For a lossless system, energy conservation implies $a_1^* a_1 + a_2^* a_2 = b_1^* b_1 + b_2^* b_2$, or

$$\left[a^*\right]^{\mathrm{T}} a = \left[b^*\right]^{\mathrm{T}} b, \tag{4.16}$$

where [ ]^{T} indicates the transposed matrix, i.e.,

$$[a]^{\mathrm{T}} = [a_1, a_2], \qquad [S]^{\mathrm{T}} := \begin{bmatrix} S_{11} & S_{21} \\ S_{12} & S_{22} \end{bmatrix}. \tag{4.17}$$

Since $[AB]^{\mathrm{T}} = [B]^{\mathrm{T}} [A]^{\mathrm{T}}$, we can conclude from Eqs. (4.16) and (4.13) that

$$\left[a^*\right]^{\mathrm{T}} a = \left[a^*\right]^{\mathrm{T}} \left[S^*\right]^{\mathrm{T}} Sa. \tag{4.18}$$

For Eq. (4.18) to be valid for arbitrary inputs, S must be unitary, $[S^*]^\mathrm{T} S = \mathbf{1}$, or

$$[S^*]^\mathrm{T} = S^{-1}. \tag{4.19}$$

We thus obtain the following equations relating the matrix components

$$S_{11}^* S_{11} + S_{21}^* S_{21} = 1 \tag{4.20}$$

$$S_{12}^* S_{12} + S_{22}^* S_{22} = 1 \tag{4.21}$$

$$S_{11}^* S_{12} + S_{21}^* S_{22} = 0 \tag{4.22}$$

$$S_{12}^* S_{11} + S_{22}^* S_{21} = 0. \tag{4.23}$$

Returning to the example of a semi-transparent mirror, the diagonal elements S_{11}, S_{22} of S represent the front and rear reflection coefficients, respectively, while the off-diagonal elements S_{12}, S_{21} are the transmission coefficients. The first two equations then simply state that the reflectance $r^* r$ and transmittance $t^* t$ must add up to 1.

A hypothetical 3 dB beam splitter with $S_{ii} = S_{ij} = \sqrt{2}/2$ obviously satisfies Eqs. (4.20) and (4.21), but violates Eq. (4.22). Such a beam splitter would produce the output Eq. (4.8) at *both* output ports of a Michelson interferometer, implying the annihilation or creation of energy, as we have remarked above. A possible choice satisfying the complete set Eqs. (4.20)–(4.23) is $S_{ii} = 1/\sqrt{2}$, $S_{ij} = \mathrm{j}/\sqrt{2}$, implying a $\pi/2$ phase shift between reflected and transmitted wave.

Another possible property of a beam splitter (or, more generally, of a system) is invariance under time reversal; in our context, time reversal swaps input and output and, as we shall see in Sect. 8.3.7, the complex amplitude of a signal into its conjugate. If the system is invariant under time reversal, it is described by the same matrix S

$$a^* = S b^*. \tag{4.24}$$

On the other hand, conjugation of Eq. (4.13) yields

$$b^* = S^* a^*. \tag{4.25}$$

Substituting Eq. (4.25) in Eq. (4.24) gives $a^* = S S^* a^*$, which is equivalent to

$$S^* = S^{-1}. \tag{4.26}$$

In combination with Eq. (4.19), we find that the scattering matrix of lossless, time reversal invariant system is symmetric

$$[S]^{\mathrm{T}} = S; \tag{4.27}$$

the transmission coefficient of such a beam splitter is independent of the propagation direction (reciprocity). A prominent example for a system that is not time reversal invariant is the Faraday rotator (Sect. 2.4.2); it allows building devices that are transparent in one direction and opaque in the opposite (Faraday isolator, Sect. 2.4.2.1). The reason for this is that a magnetic field changes its sign upon time reversal (being generated by a circulating current), so that effects that depend linearly on the magnetic field (for example, the Faraday rotation) also change sign. Another effect of this kind is the Sagnac effect that produces a phase shift in rotating systems and is the basis for the optical gyroscope (Sect. 5.3.4).

The scattering matrix formalism can be extended beyond single elements and allows us, for example, to describe an entire interferometer in a very concise way.

4.1.4 Young's Double Slit

An alternative way to produce phase correlated fields is to transmit a wave through two or more narrow slits in an opaque screen (wave front division, Fig. 4.5). The fields emanating from the slits propagate as cylindrical waves and can interfere where they overlap. This setup was used by Th. Young in 1801 to demonstrate that light is a wave phenomenon; the detector is a simple screen at a distance d from the aperture. If a is the distance between the slits, and x is the lateral coordinate on the screen, the respective propagation distances $r_{1,2}$ between slit 1,2 and the screen at

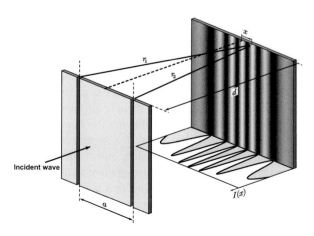

Fig. 4.5 Interference of light passing through a double slit ($d \gg a$)

coordinate x are

$$r_{1,2} = \sqrt{d^2 + (a/2 \pm x)^2} \approx d\left[1 + (a/2 \pm x)^2/2d^2\right], \qquad (4.28)$$

where the approximation is valid for $x/d \ll 1$. The path difference $\Delta r = r_2 - r_1 \approx ax/d$ results in a phase difference

$$\Delta\phi = 2\pi \frac{ax}{\lambda_0 d}, \qquad (4.29)$$

and the intensity distribution on the screen is therefore

$$I(x) \approx 2I_0\left[1 + \cos 2\pi \frac{ax}{d\lambda_0}\right] = 4I_0 \cos^2 \pi \frac{ax}{d\lambda_0}. \qquad (4.30)$$

4.1.4.1 Wave or Particle?

Young's double slit experiment is of great historical importance since it proves the wave character of light and apparently rules out the existence of light-particles, i.e., of photons. To shed some light on the nature of these well-established particles, we can perform the experiment at very low light levels, so that during the transit time τ through the apparatus there is statistically not more than one photon underway, by keeping the optical power flow well below $\hbar\omega/\tau$. The light impinging on the screen is detected with an array of sufficiently small photo detectors that ideally produce one photoelectron per incident photon. The measured histogram of photoelectron counts as a function of detector position x turns out to reproduce Eq. (4.30), provided that the integration time is long enough; in particular, at the points of zero intensity according to Eq. (4.30), no photoelectron is ever recorded. Just like other microscopic particles (electrons, neutrons, atoms, ...), a photon behaves like a wave during propagation and becomes localized when detected; the detection process converts the delocalized, wave-like photon into a localized one that excites a photoelectron. The electrodynamic intensity UU^* provides the spatial probability distribution for this process.

In this context, it is also worthwhile to note that a photon generally does not have a well defined frequency or energy. As we have seen in Sect. 3.2.1.2, a light signal of finite duration (i.e., a light pulse) cannot be monochromatic, but displays a frequency bandwidth that scales inversely with the pulse duration. If we again attenuate a light pulse to such a degree that there is not more than one photon at a time in our experimental apparatus, this photon has the same temporal and spectral properties as the original light pulse. Only if a measurement of the frequency is performed, the outcome has a certain value, and if many measurements with consecutive photons are made, we obtain a histogram of frequencies that reproduces the spectral distribution of the pulse. Stating that a photon constituting a short light pulse has a certain frequency (or energy) before a measurement is taken is as misleading as stating that a photon is transmitted through one or the other slit of Young's double slit setup. If we talk about "microscopic particles," we refer to entities that behave in this non-classical fashion as a matter of fact; there is no reason to introduce concepts such as wave–particle dualism or invoke a violation of energy conservation during temporally short interactions.

4.2 Multiple Wave Interference

A number of photonic components rely on the interference of a large (or infinite) number of partial fields $\tilde{U}_n$. Let us first study the special case of equal absolute value of the individual amplitudes and constant phase difference between the fields,

$$\tilde{U}_n = \tilde{U}_0 e^{j(n-1)\Delta\phi}, \qquad n = 1, 2, 3, \ldots, N \qquad (4.31)$$

In the complex plane, these amplitudes form a (generally open) polygon chain, with the resulting total field amplitude pointing from the origin to the final point of the chain (Fig. 4.6). The maximum possible intensity is realized if the partial amplitudes are all in phase, $\Delta\phi = 2m\pi$. The maximum intensity is given by $|N\tilde{U}_0|^2 = N^2|\tilde{U}_0|^2$, which is an N-fold enhancement over the sum of individual intensities, $N|\tilde{U}_0|^2$. The total field (and intensity) is zero whenever the polygon chain is closed; this happens if $N\Delta\phi$ is an integer multiple of 2π, or $\Delta\phi = 2m\pi/N$.

To evaluate the total field, we use $\sum_{n=1}^{N} q^{n-1} = (1 - q^N)/(1 - q)$ to obtain

$$\tilde{U}_{\text{total}} = \sum \tilde{U}_n = \tilde{U}_0 \sum_{n=1}^{N} e^{j(n-1)\Delta\phi} = \tilde{U}_0 \frac{1 - e^{jN\Delta\phi}}{1 - e^{j\Delta\phi}}. \qquad (4.32)$$

With

$$1 - e^{jN\Delta\phi} = e^{jN\Delta\phi/2} \left(e^{-jN\Delta\phi/2} - e^{jN\Delta\phi/2} \right) \qquad (4.33)$$

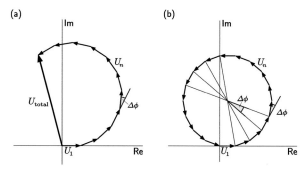

Fig. 4.6 Multiple beam interference, represented in the complex plane: the total field is the complex sum of the partial fields (**a**); if the polygon formed by the complex amplitudes is closed ($N\Delta\phi = 2m\pi$), the total field amounts to zero (**b**)

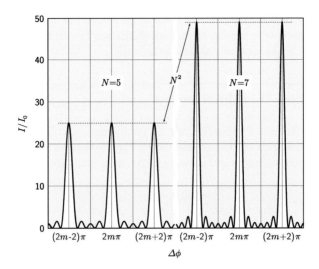

Fig. 4.7 Output intensity resulting from N-beam interference as a function of the phase difference $\Delta\phi$ for two different values of N

and $\left|\mathrm{e}^{\mathrm{j}N\Delta\phi/2}\right| = 1$ follows

$$I = \left|\tilde{U}_0\right|^2 \left|\frac{\mathrm{e}^{-\mathrm{j}N\Delta\phi/2} - \mathrm{e}^{\mathrm{j}N\Delta\phi/2}}{\mathrm{e}^{-\mathrm{j}\Delta\phi/2} - \mathrm{e}^{\mathrm{j}\Delta\phi/2}}\right|^2 = I_0 \frac{\sin^2(N\Delta\phi/2)}{\sin^2(\Delta\phi/2)} \qquad (4.34)$$

for the resulting intensity, as shown in Fig. 4.7. Compared to the result for two beam interference (Fig. 4.2), the maximum features are more pronounced and there appear $N - 2$ small intermediated peaks between the major peaks.

4.2.1   Optical Gratings

A possible realization of the multiple wave interference described above relies on an extension of the double slit experiment to an aperture with N equidistant slits, a component known as optical grating. Instead of slits, reflecting stripes on an opaque background can also be used (reflection grating). We have discussed such periodic structures in Sect. 2.1.1 and have calculated the angles under which an incident plane wave is scattered [Eq. (2.12)]. Here we want to calculate the angular dependence of the intensity of the diffracted wave, assuming an incident plane wave and a grating of finite size. The angle of incidence is θ_{in} and the slits or stripes are oriented normal to the plane of incidence (Fig. 4.8). We assume that each slit is the origin of a cylindrical wave; far away from the grating, the phase difference between waves

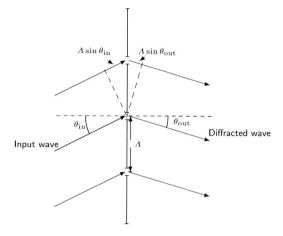

Fig. 4.8 Geometry of a transmission grating; angles are measured between the surface normal and the wave vector, in counterclockwise sense of rotation

from adjacent slits is

$$\Delta\phi = \frac{2\pi\Lambda}{\lambda_0}\left(\sin\theta_{\text{in}} - \sin\theta_{\text{out}}\right), \tag{4.35}$$

where Λ is the spatial period of the grating and θ_{out} is the angle of observation (positive or negative).

The angular intensity distribution is then essentially given by Eq. (4.34), where N is the number of slits. The condition for maximum intensity is, in agreement with Eq. (2.12), $\Delta\phi = 2m\pi$, or

$$\Lambda\left(\sin\theta_{\text{out}} - \sin\theta_{\text{in}}\right) = m\lambda_0, \tag{4.36}$$

where m is an integer denoting the order of interference. The wavelength dependence of this condition is the basis of grating spectrometers (Sect. 3.1.6.8) and monochromators that can filter a narrow frequency band out of a polychromatic input signal.

For this application, the resolving power is an important parameter. For a given set of in- and output angle θ_{in}, θ_{out}, we can, from Eq. (4.34), find the wavelength variation $\Delta\lambda_0$ that reduces the scattered amplitude from its maximum value to the first adjacent zero; according to Eq. (4.34), this requires a change of $N\Delta\phi/2$ by π, i.e., $\Delta(\Delta\phi) = 2\pi/N$. From Eq. (4.35) follows $|d\lambda_0/\lambda_0| = |d(\Delta\phi)/\Delta\phi|$, so that

$$\left|\frac{\Delta\lambda_0}{\lambda_0}\right| \approx \frac{1}{mN}; \tag{4.37}$$

the resolving power is defined as the reciprocal value and is given by

$$\left| \frac{\lambda_0}{\Delta\lambda_0} \right| = mN; \tag{4.38}$$

thus, the resolving power is proportional to the number of lines, and to the diffraction order m. Commercial gratings usually work in reflection and have about 300–1200 lines/mm; the total number of lines is typically 10^4–10^5, and the grating is operated up to the fourth order. The reflecting lines are usually grooves with triangular profile so as to maximize the reflection efficiency in a particular direction ("blazing").

4.2.2 Dielectric Multilayer Systems

Another important component relying on multiple beam interference is the dielectric multilayer mirror (see, e.g., MacLeod 2001). It consists of up to thirty dielectric layers of different propagation index and is usually designed such that the reflections from the interfaces between the layers add up constructively so as to achieve maximum reflectance in a certain wavelength range. In other applications, the opposite goal is intended, namely the reduction of the reflectance, ideally to zero (antireflection or AR coating).

According to Sect. 2.1.1, the reflection coefficient of a dielectric interface at normal incidence is given by

$$r = \frac{n_1 - n_2}{n_1 + n_2} = -\frac{1 - n_1/n_2}{1 + n_1/n_2}, \tag{4.39}$$

where $n_{1,2}$ are the propagation indices of the two media; since the propagation index of typical dielectrics in the visible is between 1 and 2.5, the corresponding reflectance $|r|^2$ is rather moderate. A stack of alternating layers of high and low index $n_{h,l}$, respectively, produces an enhanced reflected field, provided that the individual contributions add up constructively. The phase difference results from the propagation delay of the forward and backward propagating waves, and the phase jump at each interface, which is 0 or π depending on the sequence of propagation indices [Eq. (4.39)]. Considering both of these contributions, we expect maximum reflectance for a layer thickness of $\lambda/4$; the integrated reflection coefficient, however, is not a simple sum over the individual reflections, since each interface scatters the locally incident field into a reflected and transmitted wave, so that the number of partial waves is actually infinite.

This highly complex problem can be solved in an elegant way by resorting to boundary conditions at the individual interfaces. We assume a sequence of N dielectric layers with propagation index n_i and thickness d_i, supported by a substrate of index n_s (Fig. 4.9); light is incident onto the stack from the left, denoting the forward direction in the following. To establish boundary conditions at the two

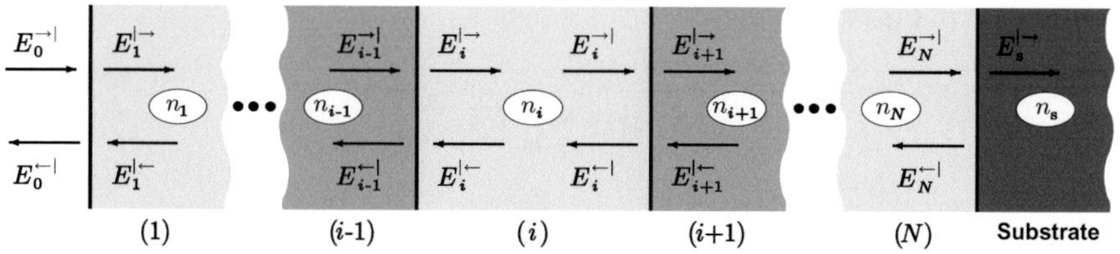

Fig. 4.9 Electric fields at the interfaces of a dielectric multilayer system

interfaces of layer i, we consider four electromagnetic fields $E_i^{|\rightarrow}, E_i^{|\leftarrow}, E_i^{\rightarrow|}, E_i^{\leftarrow|}$, forward and backward propagating, respectively, with the bar indicating the position of interface.

As in Sect. 2.1.1, we employ the continuity of the tangential component of the electric and magnetic field as boundary condition. To simplify matters, we assume normal incidence, so that the fields are equal to their tangential components. At the interface $(i-1, i)$ the electric and magnetic field, respectively, must satisfy

$$E_{(i-1,i)} = E_{i-1}^{\rightarrow|} + E_{i-1}^{\leftarrow|} = E_i^{|\rightarrow} + E_i^{|\leftarrow} \tag{4.40}$$

$$H_{(i-1,i)} = H_{i-1}^{\rightarrow|} + H_{i-1}^{\leftarrow|} = H_i^{|\rightarrow} + H_i^{|\leftarrow}. \tag{4.41}$$

According to Eqs. (1.66) and (1.68), the relation between H and E is given by

$$H = \pm \frac{n}{Z_0} E, \tag{4.42}$$

where the positive (negative) sign applies to forward (backward) propagating waves, and Z_0 is the vacuum impedance Eq. (1.69). Equation (4.41) can thus be cast in the form

$$H_{(i-1,i)} = \frac{n_{i-1}}{Z_0} \left(E_{i-1}^{\rightarrow|} - E_{i-1}^{\leftarrow|} \right) = \frac{n_i}{Z_0} \left(E_i^{|\rightarrow} - E_i^{|\leftarrow} \right). \tag{4.43}$$

Analog equations hold at interface $(i, i+1)$

$$E_{(i,i+1)} = E_i^{\rightarrow|} + E_i^{\leftarrow|}$$

$$H_{(i,i+1)} = \frac{n_i}{Z_0} \left(E_i^{\rightarrow|} - E_i^{\leftarrow|} \right). \tag{4.44}$$

The electric fields at the two interfaces differ by a phase factor

$$E_i^{\rightarrow|} = E_i^{|\rightarrow} e^{-jk_0 n_i d_i}$$

$$E_i^{\leftarrow|} = E_i^{|\leftarrow} e^{jk_0 n_i d_i}; \tag{4.45}$$

substitution in Eq. (4.44) yields

$$E_i^{|\rightarrow} = \frac{1}{2}\left(E_{(i,i+1)} + \frac{Z_0}{n_i}H_{(i,i+1)}\right)e^{jk_0 n_i d_i}$$

$$E_i^{|\leftarrow} = \frac{1}{2}\left(E_{(i,i+1)} - \frac{Z_0}{n_i}H_{(i,i+1)}\right)e^{-jk_0 n_i d_i} \tag{4.46}$$

so that Eqs. (4.40) and (4.43) can be cast in the form

$$E_{(i-1,i)} = E_{(i,i+1)}\cos k_0 n_i d_i + j\frac{Z_0}{n_i}H_{(i,i+1)}\sin k_0 n_i d_i$$

$$H_{(i-1,i)} = j\frac{n_i}{Z_0}E_{(i,i+1)}\sin k_0 n_i d_i + H_{(i,i+1)}\cos k_0 n_i d_i. \tag{4.47}$$

Thus, the fields at the two interfaces of layer (i) are related by

$$\begin{bmatrix} E_{(i-1,i)} \\ H_{(i-1,i)} \end{bmatrix} = M_i \begin{bmatrix} E_{(i,i+1)} \\ H_{(i,i+1)} \end{bmatrix} \tag{4.48}$$

where

$$M_i = \begin{bmatrix} \cos k_0 n_i d_i & j\frac{Z_0}{n_i}\sin k_0 n_i d_i \\ j\frac{n_i}{Z_0}\sin k_0 n_i d_i & \cos k_0 n_i d_i \end{bmatrix} \tag{4.49}$$

is the characteristic matrix of the layer, that accounts for all multiple reflections and transmissions within the multilayer system.

Starting from the substrate, we now can calculate the fields step by step, simply multiplying the characteristic matrices from the left:

$$\begin{bmatrix} E_{(0,1)} \\ H_{(0,1)} \end{bmatrix} = M_1 M_2 \dots M_{N-1} M_N \begin{bmatrix} E_{(N,s)} \\ H_{(N,s)} \end{bmatrix} = M_{tot}\begin{bmatrix} E_{(N,s)} \\ H_{(N,s)} \end{bmatrix}. \tag{4.50}$$

4.2.2.1 Reflection and Transmission Coefficient

The integrated reflection and transmission coefficients, respectively, are given by

$$r = \frac{E_0^{\leftarrow|}}{E_0^{\rightarrow|}}, \quad t = \frac{E_s^{|\rightarrow}}{E_0^{\rightarrow|}}. \tag{4.51}$$

According to Eqs. (4.40) and (4.43), boundary conditions at the front surface require $E_{(0,1)} = (1+r)E_0^{\rightarrow|}$ and $H_{(0,1)} = (1-r)(n_0/Z_0)E_0^{\rightarrow|}$. At the substrate interface, there is no backward wave so that $E_s^{|\rightarrow} = E_{(N,s)} = t E_0^{\rightarrow|}$ and $H_s^{|\rightarrow} = (n_s/Z_0) E_{(N,s)} = t(n_s/Z_0) E_0^{\rightarrow|}$. Equation (4.50) can therefore be written as

$$\begin{bmatrix} 1+r \\ (1-r)n_0/Z_0 \end{bmatrix} E_0^{\rightarrow|} = M_{\text{tot}} \begin{bmatrix} t \\ tn_s/Z_0 \end{bmatrix} E_0^{\rightarrow|}. \tag{4.52}$$

Solving for r and t yields

$$r = \frac{n_0 Z_0 M_{11} + n_0 n_s M_{12} - Z_0^2 M_{21} - n_s Z_0 M_{22}}{n_0 Z_0 M_{11} + n_0 n_s M_{12} + Z_0^2 M_{21} + n_s Z_0 M_{22}} \tag{4.53}$$

$$t = \frac{2 n_0 Z_0}{n_0 Z_0 M_{11} + n_0 n_s M_{12} + Z_0^2 M_{21} + n_s Z_0 M_{22}}. \tag{4.54}$$

Simple phase considerations at the beginning of this section led us to assumption that maximum reflectance is obtained by stacking quarter wavelength thick layers of alternating high and low index $n_{h,l}$ on top of each other. Since $k_0 n_i d_i = (\omega_0/c_0) n_i d_i = \pi/2$ in this case, the diagonal elements of M_i vanish and the characteristic matrix of an n_h-n_l double layer is given by

$$M_{|l|h|} = \begin{bmatrix} 0 & j\frac{Z_0}{n_l} \\ j\frac{n_l}{Z_0} & 0 \end{bmatrix} \begin{bmatrix} 0 & j\frac{Z_0}{n_h} \\ j\frac{n_h}{Z_0} & 0 \end{bmatrix} = \begin{bmatrix} -\frac{n_h}{n_l} & 0 \\ 0 & -\frac{n_l}{n_h} \end{bmatrix}. \tag{4.55}$$

A system of m such pairs has the non-vanishing components $M_{11} = (-n_h/n_l)^m$ and $M_{22} = (-n_l/n_h)^m$, so that the integrated reflection coefficient according to Eq. (4.53) is given, at the design frequency ω_0, by

$$r = \frac{M_{11} - M_{22}}{M_{11} + M_{22}} = \frac{1 - (n_l/n_h)^{2m}}{1 + (n_l/n_h)^{2m}}, \tag{4.56}$$

where we have assumed, for simplicity, $n_0 = n_s = 1$.

With 10 ($m = 5$) alternating layers of ZnS ($n_h = 2.3$) and MgF ($n_l = 1.38$), a reflectance $R = |r|^2$ of 0.976 is obtained; 20 layers yield 0.99993, which can be hardly reached in practice because of absorption and scattering losses. By comparison, the reflectance of high quality metal mirrors is limited to values below 0.098.

4.2.2.2 Bandwidth

Figure 4.10 shows the calculated reflectance of a high reflectance (HR) multilayer mirror, designed for maximum reflectance at 800 nm (optical layer thickness of 200 nm). Surprisingly, the range of high reflectance extends far beyond this central

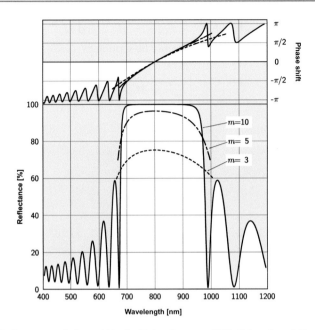

Fig. 4.10 Reflectance and phase shift of a high reflectance (HR) dielectric multilayer mirror as a function of wavelength

wavelength. For a larger number of layers, the reflection spectrum gets more and more rectangular, while the width of the reflection band turns out to depend on (and to grow with) the ratio n_h/n_l only.

To obtain an estimate of the width of the reflection band, we consider an *infinite* series of identical dielectric double layers and exploit the invariance of the stack under translation by one such pair. Let the center frequency be ω_0, so that $(\omega_0/c_0)n_i d_i = \pi/2$ and the phase term in Eq. (4.49) can be written as $k_0 n_i d_i = (\pi/2)(\omega/\omega_0)$. The fields at the interfaces of a double layer are then related by the matrix

$$
M_{|l|h|} = \begin{bmatrix} \cos\left(\frac{\pi}{2}\frac{\omega}{\omega_0}\right) & j\frac{Z_0}{n_l}\sin\left(\frac{\pi}{2}\frac{\omega}{\omega_0}\right) \\ j\frac{n_l}{Z_0}\sin\left(\frac{\pi}{2}\frac{\omega}{\omega_0}\right) & \cos\left(\frac{\pi}{2}\frac{\omega}{\omega_0}\right) \end{bmatrix} \begin{bmatrix} \cos\left(\frac{\pi}{2}\frac{\omega}{\omega_0}\right) & j\frac{Z_0}{n_h}\sin\left(\frac{\pi}{2}\frac{\omega}{\omega_0}\right) \\ j\frac{n_h}{Z_0}\sin\left(\frac{\pi}{2}\frac{\omega}{\omega_0}\right) & \cos\left(\frac{\pi}{2}\frac{\omega}{\omega_0}\right) \end{bmatrix}
$$

$$
= \begin{bmatrix} \cos^2\left(\frac{\pi}{2}\frac{\omega}{\omega_0}\right) - \frac{n_h}{n_l}\sin^2\left(\frac{\pi}{2}\frac{\omega}{\omega_0}\right) & \cdots \\ \cdots & \cos^2\left(\frac{\pi}{2}\frac{\omega}{\omega_0}\right) - \frac{n_l}{n_h}\sin^2\left(\frac{\pi}{2}\frac{\omega}{\omega_0}\right) \end{bmatrix}.
$$

$$(4.57)$$

Because of the translational invariance, the two fields must be related by the same factor $\beta_{|1|h|}$, no matter which double layer i is selected, so that

$$M_{|1|h|} \begin{bmatrix} E_{(i,i+1)} \\ H_{(i,i+1)} \end{bmatrix} = \beta_{|1|h|} \begin{bmatrix} E_{(i,i+1)} \\ H_{(i,i+1)} \end{bmatrix}; \qquad (4.58)$$

to allow for non-trivial solutions, $\beta_{|1|h|}$ must be the root of the characteristic equation $\det(M_{|1|h|} - \beta_{|1|h|}\mathbf{1}) = 0$, or

$$\beta_{|1|h|}^2 - \beta_{|1|h|}(M_{11} + M_{22}) + 1 = 0, \qquad (4.59)$$

where M_{11} and M_{22} are the diagonal components of $M_{|1|h|}$ and $\det M_i = 1$ was used. The solutions

$$\beta_{|1|h|} = \frac{M_{11} + M_{22}}{2} \pm \sqrt{\left(\frac{M_{11} + M_{22}}{2}\right)^2 - 1} \qquad (4.60)$$

can be either complex of the form $e^{\pm j\phi}$, resulting in an interfacial amplitude that oscillates along the propagation direction. For the structure to be a perfect mirror, however, the amplitude must decay exponentially in the forward direction, which requires $\beta_{|1|h|}$ to be real, implying

$$(M_{11} + M_{22})^2 \equiv \left[2\cos^2\left(\frac{\pi}{2}\frac{\omega}{\omega_0}\right) - \left(\frac{n_h}{n_l} + \frac{n_l}{n_h}\right)\sin^2\left(\frac{\pi}{2}\frac{\omega}{\omega_0}\right)\right]^2 > 4. \qquad (4.61)$$

This condition defines the reflection band

$$\cos^2\left(\frac{\pi}{2}\frac{\omega}{\omega_0}\right) < \left(\frac{n_h - n_l}{n_h + n_l}\right)^2, \qquad (4.62)$$

the borders $\omega_0 \pm \Delta\omega/2$ of which are obtained from the equation

$$\cos^2\left[\frac{\pi}{2}\left(1 \pm \frac{\Delta\omega}{2\omega_0}\right)\right] \equiv \sin^2\left(\frac{\pi}{2}\frac{\Delta\omega}{2\omega_0}\right) = \left(\frac{n_h - n_l}{n_h + n_l}\right)^2, \qquad (4.63)$$

yielding the normalized bandwidth

$$\frac{\Delta\omega}{\omega_0} = \frac{4}{\pi}\arcsin\frac{n_h - n_l}{n_h + n_l}. \qquad (4.64)$$

As can be seen, the width of the reflection band (also called stop band) increases with the propagation index contrast $(n_h - n_l)/(n_h + n_l)$; the additional peaks in the reflection spectrum Fig. 4.10 originate from the oscillatory solutions outside the stop band.

4.2.2.3 Antireflection Coatings

Dielectric layer structures can also be designed to minimize reflection by exploiting destructive interference. A particularly simple AR coating is a single dielectric layer applied to an optical surface. The propagation index n_1 of the layer is chosen to be intermediate between the indices n_0 and n_s of the adjacent media (air and glass) so that the reflection coefficient is negative at both interfaces, and the layer thickness must be $\lambda/4$ to provide destructive interference. According to Eq. (4.53), the resulting reflection coefficient is

$$r = \frac{n_0 n_s - n_1^2}{n_0 n_s + n_1^2} \tag{4.65}$$

and zero, if the index of the layer is the geometric mean value

$$n_1 = \sqrt{n_0 n_s} \tag{4.66}$$

of the adjacent media. As shown in Fig. 4.11, the reflectance vanishes at the design wavelength $\lambda_0 = 4n_1 d$ and lies significantly below the reflectance of the uncoated surface ($\approx 4\%$) over a wide wavelength range. Multilayer AR coatings provide better performance and allow for more flexibility in the choice of the layer materials.

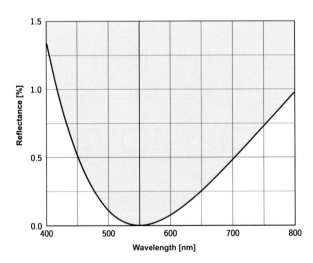

Fig. 4.11 Reflectance of a single layer antireflection coating with $n_1 = \sqrt{n_0 n_s}$

Fig. 4.12 Partial waves in a Fabry–Perot interferometer; for the meaning of r, t, and t', see text

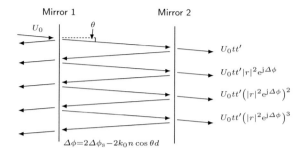

4.2.3 Fabry–Perot Interferometer

A Fabry–Perot interferometer Fig. 4.12 is a structure consisting of two parallel, partially transmitting mirrors spaced by a distance d; the enclosed space (also called cavity) can be empty or filled with a dielectric. Light incident on the input mirror is partially transmitted into the cavity and then partially reflected between the mirrors, forming an infinite series of partial waves in the forward and backward direction, respectively. The following analysis assumes a monochromatic, plane input wave and identical mirrors with the (complex) reflection coefficient $r = |r|e^{j\Delta\phi_s}$ and the transmission coefficient t. Any two consecutive partial waves impinging on the output mirror differ by the complex factor r^2 and the phase delay due to propagation over twice the distance d; the phase difference $\Delta\phi$ is therefore

$$\Delta\phi = 2\Delta\phi_s - 2k_0 nd\cos\theta = 2\Delta\phi_s - 2\frac{\omega}{c_0}nd\cos\theta, \qquad (4.67)$$

where θ is the angle of incidence and $k_0 n\cos\theta$ is the axial component of the wave vector. The total transmitted field amplitude is then

$$\tilde{U}^t = \tilde{U}_0 tt'(1 + |r|^2 e^{j\Delta\phi} + |r|^4 e^{j2\Delta\phi} + \ldots), \qquad (4.68)$$

where $t' := te^{-jk_0 nd\cos\theta}$ includes the phase shift due to the first transition through the cavity. This infinite geometric series is evaluated using $\sum_{n=1}^{\infty} q^{n-1} = 1/(1-q)$ with $q = |r|^2 e^{j\Delta\phi}$:

$$\tilde{U}^t = \tilde{U}_0 tt' \frac{1}{1 - |r|^2 e^{j\Delta\phi}}, \qquad (4.69)$$

which allows us to express the transmitted intensity by

$$\frac{I^t}{I_0} = \frac{|\tilde{U}^t|^2}{|\tilde{U}_0|^2} = \frac{T^2}{|1 - Re^{j\Delta\phi}|^2} = \frac{T^2}{1 - 2R\cos\Delta\phi + R^2}, \qquad (4.70)$$

where $R = |r|^2$ and $T = |t|^2 = |t'|^2$. Using $\cos x = 1 - 2\sin^2(x/2)$, the denominator of this expression can be cast in the form

$$1 - 2R\cos\Delta\phi + R^2 = (1-R)^2\left[1 + \frac{4R\sin^2(\Delta\phi/2)}{(1-R)^2}\right],\qquad(4.71)$$

so that we obtain

$$\frac{I^t}{I_0} = \frac{T^2}{(1-R)^2}\left[1 + \frac{4R\sin^2(\Delta\phi/2)}{(1-R)^2}\right]^{-1}.\qquad(4.72)$$

This periodic transmission function is shown in Fig. 4.13; in contrast to Fig. 4.7, there are no side maxima, which is due to the summation over an infinite number of decreasing partial fields. A maximum transmittance of $T^2/(1-R)^2$, which is equal to one if the mirrors are lossless, is obtained if the partial waves are in phase, $\Delta\phi = -2m\pi$, a situation that is called resonance.

For a given cavity length d, the resonance condition yields a series of resonance frequencies

$$\omega_m = \frac{c_0}{nd\cos\theta}(\Delta\phi_s + m\pi) =: \Delta\omega_s + m\Delta\omega_r,\qquad(4.73)$$

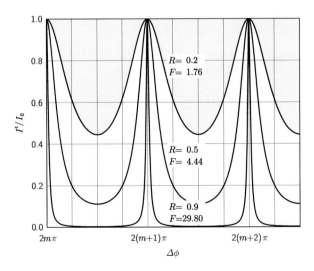

Fig. 4.13   Transmittance of a Fabry–Perot interferometer

spaced by a frequency interval $\Delta\omega_r$

$$\Delta\omega_r = \frac{c_0 \pi}{nd \cos\theta},\tag{4.74}$$

called free spectral range, that is approximately constant if the dispersion of the cavity medium can be neglected. It follows from Eq. (4.74) that the "comb" of resonance frequencies can be stretched or compressed by changing d, n, or θ. Note that the effect of the mirror induced phase shift is a frequency-offset of the entire comb by $\Delta\omega_s = c_0 \Delta\phi_s / nd \cos\theta$. At optical frequencies and macroscopic cavities ($d \gg \lambda_0$), the resonator mode index is a very large number, $m \gg 1$, so that $\Delta\phi_s$ can often be neglected.

Another characteristic parameter of a Fabry–Perot interferometer is the so-called *finesse*, defined as

$$F := \pi \frac{\sqrt{R}}{1-R};\tag{4.75}$$

with this parameter, Eq. (4.72) can be written as

$$\frac{I^t}{I_0} = \frac{1}{1 + (2F/\pi)^2 \sin^2(\Delta\phi/2)}.\tag{4.76}$$

Neglecting the phase shift introduced by the mirror, we can set $\Delta\phi = 2\pi\omega/\Delta\omega_r$, and

$$\frac{I^t}{I_0} = \frac{1}{1 + (2F/\pi)^2 \sin^2(\pi\omega/\Delta\omega_r)}.\tag{4.77}$$

The frequency deviation $\Delta\omega$ from a resonance that reduces the transmittance to 50 % is a measure of the line width of the comb filter and is given by

$$\sin^2\left(\pi\frac{\Delta\omega}{\Delta\omega_r}\right) = \left(\frac{\pi}{2F}\right)^2.\tag{4.78}$$

Provided that $R \approx 1$, $\pi/2F$ is small and we can approximate $\sin^2 x \approx x^2$ so that

$$\frac{\Delta\omega}{\Delta\omega_r} = \frac{1}{2F}.\tag{4.79}$$

The FWHM-width $\Delta\omega_{\mathrm{res}}$ of the transmission peak is then

$$\Delta\omega_{\mathrm{res}} \approx \frac{\Delta\omega_{\mathrm{r}}}{F}; \tag{4.80}$$

thus, the finesse Eq. (4.75) is the ratio of free spectral range to bandwidth.

Fabry–Perot interferometers are used as high resolution filters; frequency tuning is provided by adjusting d, θ, or n; filters with constant d and n are called etalon-filter and can be tuned by tilting. The most important aspect of the structure, however, is its application as laser resonator.

4.3   Resonators

The rather counterintuitive fact that a sequence of two mirrors transmits, at resonance, $100\,\%$ of the incident light, while each of the mirrors transmits only a fraction $T = 1 - R$, finds its explanation in the enhancement of the field inside the cavity. As can be seen in Fig. 4.12, the total field incident on the output mirror is $1/t$ times larger than the transmitted field, implying that the corresponding intensity is $1/T = 1/(1 - R)$ times larger (Fig. 4.14). Under resonance conditions, the transmitted intensity is equal to I_0, so that we have to conclude that the intensity incident on the output mirror is $I_0/(1 - R)$; if, for example, the reflectance of the mirror is $80\,\%$, the intensity of the right-propagating intracavity field is five times enhanced over the input intensity.

Energy conservation requires the reflectance of the Fabry–Perot interferometer at resonance to be zero, which implies the cancellation of the reflected field rU_0 by the transmitted fraction t of the left-propagating intracavity field; for complete

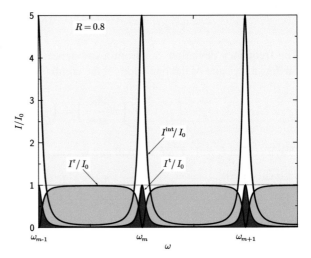

Fig. 4.14 Normalized transmitted (I^{t}), reflected (I^{r}), and right-propagating internal intensity (I^{int}) of a Fabry–Perot interferometer

cancellation, the magnitude of this field must be $-(r/t)U_0$, corresponding to a left-propagating intensity of $I_0(R/T)$. The net power flow (the difference between right- and left-propagating power flow) through the cavity is therefore $I_0[1/(1-R)-R/(1-R)] = I_0$, while the remaining power flow $I_0(R/T)$ is just circulating between the mirrors.

The resonant enhancement is due to constructive interference of the fields propagating inside the cavity. The resonance condition of the Fabry–Perot interferometer implies that the wave circulating between the mirrors exactly reproduces itself in terms of the phase; such waves are called eigenmodes or simply modes of the resonator, and the corresponding frequencies (ω_m) are its eigenfrequencies. These modes play a central role in the theory of laser oscillators (see Chap. 7). Neglecting $\Delta\phi_s$, the eigenfrequencies of the resonator are given by Eq. (4.73) with $\theta = 0$

$$\omega_m = m\frac{c_0\pi}{nd}, \tag{4.81}$$

where d is the resonator length, which, depending on the type of laser, lies between some $100\,\mu m$ (semiconductor lasers) and 1–$2\,m$ (gas lasers). Consequently, the mode index m ranges between 10^2 and several 10^6 in the VIS and NIR. The mode spacing,

$$\Delta\omega_r = \frac{c_0\pi}{nd}, \tag{4.82}$$

lies between $1000\,GHz$ (semiconductor laser) and $100\,MHz$ (gas laser).

Laser resonators are lossy, not only because of the finite reflectance of the mirrors (which serves to couple the laser light out of the cavity), but also because of various internal losses. The power loss per round trip can be described by a loss factor $e^{-\alpha_{res}2d}$ [Eq. (7.5)], while the round trip loss of an ideal, symmetric Fabry–Perot resonator is represented by R^2. To describe a lossy resonator, we can replace, in the expression (4.75) for the finesse, the term R by $e^{-\alpha_{res}d}$

$$F = \pi\frac{e^{-\alpha_{res}d/2}}{1-e^{-\alpha_{res}d}} \approx \frac{\pi}{\alpha_{res}d}, \tag{4.83}$$

where we have used the approximation $e^x \approx 1+x$. In Chap. 7 we will introduce the concept of a resonator life time τ_{res} as the time it takes a given photon number in the cavity to decay to a fraction $1/e$. As we will see [Eq. (7.7)], this life time is related to the loss coefficient α_{res} by $\alpha_{res} = 1/c\tau_{res}$, so that the bandwidth Eq. (4.80) of a resonator in the absence of gain can be expressed, with the help of Eqs. (4.82) and (4.83), by

$$\Delta\omega_{res} \approx \frac{1}{\tau_{res}}. \tag{4.84}$$

In a laser under stationary operating conditions, the gain (represented by the gain coefficient γ) provided by stimulated emission exactly compensates the losses so that the effective loss coefficient $\alpha_{res} + \gamma = 0$ [Eq. (7.9)]; consequently, the finesse tends towards infinity and the spectral width of the laser mode approaches zero (see Sect. 7.2.3).

4.3.1   Spherical Mirror Resonators

In practice, mirrors are of finite size and the quasi-plane waves circulating in a realistic plane mirror Fabry–Perot resonator experience losses at the mirrors because the wave diverges during propagation. By contrast, a Gaussian wave function with its rapidly decaying radial amplitude can be reflected by finite size mirrors very efficiently. As we have seen in Sect. 3.1.2, Gaussian beams have spherical phase fronts; if such a beam is reflected at a mirror of matching curvature, it is reflected exactly into itself; two spherical mirrors, spaced by a certain distance, allow a Gaussian beam with appropriate parameters to circulate between them without changing its shape (Fig. 4.15). The problem of finding the required beam parameters was treated in Sect. 3.1.2, where the confocal parameter Eq. (3.27) and the position of the waist Eq. (3.26) were calculated for a given set of phase front curvatures; note that these two parameters do not depend on the frequency of the wave, so that one degree of freedom for the specification of the mode is left.

4.3.1.1 Eigenfrequencies

Just like the eigenmodes of a plane mirror Fabry–Perot resonator, the modes of a spherical mirror resonator must reproduce themselves after a round trip; the longitudinal phase of a Gaussian wave function is, according to Eq. (3.17), $kz - \xi(z)$ so that the resonance condition for a resonator with spherical mirrors at z_1 and z_2 is

$$2k(z_2 - z_1) - 2[\xi(z_2) - \xi(z_1)] = 2m\pi, \qquad (4.85)$$

Fig. 4.15   Spherical mirror resonator with Gaussian mode, shown with contours of constant energy density

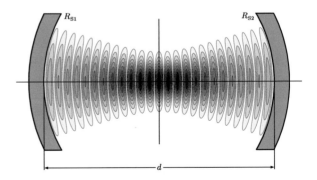

where m is the longitudinal mode index and $z_{1,2}$ is measured in respect to the waist; an additional phase introduced by the mirror has been neglected for the sake of simplicity. With $k = \omega/c$ and $z_2 - z_1 = d$, we obtain

$$\omega_m = m\frac{c\pi}{d} + \frac{c[\xi(z_2) - \xi(z_1)]}{d}. \tag{4.86}$$

The Gouy term $c[\xi(z_2) - \xi(z_1)]/d$ depends only on the position and curvature of the mirrors and shifts the comb of eigenfrequencies by a constant offset; the mode spacing is not affected and equal to that of a plane mirror Fabry–Perot resonator [Eq. (4.82)] of the same length

$$\Delta\omega_r \doteq \frac{c\pi}{d}. \tag{4.87}$$

4.3.1.2 Stability Condition

Not every configuration of two mirrors supports a Gaussian mode; a pair of convex mirrors, for example, obviously cannot be matched by the phase fronts of any Gaussian beam. The condition that the set R_{s1}, R_{s2}, d has to meet for a mode to exist is called resonator stability condition and can be derived from Eq. (3.27) by restricting z_0^2 to positive values. Resonator configurations that do not satisfy this condition are called instable and are highly lossy.

The sign of the curvature $1/R_s$ of a spherical mirror is defined in respect to the reflecting surface: by convention, it is negative (positive) for concave (convex) mirrors. The sign of the phase front curvature $1/R$ in Eq. (3.13), however, is given in respect to the orientation of the z-axis; to account for this conflicting definitions, we formulate the curvature matching condition as $R_1 = R_{s1}$ for the left mirror and $R_2 = -R_{s2}$ for the right mirror in Fig. 4.15. Equation (3.27) then assumes the form

$$z_0^2 = \frac{d(d + R_{s2})(d + R_{s1})(-R_{s2} - R_{s1} - d)}{(R_{s2} + R_{s1} + 2d)^2} > 0. \tag{4.88}$$

With the substitutions

$$g_1 := 1 + d/R_{s1}, \quad g_2 := 1 + d/R_{s2}, \tag{4.89}$$

the inequality can be cast in the form

$$d^2 \frac{g_1 g_2 (1 - g_1 g_2)}{(g_1 + g_2 - 2g_1 g_2)^2} > 0. \tag{4.90}$$

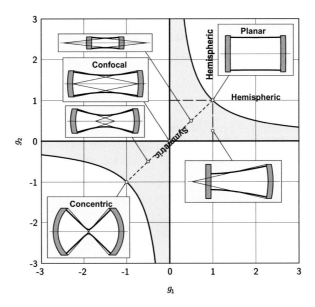

Fig. 4.16 Stability diagram of spherical mirror resonators in terms of the parameters $g_1 :=$ $1 + d/R_{s1}$ and $g_2 := 1 + d/R_{s2}$: the *shaded area* indicates the range of stable resonators; the lines $g_{1,2} = 1$ mark the border between concave and convex mirrors

Since the denominator of this expression is positive, the terms $g_1 g_2$ and $1 - g_1 g_2$ must have the same sign, yielding the stability condition

$$0 < g_1 g_2 < 1, \tag{4.91}$$

which is graphically represented by the map Fig. 4.16.

For a symmetric resonator $R_{s1} = R_{s2} = R_s$, Eq. (4.91) is reduced to

$$0 \le -\frac{d}{R_s} \le 2, \tag{4.92}$$

while a plano-concave resonator ($d/R_{s1} = 0$ and $R_{s2} = R_s$) has to satisfy

$$0 \le -\frac{d}{R_s} \le 1. \tag{4.93}$$

4.3.1.3 Mode Parameters

The position of the waist, in reference to the left mirror, is given by Eq. (3.26); for symmetric resonators, the waist is naturally in the center of the resonator. The confocal parameter follows from Eq. (3.13) with $z = d/2$,

$$z_0 = \tfrac{1}{2}\sqrt{2|R_s|d - d^2}; \tag{4.94}$$

the waist radius of the mode is given by Eq. (3.12)

$$w_0 = \sqrt{\frac{\lambda z_0}{\pi}}. \tag{4.95}$$

Equation (3.11) yields the mode radius at the mirrors (Fig. 4.17); the mirror must be several times larger than the mode radius to keep losses low.

The focal length of a spherical mirror is equal to $-R_s/2$; for a symmetric confocal resonator (Fig. 4.18), $R_s = -d$, so that $z_0 = |R_s|/2$ and $w_0 = \sqrt{\lambda d/2\pi}$. This type of

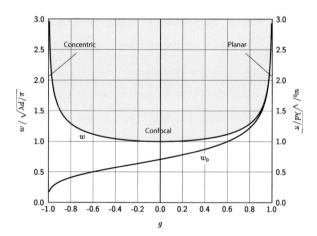

Fig. 4.17 Beam waist w_0, and mode radius w at the mirrors for a symmetric spherical mirror resonator as a function of $g = 1 + d/R_s$

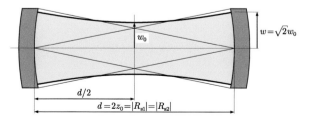

Fig. 4.18 Geometry of a confocal resonator

resonator has, at a given length, the smallest possible mode radius $w(d/2) = \sqrt{2}w_0$ at the mirrors (Fig. 4.17).

All members of the Hermite–Gaussian family of wave functions Eq. (3.95) exhibit the same spherical phase front curvature and thus are equally well suited as eigenmodes of spherical mirror resonators; it is common to refer to these modes as transverse modes. The eigenfrequencies Eq. (4.86) of the TEM_{ij}-modes are slightly up-shifted in comparison to the Gaussian TEM_{00}-mode because of the increased Gouy phase term $(i + j + 1)\xi(z)$ [Eq. (3.97)]. The intensity profile of the modes is shown in Fig. 3.14.

The superposition of left- and right-propagating waves inside the cavity results in an axial amplitude modulation $\propto e^{-jkz} + e^{jkz} = 2\cos kz$; the energy density is therefore axially modulated $\propto \cos^2 kz$ (Fig. 4.15), giving rise to a total of $2d/\lambda$ axial nodes.

4.3.1.4 ABCD Formalism for Spherical Mirror Resonators

The ABCD formalism, discussed in Sect. 3.1.4 provides a very powerful tool to analyze spherical mirror resonators. As an example, we present a generalized formulation of the stability condition: let the ABCD matrix of a resonator round trip be $\boldsymbol{M}_{\text{res}}$; for a Gaussian wave to be an eigenmode of the resonator, it must reproduce itself after one round trip, i.e., its q-parameter must remain unchanged

$$q = \frac{Aq + B}{Cq + D}, \tag{4.96}$$

where A, B, C, D are the coefficients of $\boldsymbol{M}_{\text{res}}$. The solutions

$$q_{1,2} = \frac{A - D}{2C} \pm \frac{1}{2C}\sqrt{(A - D)^2 + 4BC} \tag{4.97}$$

must have a non-vanishing imaginary part to be a meaningful q-parameter. Using $\det \boldsymbol{M}_{\text{res}} = AD - BC = 1$, we can express this condition in the form

$$\frac{(A + D)^2 - 4}{C^2} < 0 \tag{4.98}$$

or

$$-2 < A + D < 2; \tag{4.99}$$

the absolute value of the trace of the ABCD matrix must be less than 2. This is an elegant formulation of the resonator stability condition, that also applies to complex cavities containing lenses etc. The negative real part of q,

$$\frac{D - A}{2C} \tag{4.100}$$

is equal to the distance of the mode waist from the reference plane of the ABCD matrix. The mode parameters are given by

$$z_0 = \frac{1}{2|C|} \sqrt{4 - (A + D)^2}, \tag{4.101}$$

$$w_0^2 = \frac{\lambda}{\pi} \frac{1}{2|C|} \sqrt{4 - (A + D)^2}. \tag{4.102}$$

For the spherical mirror resonator discussed above, we obtain, with the reference plane at mirror 2

$$M = \begin{bmatrix} 1 & d \\ 0 & 1 \end{bmatrix} \begin{bmatrix} 1 & 0 \\ 2/R_{s1} & 1 \end{bmatrix} \begin{bmatrix} 1 & d \\ 0 & 1 \end{bmatrix} \begin{bmatrix} 1 & 0 \\ 2/R_{s2} & 1 \end{bmatrix} \tag{4.103}$$

$$= \begin{bmatrix} (1 + 2d/R_{s1})(1 + 2d/R_{s2}) + 2d/R_{s2} & d(2 + 2d/R_{s1}) \\ 2(1 + 2d/R_{s2})/R_{s1} + 2/R_{s2} & 2d/R_{s1} + 1 \end{bmatrix};$$

substituting this matrix in Eq. (4.99) immediately reproduces Eq. (4.91).

4.3.2 3D Resonators

A question of great theoretical importance is the spectral mode density of the electromagnetic field, that is the number of electromagnetic modes per unit volume in the frequency interval $\omega, \omega + d\omega$. For an estimate, we choose a rectangular, box shaped cavity of dimension $d_{x,y,z}$, with perfectly conducting (and reflecting) walls, imposing the boundary condition that the tangential component of the electric field at the walls must vanish. The (standing) waves

$$E_x(\mathbf{x}) = E_{0,x} \cos k_x x \sin k_y y \sin k_z z$$

$$E_y(\mathbf{x}) = E_{0,y} \sin k_x x \cos k_y y \sin k_z z$$

$$E_z(\mathbf{x}) = E_{0,z} \sin k_x x \sin k_y y \cos k_z z \tag{4.104}$$

are, as can be easily shown, solutions of the Helmholtz equation Eq. (1.22). The boundary conditions are satisfied if the components k_i of the wave vector assume the values

$$k_i d_i = \pi m_i, \qquad m_i = 1, 2, 3, \ldots; \tag{4.105}$$

for each wave vector, there exist two linearly independent modes with orthogonal polarization. In three-dimensional k-space, these modes are represented by equidistant points (k_x, k_y, k_z) in the positive octant. Because of the constant mode spacing, each of the modes occupies the volume $\pi^3/d_x d_y d_z$.

Before we determine the number of modes at a certain frequency, we estimate the number of modes in the interval $[k, k + dk]$, where $k = |\mathbf{k}|$ is the wave number. In k-space, this interval is represented by a spherical shell octant of radius k, thickness dk, and volume $4\pi k^2\, dk/8$. Assuming that the dimensions of the box are very large in comparison to the wavelength of interest, the volume $\pi^3/d_x d_y d_z$ is very small and we can approximate the number of modes by dividing the volume of the shell by the volume per mode. If we finally divide the resulting number by the volume $d_x d_y d_z$ of the box and multiply with 2 to account for the two polarization states, we obtain the number $N(k)\, dk$ of modes per volume in the interval $[k, k + dk]$[1]

$$N(k)\, dk = \frac{k^2}{\pi^2}\, dk. \tag{4.106}$$

The dispersion relation Eq. (1.28) $\omega = k c_0/n$ allows us to evaluate the spectral density $N(\omega)$ of modes as a function of frequency: with $N(\omega)\, d\omega = N(k)\, dk$ we obtain

$$N(\omega) = \frac{\omega^2 n^3}{\pi^2 c_0^3}. \tag{4.107}$$

For large resonators, the mode density is independent of the shape and size of the resonator, and proportional to the square of the frequency.

---

[1]In the same way and with the same result, one can calculate the density of states $\rho_{\mathrm{B}}(k)$ of electron Bloch waves in a semiconductor [see Eq. (6.107)].

4.4 Coherence*

4.4.1 Temporal Coherence

Up to this point, we have treated interference effects of purely monochromatic waves that are completely coherent. We now extend the discussion to light fields of constant intensity (stationary fields) that have statistical fluctuations of the phase and a spectral density that extends over a finite, narrow bandwidth. Such light is emitted, for example, by luminescence diodes or by thermal light sources with a narrow-band transmission filter. Since the visibility of interference phenomena depends on the stability of phase relations, the coherence properties of such light can be analyzed with interferometers (see, e.g., Goodman 2015).

4.4.1.1 Complex Analytic Signal
We assume that the light field is given by the scalar function $u(t)$ with the Fourier transform pair

$$U_\omega = \int_{-\infty}^{\infty} u(t) e^{-j\omega t}\, dt, \qquad (4.108)$$

$$u(t) = \frac{1}{2\pi} \int_{-\infty}^{\infty} U_\omega e^{j\omega t}\, d\omega. \qquad (4.109)$$

Since $u(t)$ is real, $U_{-\omega} = U_\omega^*$, and Eq. (4.109) can be cast in the form

$$u(t) = \frac{1}{2}\left[\underbrace{\frac{1}{\pi}\int_0^\infty U_\omega e^{j\omega t}\, d\omega}_{\hat{U}(t)} + \underbrace{\frac{1}{\pi}\int_0^\infty U_\omega^* e^{-j\omega t}\, d\omega}_{\hat{U}^*(t)} \right], \qquad (4.110)$$

defining the so-called analytical $\hat{U}(t)$

$$\hat{U}(t) = \frac{1}{\pi}\int_0^\infty U_\omega e^{j\omega t}\, d\omega \qquad (4.111)$$

with the property

$$u(t) = \mathrm{Re}\left[\hat{U}(t)\right]; \qquad (4.112)$$

if $u(t)$ is normalized so that $I = 2\langle u(t)u(t)\rangle$, then

$$I = \left\langle \hat{U}(t)\hat{U}^*(t) \right\rangle \qquad (4.113)$$

[compare Eq. (1.59)]. A superposition of two fields $u_{1,2}(t)$, with the corresponding analytic signals $\hat{U}_{1,2}(t)$, produces the intensity

$$I = \left\langle \left[\hat{U}_1(t) + \hat{U}_2(t) \right] \left[\hat{U}_1(t) + \hat{U}_2(t) \right]^* \right\rangle$$

$$= I_1 + I_2 + 2\mathrm{Re}\left[\left\langle \hat{U}_1(t)\hat{U}_2^*(t) \right\rangle \right]. \qquad (4.114)$$

Thus, we can treat partially coherent signals in the same way as coherent signals by using the analytic signal instead of the complex amplitude.

4.4.1.2 Correlation Functions

If we launch a polychromatic field $u(t)$ of constant intensity I_0 into a Michelson interferometer, we obtain, at output port 1 (Fig. 4.1(a)), the superposition $rt[u(t) + u(t + \tau)]$, where r, t are the reflection and transmission coefficients of the beam splitter ($rr^* = tt^* = 1/2$), and $\tau = 2\Delta s/c$ is the delay introduced between the two partial fields by the length difference Δs of the interferometer legs. The intensity at the detector is then, according to Eq. (4.114),

$$I(\tau) = \frac{1}{2}\left(I_0 + \mathrm{Re}\left[\left\langle \hat{U}(t)\hat{U}^*(t + \tau) \right\rangle \right] \right), \qquad (4.115)$$

which is the sum of a constant background and the real part of the function $\left\langle \hat{U}(t)\hat{U}^*(t + \tau) \right\rangle$. Obviously, if $\hat{U}(t)$ describes a coherent wave with time dependence $e^{j\omega t}$, the real part of this function lies between $\pm I_0$, and the output of the Michelson interferometer varies between 0 and I_0; on the other hand, if $\hat{U}(t)$ and $\hat{U}(t + \tau)$ are completely uncorrelated, this function is zero; interferometers are therefore well suited to analyze the statistical properties of light.

The correlation of two complex analytic signals $\hat{U}_i(t)$ and $\hat{U}_j(t)$ can be characterized by the time averaged correlation function

$$\Gamma_{ij}(\tau) := \left\langle \hat{U}_i(t)\hat{U}_j^*(t + \tau) \right\rangle; \qquad (4.116)$$

for $i \neq j$ this is the averaged cross correlation which in optics usually refers to the field at two different points $\mathbf{x}_{1,2}$

$$\Gamma_{ij}(\tau) = \Gamma(\mathbf{x}_1, \mathbf{x}_2, \tau) = \left\langle \hat{U}(\mathbf{x}_1, t)\hat{U}^*(\mathbf{x}_2, t + \tau) \right\rangle; \qquad (4.117)$$

if $i = j$, the function is the averaged autocorrelation. Since $\Gamma_{ii}(0) = I_i$, the local intensity can be used to introduce the normalized correlation function

$$\gamma_{ij}(\tau) := \frac{\Gamma(\mathbf{x}_1, \mathbf{x}_2, \tau)}{\sqrt{I(\mathbf{x}_1)I(\mathbf{x}_2)}}, \quad 0 \leq |\gamma_{ii}(\tau)| \leq 1, \tag{4.118}$$

which is called the mutual coherence function, while $\gamma_{ii}(\tau)$ is called complex degree of temporal coherence.

The output intensity Eq. (4.115) of our Michelson interferometer can therefore be written as

$$I(\tau) = \tfrac{1}{2} \left(1 + \mathrm{Re}\left[\gamma_{ii}(\tau)\right] \right) I_0, \tag{4.119}$$

which varies between the values $I_0 \left(1 \pm |\gamma_{11}| \right)/2$. The visibility of the interference is defined as the contrast ratio $(I_{\max} - I_{\min})/(I_{\max} + I_{\min})$ and is equal to the absolute value of the degree of coherence,

$$\frac{I_{\max} - I_{\min}}{I_{\max} + I_{\min}} = |\gamma_{11}(\tau)|. \tag{4.120}$$

For light with statistically distributed phase, the visibility usually decreases with growing delay τ (Fig. 4.19). The delay for which the visibility drops to $1/e$ is called coherence time τ_{coh}; it corresponds to the longitudinal coherence length $l_{\mathrm{coh,long}} := c_0 \tau_{\mathrm{coh}}$, that is a measure for the distance in propagation direction

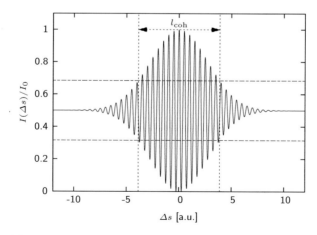

Fig. 4.19 Michelson interferometer output of a narrow-band polychromatic signal; Δs is the displacement of the scanning mirror, λ_0 is the central wavelength of the signal the *dashed lines* mark the $1/e$-points of the visibility

over which a significant phase correlation is maintained. The complex degree of coherence of monochromatic light $\hat{U} \propto e^{j\omega t}$ is given by $\gamma_{11}(\tau) = e^{-j\omega\tau}$, with $|\gamma_{11}(\tau)| = 1$ for arbitrary times, implying an infinite coherence time and length.

4.4.1.3 Coherence and Spectral Width

To spectrally characterize a polychromatic signal of constant intensity, we start from the Fourier transform V_ω of the analytic signal

$$V_\omega = \int_{-\infty}^{\infty} \hat{U}(t) e^{-j\omega t}\, dt; \qquad (4.121)$$

$|V_\omega|^2$ can be interpreted as the energy content of the field in the frequency interval $[\omega, \omega + d\omega]$. For a stationary signal, however, the energy content is infinity, and we use instead the truncated Fourier transform $V_T(\omega)$

$$V_T(\omega) := \int_{-T/2}^{T/2} \hat{U}(t) e^{-j\omega t}\, dt, \qquad (4.122)$$

and define the power spectral density $S(\omega)$ of $\hat{U}(t)$ as

$$S(\omega) := \lim_{T\to\infty} \frac{\langle |V_T(\omega)|^2 \rangle}{T}. \qquad (4.123)$$

According to the Wiener–Khinchin theorem, $S(\omega)$ is the Fourier transform of the autocorrelation function $\Gamma = \Gamma_{ii}$ [Eq. (4.116)],

$$S(\omega) = \int_{-\infty}^{\infty} \Gamma(\tau) e^{-j\omega\tau}\, d\tau, \qquad (4.124)$$

and

$$\Gamma(\tau) = \frac{1}{2\pi} \int_{-\infty}^{\infty} S(\omega) e^{j\omega\tau}\, d\omega. \qquad (4.125)$$

This is the basis of Fourier transform spectroscopy that determines the power spectrum by numerical Fourier transformation of the autocorrelation function, measured with a Michelson interferometer. The spectrum is thus obtained without any dispersive element (grating) and relies only on a power detector.

Quite generally, the widths of a Fourier transform pair such a Γ and S are reciprocal to each other. We therefore can conclude that the width of the power spectrum and the coherence time are related by

$$\tau_{\text{coh}} \Delta\omega = 2\pi C, \tag{4.126}$$

where C is a constant (of order 1) that depends on the shape of the spectrum and the definition of the spectral width. With $\omega = 2\pi c_0/\lambda_0$, and $\Delta\omega/\Delta\lambda_0 \approx -2\pi c_0/\lambda_0^2$, we can extract a useful relation between coherence length and spectral width $\Delta\lambda_0$ from Eq. (4.126)

$$l_{\text{coh,long}} = c_0 \tau_{\text{coh}} \approx \frac{\lambda_0^2}{|\Delta\lambda_0|}. \tag{4.127}$$

A typical thermal white light source has a power spectrum centered around 500 nm with a bandwidth of several hundred nm; accordingly, the coherence length is a few μm, which is still sufficient to observe interference patterns from a thin oil film on water, for example. A Helium–Neon laser ($\lambda_0 = 632$ nm) emits light with a bandwidth of about 1 MHz, corresponding to a coherence time of 1 μs and a coherence length of 300 m. With appropriate filters, light of 1 MHz bandwidth can also be obtained from a thermal white light source. The power of such a signal is a very small fraction $\approx 10^6$ Hz$/10^{15}$ Hz $= 10^{-9}$ of the power of thermal source. Assuming a lamp emitting 1 W of visible light, only 1 nW would be in a 1 MHz spectral window, which has to be compared to the typical 10 mW output power of a HeNe laser. The temporal coherence properties, however, would be the same. Thermal light, however, is also characterized by low spatial coherence (see following section) and differs also in terms of noise statistics from laser light (Sect. 9.3).

4.4.2 Spatial Coherence

Temporal coherence refers to the phase correlation of a light wave at a selected point in space at different times. Phase correlations can also be measured at different points in space. For this purpose, the mutual complex degree of coherence

$$\gamma(\mathbf{x}_1, \mathbf{x}_2, \tau) = \frac{\Gamma(\mathbf{x}_2, \mathbf{x}_2, \tau)}{\sqrt{I(\mathbf{x}_1)I(\mathbf{x}_2)}} \tag{4.128}$$

is used; it can be measured by a scheme similar to Young's double slit interferometer, using two pinholes of variable distance. If the optical field is a beam, the two points can be chosen in a common plane normal to the propagation direction. The distance $|\mathbf{x}_2 - \mathbf{x}_1|$ at which the visibility of the resulting interference pattern on a screen drops below a certain value defines the *transverse* coherence length $l_{coh,trans}$. Spatial coherence is relevant for the degree to which a light field can be focused or collimated. A Gaussian beam with a perfectly coherent phase front has a divergence angle of $2\theta = 2\lambda/\pi w_0$ [Eq. (3.19)]; the transverse coherence length is given by the beam diameter $2w_0$ in this case. We therefore can estimate the divergence of an arbitrary light beam with the transverse coherence length $l_{coh,trans}$ to be

$$2\theta \approx \frac{4\lambda}{\pi l_{coh,trans}}, \tag{4.129}$$

since in terms of coherence, the beam can be treated as being composed of independent coherent beams of diameter $l_{coh,trans}$.

4.5   Summary

Interference is a universal phenomenon in optics: any wave function can be understood as the result of an interference of elementary waves radiated by the electrons of the emitter. In a more applied sense, interference refers to an optical design where light waves are superimposed to reach a certain goal: high reflectance of dielectric multilayer mirrors, controlled spatial intensity patterns, field enhancement in resonators, spectral analysis by gratings, sensing capability of interferometers, etc.

The partial fields in these devices are usually generated using beam splitters, or by selecting different parts of an original phase front. The properties of beam splitters are very interesting mathematically; the amplitude of the fields reflected by and transmitted through a 50% beam splitter, for example, has an absolute value of about 70% of the original field. A scheme that would allow adding these fields would produce a field amplitude with an absolute value of 140%—in violation of energy conservation. The restrictions imposed by energy conservation, reciprocity or time reversal invariance reduce the number and values of independent (complex) parameters of a beam splitter accordingly. The representation of beam splitters by a scattering matrix provides an elegant and stringent formalism that can be extended to interferometers and more complex systems.

The central parameter determining interference is the phase of the participating fields, and a convenient way to analyze related effects is to represent the fields in the complex amplitude plane. The reader is advised to visualize, as a valuable exercise, effects such as multiple beam interference with adequate software in this manner.

The crucial role of phase in interference renders interferometers also ideal tools to analyze the (statistical) coherence properties of light. Since a Michelson interferometer provides the autocorrelation function of the input signal, it can be used to determine the coherence length of partially coherent light. A Michelson interferometer can also be used to measure the power spectrum of a signal, which is the Fourier transform of the autocorrelation function; this technique, which requires a computer to perform the Fourier transform, is an important tool used in infrared spectroscopy.

4.6 Problems

1. Design a high reflecting multilayer mirror with 10 pairs of layers, $n_l = 1.3$, $n_h = 1.8$ on glass ($n_{\text{glass}} = 1.5$) with a central wavelength of 660 nm. Calculate numerically the complex reflection coefficient and the reflectance of the mirror as a function of the frequency, and display them in a suitable plot. Calculate the electric field (normalized to the input field) at each interface and plot it as a function of the layer index, (a) for a wavelength within the "stop band," (b) for a wavelength outside the stop band.

2. Same as problem 1 but with a "defect layer" (=missing single layer in the middle of the stack) that produces a narrow high transmission line in the center of the "stop band."

3. Same as problem 1 but with gradually increasing layer thickness ("chirped mirror"); the center layer pair is designed for 660 nm, the first and last layer pair is designed for 660 $\pm$20 %, respectively. For simplicity, omit the substrate. What happens if the propagation direction is reversed? Assume a Fourier limited Gaussian pulse with a bandwidth equal to that of the mirror; calculate numerically the pulse shape after (multiple) reflection.

4. Calculate the reflectance R of a silver mirror ($n = 0.050 - 3.13 j$) at normal incidence. Can one increase R by coating the silver layer with a single dielectric layer of appropriate n and thickness? For the analysis, apply either the theory of the Fabry–Perot interferometer or the multilayer formalism.

5. A hypothetical 1:1 beam splitter (angle of incidence 45°) has reflection and transmission coefficients, respectively, of $r = t = 1/\sqrt{2}$, so that $R = T = 1/2$ and $R + T = 1$. With two of these beam splitters, build a Mach–Zehnder interferometer and calculate the output power at the two outputs as a function of the phase difference in the two interferometer branches. What follows from the result? Propose a (more) realistic beam splitter.

6. Calculate the reflection and transmission coefficient of a symmetric Fabry–Perot interferometer and construct its S-matrix. Confirm that it fulfills condition Eq. (4.27). Repeat for an asymmetric Fabry–Perot interferometer, where the reflectance of the two mirrors is not identical.

7. Assume an optical attenuation filter with a given complex refractive index and thickness. Is it possible to apply a single layer antireflection coating at the front face of the filter so that the reflectance of the filter at a certain wavelength is zero? Take both surfaces of the filter into account. Formulate the S-matrix of the filter and compare with Eq. (4.27). What happens if the propagation direction is reversed?

8. A Michelson interferometer, in its basic form, is not well suited for the measure of distances because of the cosinusoidal output characteristic [Eq. (4.8)]. To overcome this problem, assume that the input beam of the interferometer is

circularly polarized. Insert a $\lambda/8$ wave plate in the reference arm with one of its axes parallel to the plane formed by the interferometer legs. Calculate the output power as a function of the length difference Δs. Next, insert a polarization beam splitter into the output beam producing σ and π polarized outputs. Show that one of the output powers has a $\cos 2k_0\Delta s$ dependence, the other a $-\sin 2k_0\Delta s$ dependence. Denoting with P_π and P_σ the respective output powers normalized such that they vary between 0 and 1, show that $\int_{t_1}^{t_2}[(P_\pi-0.5)\dot{P}_\sigma-\dot{P}_\pi(P_\sigma-0.5)]\,dt$ is proportional to the distance travelled by the object mirror between t_1 and t_2.

9. Assume a semiconductor laser emitting two equally strong modes at a wavelength of $\approx 1\mu$m with a linewidth of 10 MHz each, separated by 1 THz. Determine the autocorrelation function and discuss visibility and coherence length.

References and Suggested Reading

Goodman, J. W. (2015). *Statistical optics*. New York: John Wiley.
Haus, H. A. (1984). *Waves and fields in optoelectronics*. Englewood Cliffs, NJ: Prentice Hall.
Hecht, E., & Zajac, A. (1987). *Optics*. San Francisco, CA: Addison-Wesley.
Klein, M. V., & Furtak, T. E. (1986). *Optics*. New York: John Wiley.
Lipson, S. G., & Lipson, H. (1969). *Optical physics*. London: Cambridge University Press.
MacLeod, H. A. (2001). *Thin-film optical filters*. Abingdon: Taylor & Francis.
Sakoda, K. (2005). *Optical properties of photonic crystals*. New York: Springer.
Saleh, B. E., & Teich, M. C. (2007). *Fundamentals of photonics*. New York: Wiley.
Sibilia, C., & Benson, T. (2008). *Photonic crystals*. New York: Springer.

Dielectric Waveguides

5

Dielectric waveguides are key components of photonics; the success of optical communications relies to a great degree on the availability of glass fibers with extremely low losses. In contrast to (metallic) radio frequency waveguides that are bulky and lossy, photonic waveguides rely on total internal reflection in dielectrics, are very small in diameter and can transport optical fields over tens of kilometers before signal regeneration is necessary.

Electromagnetic fields in waveguides are called modes; of particular importance are guided modes, having a field distribution that is essentially confined to the core of the waveguide over the entire propagation distance. Guided modes that do not change their tranverse amplitude profile during propagation are called eigenmodes, sometimes also simply waveguide modes. The number of (guided) eigenmodes is finite, growing with the radius of the waveguide core in relation to the wavelength. Different eigenmodes usually have different propagation constants (eigenvalues) and thus different propagation velocity, which renders multimode-waveguides not very well suited for optical long distance communications or interferometric sensor applications. As we shall see, however, it is possible to design waveguides such that they support not more than one mode at a given wavelength.

Apart from cylindrical waveguides (fibers) there are also planar waveguide structures (integrated optics). Beyond the waveguide as a means for light transportation, there is a host of waveguide components such as couplers, mirrors, filters, sensors, modulators, amplifiers, and oscillators. The integration of these components allows, for example, the setup of all optical data networks.

5.1 Planar Waveguides

Planar waveguides are layered dielectric structures, with a guiding layer of elevated propagation index n_g (the core) confined between layers of lower propagation index n_s and n_c, usually called substrate and cladding, respectively (Fig. 5.1). Such structures can be produced, for example, by applying a polymer layer on top of

© Springer International Publishing Switzerland 2016
G.A. Reider, *Photonics*, DOI 10.1007/978-3-319-26076-1_5

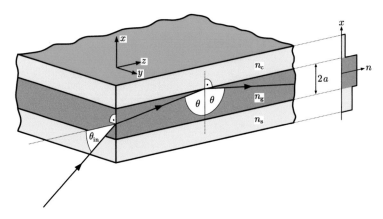

Fig. 5.1 Geometry of a planar waveguide

a substrate, or by ion diffusion of different dopants (such as H^+) into the surface region of a glass or crystal substrate. In lithium niobate, which is a popular substrate for electro-optic waveguide structures, the waveguide is produced by diffusion of Ti-ions into the surface region.

Total internal reflection as the basic guiding mechanism requires the tangential component of the wave vector (denoted as β) to be larger than the wave number in the adjacent media (Sect. 2.1.3),

$$\beta := k_\| = n_g k_0 \sin \theta > n_{s,c} k_0, \tag{5.1}$$

which is possible only if $n_g > n_{s,c}$. Assuming $n_s \geq n_c$, we find the condition

$$\sin \theta > \sin \theta_{crit} := \frac{n_s}{n_g}. \tag{5.2}$$

Light is usually launched into the waveguide from the front face of the structure (Fig. 5.1). Taking refraction at the air/guide interface into account, condition (5.2) requires that the angle of incidence θ_{in} must fulfill $\sin \theta_{in} < n_g \cos \theta_{crit}$; with $\sin \theta_{in} \approx \theta_{in}$, this can be expressed as

$$\theta_{in} < n_g \sqrt{1 - \sin^2 \theta_{crit}} = \sqrt{n_g^2 - n_s^2} =: NA. \tag{5.3}$$

The so-called numerical aperture NA is equal to the angle of acceptance of the waveguide and scales with the propagation index difference between guiding layer and substrate.

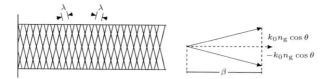

Fig. 5.2 Waveguide mode as superposition of two plane waves

5.1.1 Eigenmodes

We consider an infinite planar layered structure with a propagation index profile as shown in Fig. 5.1. For geometric reasons, plane waves are natural candidates for the construction of the eigenmodes of such a structure. Let $[k_\perp, 0, k_\parallel]$ be the wave vector of such a wave, where

$$k_\perp = n_\mathrm{g} k_0 \cos\theta, \quad k_\parallel = n_\mathrm{g} k_0 \sin\theta = \beta; \tag{5.4}$$

by reflection at the upper interfaces ($n_\mathrm{g}/n_\mathrm{c}$) this wave is converted into another plane wave with the wave vector $[-k_\perp, 0, k_\parallel]$, which is the second wave component of the eigenmode (Fig. 5.2). For reasons of self-consistency, this wave, after reflection at the second interface ($n_\mathrm{g}/n_\mathrm{s}$), must be indistinguishable from the original plane wave (Fig. 5.3).

During its "round trip" between the interfaces, the wave acquires a phase of $-4ak_\perp$, where $2a$ is the distance between the interfaces. According to Sect. 2.1.3, the reflection coefficient has the form $e^{i\phi^\mathrm{s,c}}$, implying that the wave experiences an additional phase shift of $\phi^\mathrm{s} + \phi^\mathrm{c}$ due to reflection at the two interfaces. Self-consistency requires that the total phase is an integer multiple of 2π

$$-4ak_\perp + \phi^\mathrm{s}_{\sigma,\pi} + \phi^\mathrm{c}_{\sigma,\pi} = -2m\pi, \quad m = 0, 1, 2, \ldots \tag{5.5}$$

For σ-polarized light, the phase shift according to Eq. (2.46) is

$$\tan\frac{\phi^\mathrm{s,c}_\sigma(\theta)}{2} = \frac{\sqrt{n_\mathrm{g}^2 \sin^2\theta - n_\mathrm{s,c}^2}}{n_\mathrm{g} \cos\theta}. \tag{5.6}$$

For a given set a, k_0, and $n_\mathrm{g,s,c}$, Eq. (5.5) has a finite number of solutions $\theta^{(m)}$, corresponding to modes with the propagation constants

$$\beta^{(m)} = n_\mathrm{g} k_0 \sin\theta^{(m)}, \tag{5.7}$$

Fig. 5.3 Self-consistency condition in a planar waveguide: (**a**) partial wave with $\mathbf{k} = [k_\perp, 0, k_\parallel]$; (**b**) reflected partial wave $([-k_\perp, 0, k_\parallel])$, shown in the *inverted* coordinate system; (**c**) doubly reflected partial wave with an accumulated phase shift of $-4ak_\perp + \phi^s + \phi^c = -2m\pi$

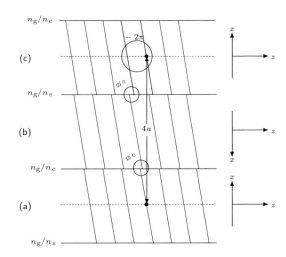

and the phase velocity

$$v_{\text{ph}}^{(m)} = \frac{\omega}{\beta^{(m)}} =: \frac{c_0}{n_{\text{eff}}^{(m)}}, \tag{5.8}$$

where $n_{\text{eff}}^{(m)}$ is the effective propagation index

$$n_{\text{eff}}^{(m)} = n_g \sin \theta^{(m)} \quad n_{\text{s,c}} < n_{\text{eff}}^{(m)} < n_g. \tag{5.9}$$

In the following, we restrict the discussion to symmetric waveguides $n_c = n_s$, which simplifies the treatment. The mode condition Eq. (5.5) scales with $ak_0 = 2\pi a/\lambda_0$, the ratio of waveguide width to wavelength. It is therefore common to introduce a normalized parameter V

$$V := \frac{2\pi a}{\lambda_0}\sqrt{n_g^2 - n_s^2} = ak_0\text{NA} = a\frac{\omega}{c_0}\text{NA} \tag{5.10}$$

that is also called normalized frequency because it is proportional to the frequency of the mode; V comprises all relevant properties of the light field and the waveguide. In addition, the normalized parameters

$$u := ak_\perp = ak_0\sqrt{n_g^2 - n_{\text{eff}}^2} = ak_0 n_g \cos\theta = a\sqrt{k_0^2 n_g^2 - \beta^2} \tag{5.11}$$

$$w := ak_0\sqrt{n_{\text{eff}}^2 - n_s^2} = a\sqrt{\beta^2 - k_0^2 n_s^2} \tag{5.12}$$

are introduced that are related to V by

$$u^2 + w^2 = V^2. \tag{5.13}$$

Obviously, $-2u$ is the transverse phase difference of the mode between the two interfaces, while $1/w$ is the normalized penetration depth of the evanescent field into the substrate or cladding, as we shall see shortly; note that u and w are both functions of θ. Since $u(\theta_{\mathrm{crit}}) = V$, total reflection requires $u < V$.

With these parameters, Eq. (5.6) can be expressed as $\tan(\phi/2) = w/u$ and the mode condition Eq. (5.5) assumes the form

$$\tan\left(u - m\frac{\pi}{2}\right) = \frac{w}{u} \tag{5.14}$$

or, using Eq. (5.13),

$$\tan\left(u - m\frac{\pi}{2}\right) = \frac{\sqrt{V^2 - u^2}}{u}. \tag{5.15}$$

Figure 5.4 shows the graphical representation of the two sides of this transcendental equation as a function of $u = ak_\perp$. The left-hand side is a series of tangens branches (for $m = 0, 2, 4, \ldots$), interleaved with negative co-tangens branches (for $m = 1, 3, 5, \ldots$). The points of intersection with the right-hand side yield solutions $u^{(m)}$, and, with Eq. (5.11), $\theta^{(m)}$ and $\beta^{(m)}$.

As already mentioned, u cannot exceed V; consequently, the right-hand side of Eq. (5.15) is defined only for $0 < u \leq V$ (Fig. 5.4). Since the branches of the left-hand side are separated by $\pi/2$, the number of solutions is

$$M = \left[\frac{V}{\pi/2}\right] + 1 \tag{5.16}$$

(the square brackets denote the maximum integer contained in the argument). The condition $u < V$ (that is equivalent to $\theta < \theta_{\mathrm{crit}}$) is called cutoff condition. In many applications, the existence of more than one mode is not desired and one designs the waveguide such that

$$V < V_{\mathrm{c}} = \pi/2. \tag{5.17}$$

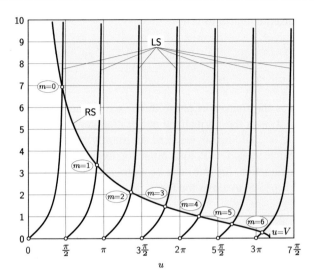

Fig. 5.4 Plot of the left-hand (LS) and right-hand side (RH) of Eq. (5.15) for $V = 10$

With Eq. (5.10), this monomode or single mode condition can be expressed in terms of the vacuum wavelength λ_0

$$\lambda_0 > \lambda_{0,c} = 4a\mathrm{NA}; \qquad (5.18)$$

$\lambda_{0,c}$ is called the monomode cutoff wavelength of the waveguide. For practical reasons, NA is usually on the order of 0.1, which implies that the thickness $2a$ of the guiding layer of a typical single mode waveguide is a few times the wavelength.

The above discussion refers to σ-polarized light; the electric field in this case has no longitudinal component and the modes are consequently called transverse electric (TE). To adapt the results for π-polarized light, we only have to replace the phase shift at reflection according to (2.48) by multiplying the right-hand side of Eq. (5.15) with $(n_g/n_s)^2 > 1$. This results in somewhat larger values of $u^{(m)}$ and smaller values of $\beta^{(m)}$; under weakly guiding conditions $(n_g - n_s)/n_g \ll 1$, this difference is very small, however. While the electric field of these modes has a longitudinal component, the magnetic field is purely transverse, and this set of modes is called transverse magnetic (TM).

The cutoff condition and thus the number of modes is the same for both polarizations. A so-called monomode waveguide therefore supports actually two modes of different polarization.

5.1.2 Transverse Mode Profile

With $\theta^{(m)}$ and the related parameters $\beta^{(m)}$, $u^{(m)}$, $w^{(m)}$ given, we can now construct the wave function of the modes by combining two plane waves with the wave vectors $\mathbf{k} = (\pm u/a, 0, \beta)$. We set the origin of the transverse coordinate in the central plane of the guiding layer so that the interfaces are at $x = \pm a$, respectively. In the central plane $x = 0$, the two plane waves constituting the mode have a phase difference of $m\pi$, so that the field in the guiding layer is

$$E^{\mathrm{g}} \propto \left[e^{j(u/a)x} + e^{-j(u/a)x} e^{-jm\pi} \right] e^{-j\beta z}. \qquad (5.19)$$

Introducing the normalized transverse coordinate $x' := x/a$, and neglecting a prefactor j as irrelevant, we obtain, for $|x'| \leq 1$

$$E^{\mathrm{g}} = E_0^{\mathrm{g}} \cos(ux') e^{-j\beta z} \quad \text{for} \quad m = 0, 2, 4 \ldots,$$
$$= E_0^{\mathrm{g}} \sin(ux') e^{-j\beta z} \quad \text{for} \quad m = 1, 3, 5 \ldots. \qquad (5.20)$$

Modes of even (odd) order are (anti)symmetric with respect to the central plane (Fig. 5.5). The mode order m is equal to the number of nodal planes, where the electric field is zero.

In the adjacent media $|x'| > 1$, the tangential component of the wave vector is again β, while the normal component is imaginary, $k_\perp = \pm j \sqrt{\beta^2 - n_{\mathrm{s}}^2 k_0^2} =$

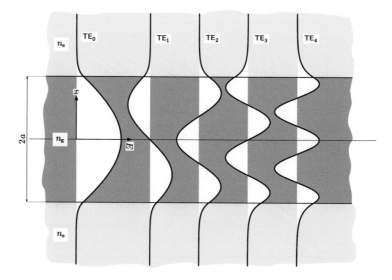

Fig. 5.5 Transverse mode profile for 5 TE modes of a symmetric waveguide ($n_{\mathrm{s}} = n_{\mathrm{c}}$)

$\pm j(w/a)$; the wave vector is therefore complex, $\mathbf{k} = [\pm j(w/a), 0, \beta]$, and the field for $|x'| > 1$ is given by

$$E^s = E_0^s e^{-w|x'|} e^{-j\beta z}; \qquad (5.21)$$

as stated above, w is the spatial decay constant of the field in normalized coordinates. From the continuity of the (transverse) electric field at $|x'| = 1$ follows $E_0^s = E_0^g \cos u$ (even mode order) and $E_0^s = \pm E_0^g \sin u$ (odd mode order).

The set of eigenmodes constitutes a complete base of orthogonal wavefunctions; any TE field guided by the waveguide can be written as a linear combination of these modes.

5.1.3   Waveguide Dispersion

An inspection of Fig. 5.4 shows that not only the number of modes, but also the propagation constant of a mode of given order depends on the frequency. This is a consequence of the fact that the mode condition Eq. (5.15) depends on the ratio a/λ_0 of the waveguide; one and the same waveguide appears to be wider for light of shorter wavelength. This purely geometric contribution to the dispersion $\beta^{(m)}(\omega)$ is called waveguide dispersion and has to be taken into account in addition to the material dispersion $n_{g,s,c}(\omega)$ of the waveguide materials. The waveguide dispersion can be obtained from Eq. (5.15) assuming constant $n_{g,s,c}(\omega)$ and is shown in Fig. 5.6 for a typical waveguide. The dispersion functions $\beta^{(m)}(\omega)$ are confined between the dispersion lines $\beta = k = (n_g/c_0)\omega$ and $\beta = k = (n_s/c_0)\omega$ for free wave

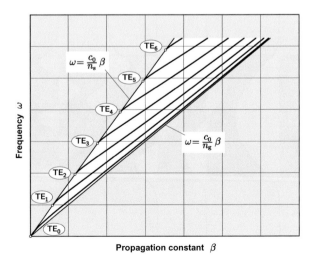

Fig. 5.6 Dispersion diagram of a planar waveguide

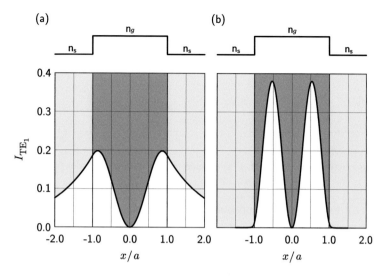

Fig. 5.7 Transverse intensity profile of the TE_1-mode at (**a**) $V \approx V_c$, (**b**) $V \gg V_c$

propagation in the respective medium (guide or substrate/cladding). With decreasing frequency, each dispersion curve approaches the dispersion line of the substrate until it terminates on this line (cutoff). This is to be expected from Eqs. (5.10) and (5.11), according to which $n_{eff} \rightarrow n_s$ near the cutoff. Far above the cutoff frequency, the effective propagation index approaches the free propagation index of the guiding layer. The physical reason for this becomes obvious from an inspection of Fig. 5.7: close to the cutoff frequency, the penetration depth $1/w$ of the evanescent field increases [Eq. (5.12)], so that a large fraction of the mode profile lies in the low index substrate/cladding. Far above the cutoff, the penetration depth is small and the mode profile is concentrated in the guiding layer.

The combined waveguide and material dispersion is called chromatic dispersion; the chromatic dispersion coefficient Eq. (3.159) of a waveguide can, in very good approximation, be calculated as the sum of the waveguide and material dispersion coefficients, provided that the latter is the same for guiding layer and substrate.

The existence of more than one mode at a given frequency is often referred to as mode dispersion, and the dependence of the propagation index on the polarization as polarization dispersion.

5.2 Fiber Waveguides

So far we have discussed the confinement of a light field in one dimension. In most applications, confinement in both lateral dimensions is required. This can be obtained by designing the guiding medium in the shape of a rectangular channel that is either embedded in a medium of lower propagation index (channel waveguide) or

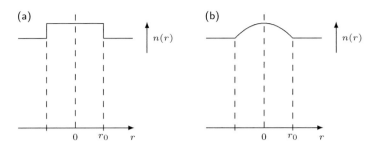

Fig. 5.8 Refractive index profile of (**a**) a step index fiber, (**b**) a gradient index fiber

is placed on top of a planar substrate layer (ridge waveguide). The mathematical treatment is similar to that of a planar waveguide, with the main difference that for each of the two transverse dimensions, a separate mode condition is established. Such waveguides are of great importance for integrated optical devices such as modulators, sensors, couplers, and multiplexers that will be discussed later.

For the transportation of light signals over large distances, cylindrical waveguides (fibers) are employed. They can be produced in virtually arbitrary length and support data rates above 100 Gbit/s; under optimized conditions, the signal loss is as low as 0.16 dB/km.

The operating principle of such waveguides is the same as that of planar dielectric waveguides. A guiding core of elevated propagation index n_c is surrounded by an optically thinner cladding. For protection purposes, this fiber, which is usually made out of silica glass, is coated by plastic layer that has no optical function. The transition between core and cladding can be step-like (step index fibers) or continuous (gradient index fibers), as shown schematically in Fig. 5.8. Gradient fibers will be discussed later; the following treatment of step index fibers is very similar to that of symmetric planar waveguides.

5.2.1 Step Index Fibers

Mathematically, discontinuous structures such as propagation index steps can be treated by imposing boundary conditions on the wave equations. The description of planar waveguides was particularly simple because the boundary condition problem at a planar interface has already been solved in Sect. 2.1.3, yielding the Fresnel coefficients. For a cylindrical geometry, we have to start from the Helmholtz equation Eq. (1.22) in cylindrical coordinates z, r, φ

$$\frac{\partial^2 U}{\partial r^2} + \frac{1}{r}\frac{\partial U}{\partial r} + \frac{1}{r^2}\frac{\partial^2 U}{\partial \varphi^2} + \frac{\partial^2 U}{\partial z^2} + n^2 k_0^2 U = 0, \tag{5.22}$$

where U is a cartesian component of the $\mathbf{E}$ or $\mathbf{H}$-field. With the separation-ansatz

$$U(r, \varphi, z) = R(r)\Phi(\varphi)\mathrm{e}^{-\mathrm{j}\beta z}, \tag{5.23}$$

Eq. (5.22) yields the azimuthal differential equation

$$\frac{\mathrm{d}^2\Phi}{\mathrm{d}\varphi^2} + l^2\Phi = 0 \tag{5.24}$$

with the two independent solutions

$$\Phi = \mathrm{e}^{\pm \mathrm{j}l\varphi}, \tag{5.25}$$

which can be combined to $\cos l\varphi$ or $\sin l\varphi$, respectively, where $l = 0, 1, 2, \ldots$ to meet the self-consistency condition $U(\varphi) = U(\varphi + 2\pi)$.

For a given value of l, the radial differential equation is then

$$\frac{\mathrm{d}^2R}{\mathrm{d}r^2} + \frac{1}{r}\frac{\mathrm{d}R}{\mathrm{d}r} + \left[n^2(r)k_0^2 - \beta^2 - \frac{l^2}{r^2}\right]R = 0. \tag{5.26}$$

We introduce the normalized radius $\rho := r/r_0$, where r_0 is the core radius, so that

$$n(\rho) = n_\mathrm{g} \quad \text{for} \quad \rho \le 1$$
$$n(\rho) = n_\mathrm{c} \quad \text{for} \quad \rho > 1. \tag{5.27}$$

With this propagation index profile, Eq. (5.26) becomes

$$\frac{\mathrm{d}^2R}{\mathrm{d}\rho^2} + \frac{1}{\rho}\frac{\mathrm{d}R}{\mathrm{d}\rho} + \left(u^2 - \frac{l^2}{\rho^2}\right)R = 0 \quad \text{for} \quad \rho \le 1 \tag{5.28}$$

$$\frac{\mathrm{d}^2R}{\mathrm{d}\rho^2} + \frac{1}{\rho}\frac{\mathrm{d}R}{\mathrm{d}\rho} - \left(w^2 + \frac{l^2}{\rho^2}\right)R = 0 \quad \text{for} \quad \rho > 1, \tag{5.29}$$

where, in analogy to Eqs. (5.10)–(5.12), the normalized frequency V and the parameters u and w, defined as

$$V = r_0 k_0 \sqrt{n_\mathrm{g}^2 - n_\mathrm{c}^2} =: r_0 k_0 \mathrm{NA} \tag{5.30}$$

$$u := r_0 \sqrt{n_\mathrm{g}^2 k_0^2 - \beta^2} \tag{5.31}$$

$$w := r_0 \sqrt{\beta^2 - n_\mathrm{c}^2 k_0^2} \tag{5.32}$$

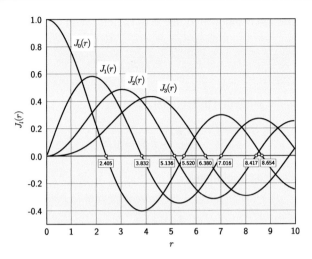

Fig. 5.9 Bessel functions of first kind

have been used; the numerical aperture is defined in the same way as in Eq. (5.3). Again, $u^2 + w^2 = V^2$, and u cannot exceed V.

The solutions of Eqs. (5.28) and (5.29) are Bessel functions

$$R(\rho) = A_g J_l(u\rho) \quad \text{for} \quad \rho \le 1$$
$$R(\rho) = A_c K_l(w\rho) \quad \text{for} \quad \rho > 1, \tag{5.33}$$

where the Bessel functions J_l of first kind and l-th order resemble sine- and cosine functions with radially decaying amplitude (Fig. 5.9), and the modified Bessel functions K_l of second kind and l-th order resemble decaying exponential functions (Fig. 5.10); the amplitudes $A_{g,c}$ are determined by the boundary conditions.

Assuming weak guiding $(n_g - n_c)/n_g \ll 1$, total internal reflection requires grazing incidence of the field at the core/cladding interface, so that the field has only very small longitudinal components; accordingly it can be treated as approximately transverse electromagnetic (TEM) and the boundary conditions require the continuity of $R(\rho)$ and its derivative $dR(\rho)/d\rho$ at $\rho = 1$

$$A_g J_l(u) - A_c K_l(w) = 0$$
$$A_g u J_l'(u) - A_c w K_l'(w) = 0, \tag{5.34}$$

where the prime denotes the derivative in respect to ρ. For this system of equations to have non-trivial (i.e., non-zero) solutions, the system determinant must vanish

$$J_l(u) w K_l'(w) - K_l(w) u J_l'(u) = 0. \tag{5.35}$$

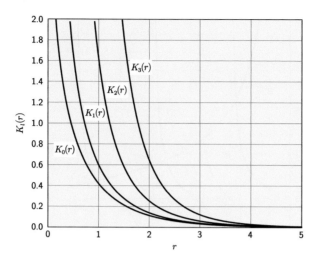

Fig. 5.10 Modified Bessel functions of second kind

Employing the identities

$$J_l'(x) = -(l/x)J_l(x) + J_{l-1}(x)$$

$$K_l'(x) = -(l/x)K_l(x) - K_{l-1}(x) \tag{5.36}$$

(see, e.g., Abramowitz and Stegun 2014), we can express Eq. (5.35) by

$$-\frac{J_{l-1}(u)K_l(w)}{J_l(u)K_{l-1}(w)} = \frac{w}{u}; \tag{5.37}$$

this is the mode condition for cylindrical step index waveguides under the weakly guided mode approximation.

Figure 5.11 shows both sides of Eq. (5.37) for $l = 0, 1$ as a function of u (note the similarity with Fig. 5.4), where the identities $J_{-1}(u) = -J_1(u)$ and $K_{-1}(u) = K_1(u)$ have been used. For a given value of l, Eq. (5.37) can have one or more solutions $u^{(lm)}$, denoted by the radial mode index $m = 1, 2, \ldots$ (Fig. 5.12), with corresponding propagation constants $\beta^{(lm)}$ given by Eq. (5.31), and wave functions according to Eq. (5.23). A mode is therefore characterized by the two indices l and m; according to Eq. (5.25), for $l \geq 1$ each set (l, m) can be represented by two azimuthal field distributions that are offset by 90° and are, for symmetry reasons, degenerate (i.e., they have identical propagation constants). Moreover, each set (l, m) allows for two orthogonal polarization states that are also degenerate. It is therefore common to classify these degenerate modes as one, linearly polarized mode LP_{lm}; while l denotes the number of azimuthal nodes, m gives the number of radial intensity peaks (Fig. 5.13). The mode profile of LP_{01} is similar to a Gaussian

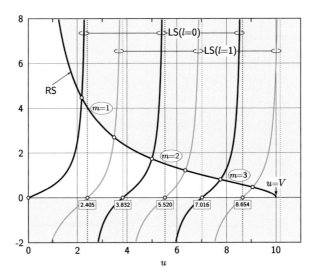

Fig. 5.11 Plot of the left-hand (LS) and right-hand (RS) sides of the mode equation (5.37) of a step index fiber for $l = 0$ and $l = 1$, $V = 10$

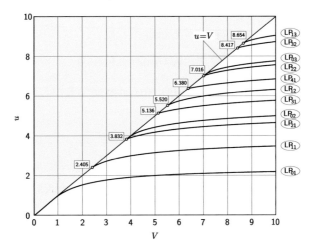

Fig. 5.12 Solutions of Eq. (5.37) as a function of the normalized frequency V

profile, so that a Gaussian beam of appropriate waist diameter is well suited to excite this mode in a fiber.

We have mentioned before that $u < V$, and this condition also shows up in Fig. 5.11, where the right-hand side vanishes at $u = V$. For a given waveguide, V can be varied by changing the frequency of the light field; with decreasing (normalized) frequency, there are less and less intersections and thus solution of Eq. (5.37). If V falls below the value 2.405 (the first root of the Bessel function J_0, see Fig. 5.9), there is only one solution left, LP_{01}. The upper frequency limit for single mode

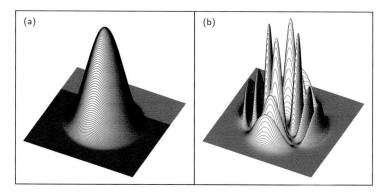

Fig. 5.13 Intensity profiles of two modes in a fiber: (**a**) LP_{01} (**b**) LP_{23}

operation of a cylindrical waveguide is therefore given by

$$V < V_c = 2.405; \tag{5.38}$$

in terms of wavelength, the single mode limit is

$$\lambda_0 > \lambda_{0c} = r_0 NA \frac{2\pi}{2.405}. \tag{5.39}$$

As an example, the maximum core radius of a single mode fiber with $NA = 0.1$ and $\lambda_{0c} = 1\,\mu m$ is $3.8\,\mu m$.

Figure 5.14 shows the effective propagation index n_{eff} of the various modes; it increases with frequency and approaches n_g far above the cutoff frequency of the respective mode. Close to the cutoff, n_{eff} tends toward the cladding index n_c, just as in the case of planar waveguides, and for the same reason: as illustrated in Fig. 5.15, the fraction of the mode that is transported in the core decreases with decreasing frequency and approaches zero at cutoff. To prevent excessive losses, the cladding has to be thick enough to accommodate the evanescent field; typical values are around $50\,\mu m$, so that the total diameter of a single mode fiber for $1\,\mu m$-wavelength light is about $100\,\mu m$.

The dispersion of n_{eff} also results in a group velocity dispersion (Fig. 5.16) due to the waveguide structure: different modes have different group velocities, and the group velocity of a given mode depends on the frequency.

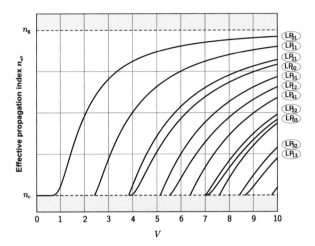

Fig. 5.14 Effective propagation index n_{eff} of the modes in a step index fiber; the phase velocity is $v_{\text{ph}} = c_0/n_{\text{eff}}$

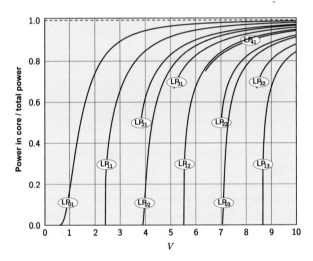

Fig. 5.15 Fraction of the energy flux transported inside the core

5.2.2 Fiber Losses and Dispersion

High quality fiber waveguides are usually made of quartz glass (SiO_2) whose propagation index is modified by controlled doping with GeO_2 and other dopants that increase (Ge, P) or decrease (B) the refractive index. To avoid contamination with absorbing impurities, SiO_2 is grown by modified chemical vapor deposition (MCVD) from a gas phase reaction of $SiCl_4$ and O_2

$$SiCl_4 + O_2 \rightarrow SiO_2 + 2Cl_2.$$

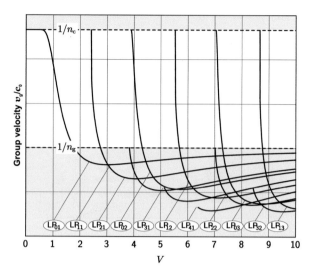

Fig. 5.16 Group velocity of modes in a step index fiber as a function of the normalized frequency V

Dopants are added by admitting fractions of $GeCl_4$, $POCl_3$, or BCl_3 to the reaction (10–20 mol% are required to change the index by 1 %). The reactor is a rotating quartz glass tube of some 10 mm diameter. The silicon oxide including the dopant oxide is deposited on the inner wall of the tube and fused to glass at about 1000 °C. The resulting tube is collapsed under vacuum into a so-called preform that shows the refractive index profile of choice. The preform is then heated to 2000 °C and the glass fiber is drawn from it in a vertical tower; immediately after cooling it is coated by a polymer film to protect it from diffusive impurities such as hydrogen.

Transmission distance and channel data transmission capacity is limited by losses and group velocity dispersion. Losses are specified by a loss coefficient $[10 \lg P(0)/P(l)]/l$ in decibel per kilometer (dB/km), where $P(0)$ is the optical power fed into the fiber and $P(l)$ is the output power after the distance l. Silica glass fibers can have loss coefficients as low as 0.16 dB/km, which corresponds to a transmission of 10 % for a fiber length of 62.5 km or an attenuation by a factor of 40 for a 100 km long fiber. Figure 5.17 shows the various loss contributions and their wavelength dependence. The global loss minimum is found at a wavelength of 1.55 μm; in the visible, the losses are much higher, resulting from the wings of electronic resonances in the UV and from Rayleigh scattering that scales with $1/\lambda_0^4 \propto \omega^4$ and originates from density fluctuations in the glass that depend on the melting point of the glass.[1] In the near infrared, there are two vibrational absorption

---

[1] The ω^4-dependence can be understood from an inspection of Eq. (2.1): the field scattered from an inhomogeneity scales with the second time derivative $\partial^2 P/\partial t^2 \propto \omega^2$; the radiated power is therefore proportional to ω^4.

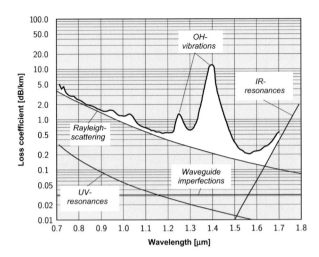

Fig. 5.17 Fiber losses and loss mechanisms of a quartz glass fiber as a function of wavelength

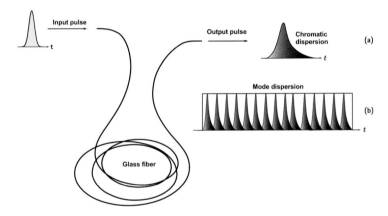

Fig. 5.18 Schematic illustration of dispersive effects on pulse propagation in fibers (**a**) waveguide and material dispersion, (**b**) mode dispersion

lines at 0.95 and 1.39 μm that are overtones of the hydroxyl vibration at 2.8 μm; a OH$^-$-concentration of 10^{-6} results in a loss of 30 dB/km at 1.39 μm. For this reason, the in-diffusion of hydrogen into the silica matrix must be kept as low as possible ($<10^{-8}$). The absorption "valleys" at 1.3 and 1.55 μm define the operating wavelengths of optical communications.

Group velocity dispersion (GVD) is the other limiting factor for the transmission capacity of fibers. Optical data are transmitted in the form of pulses, and the pulse transmission rate determines the data rate. Mode dispersion results in the splitting of a single input pulse into multiple pulses because of the different group velocities of the modes (Fig. 5.18). This problem (the group delay differences can be as large as 10 ns/km) is avoided in single mode fibers.

The combined waveguide and material dispersion results, as discussed in Sect. 3.2, in pulse broadening during propagation. If the broadening exceeds the temporal separation of the pulses, they start to overlap, rendering their identification at the detector impossible. The relevant measure of group velocity dispersion is the dispersion coefficient D_ω or D_λ that denotes the differential change of the group delay as a function of frequency. The treatment in Sect. 3.2 is independent on the underlying dispersion mechanism and its results can be directly applied to the chromatic dispersion of a waveguide. The only formal difference is the replacement of the wave number k by β, or n by n_{eff}. Thus, Eq. (3.162) assumes the form

$$D_\lambda = -\frac{\lambda_0}{c_0}\frac{\mathrm{d}^2 n_{\text{eff}}}{\mathrm{d}\lambda_0^2}; \qquad (5.40)$$

the pulse broadening Eq. (3.175) over a distance l is qualitatively given by

$$\Delta\tau \approx l|D_\omega|\Delta\omega = l|D_\lambda|\Delta\lambda_0, \qquad (5.41)$$

where $\Delta\omega$ and $\Delta\lambda_0$, respectively, are the spectral bandwidth of the pulse. Typical values of $|D_\lambda|$ are between 0 and 100 ps/nm km, so that dispersive pulse broadening plays a significant role only for sub-ns pulses.

As an inspection of Fig. 5.19 shows, the waveguide dispersion coefficient is strongly frequency dependent and may actually change its sign at a certain

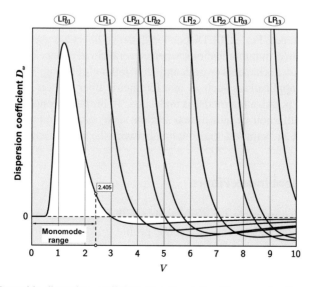

Fig. 5.19 Waveguide dispersion coefficient of a step index fiber as a function of normalized frequency; for $V < 2.405$, the coefficient D_ω is positive and can be compensated only by *negative* (anomalous) material dispersion

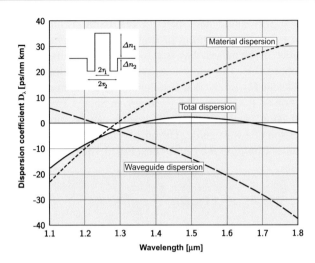

Fig. 5.20 Dispersion flattened fiber waveguide: the refractive index profile is designed such that the group delay dispersion is close to zero over a broad spectral range

frequency. Within the single mode range, however, the waveguide dispersion of an ordinary step index fiber is positive and requires negative material dispersion to be compensated. As can be seen from Fig. 3.17, the material dispersion coefficient of silica glass goes to zero (and changes sign) at about $1.27\,\mu$m. By sophisticated choice of the index profile (exploiting the frequency dependence of the mode diameter), the waveguide dispersion can be modified such that the combined chromatic dispersion is zero at a wavelength of choice (dispersion shifted), or close to zero over a selected wavelength range, such as the attenuation valley at $1.55\,\mu$m (dispersion flattened, Fig. 5.20). Dispersion flattened fibers are used for wavelength-multiplexed optical communications, where several closely spaced carrier waves act as independent data channels (wavelength division multiplexing, WDM).

For certain applications such as interferometric fiber sensors, the existence of two degenerate polarization modes is undesired. To lift the degeneracy, the fiber can be produced with a non-cylindric cross section; such polarization maintaining fibers are birefringent and suppress the coupling between the two polarization states.

5.2.3   Gradient Index Fibers

As an instructive example for a waveguide with a continuous propagation index profile, we assume a fiber with a parabolic profile

$$n(r) = n_0\left(1 - \tfrac{1}{2}\alpha_\mathrm{g}^2 r^2\right). \tag{5.42}$$

We have encountered such a structure already in the context of Gaussian beams in Sect. 3.1.3, where a slice of glass with such an index profile was employed as (GRIN)-lens. In a sense, a gradient index fiber (also known as graded index fiber) is just a very long GRIN-lens.

As we have seen in Sect. 3.1.4, the q-parameter of a Gaussian beam propagating along the axis of such a structure develops from q at input to q' at output according to the ABCD-transformation Eq. (3.65)

$$q' = \frac{Aq + B}{Cq + D} \tag{5.43}$$

with the ABCD matrix Eq. (3.94)

$$M_{\mathrm{GRIN}} = \begin{bmatrix} \cos \alpha_g d & (1/n_0 \alpha_g) \sin \alpha_g d \\ -n_0 \alpha_g \sin \alpha_g d & \cos \alpha_g d \end{bmatrix}. \tag{5.44}$$

We now search for a Gaussian field distribution that conserves its q-parameter throughout the propagation, and is therefore an eigenmode of the fiber, by solving the equation $q' = q$ for arbitrary d. The solutions

$$q_{1,2} = \frac{A - D}{2C} \pm \frac{1}{2C} \sqrt{(A - D)^2 + 4BC} \tag{5.45}$$

with the ABCD matrix Eq. (5.44) are purely imaginary

$$q_{1,2} = \pm j \frac{1}{n_0 \alpha_g}; \tag{5.46}$$

from the relations given in Table 3.1 we obtain

$$z_0 = \mathrm{Im}\,[q] = \frac{1}{n_0 \alpha_g} = \frac{n_0 \pi w_0^2}{\lambda_0} \tag{5.47}$$

and

$$w_0^2 = \frac{\lambda_0}{\pi n_0^2 \alpha_g}. \tag{5.48}$$

The radial field distribution $E(r) \propto e^{-r^2/w_0^2}$ represents an eigenmode of the parabolic gradient index fiber because the distributed GRIN-lens compensates the tendency of the field to diverge. Since the ABCD formalism is also applicable to Hermite–Gaussian beams (Sect. 3.1.5), Hermite–Gaussian field profiles with a w_0-parameter given by Eq. (5.48) also represent eigenmodes of such a fiber. The intensity profile of these modes is given by Eq. (3.98) and shown in Fig. 3.14.

The axial dependence of the wave function follows from Eqs. (3.95) and (3.97) to be $e^{-j[kz-(1+l+m)\xi(z)-\omega t]}$, where l and m denote the order of the Hermite–Gaussian mode. The propagation constant $\beta^{(lm)}$ is obtained by taking the z-derivative of the phase $[kz - (1+l+m)\xi(z) - \omega t]$ at $z = 0$. With Eq. (3.16) we find

$$\beta^{(lm)} = k - (1+l+m)/z_0 = n_0[\omega/c_0 - (1+l+m)\alpha_g]. \tag{5.49}$$

Modes with the same value $l + m$ are degenerate in the sense that they have the same propagation constant. Note that any linear combination of mutually degenerate eigenmodes is again an eigenmode. The group velocity of the mode (lm) follows from $1/v_g = d\beta/d\omega$ to be $v_g = c_0/n_0$, independent of the mode order. Different from step index fibers (Fig. 5.16), the mode dispersion of parabolic gradient index fibers vanishes.

In practice, the index profile is parabolic only up to a certain radius r_0 and remains constant in the cladding region $r > r_0$ (Fig. 5.8b). Guiding requires that the effective mode propagation index β/k_0 is larger than the cladding index, $\beta/k_0 > n_0\left(1 - \frac{1}{2}\alpha_g^2 r_0^2\right)$. With Eq. (5.49) we obtain

$$n_0\left[1 - \frac{(1+l+m)\alpha_g}{k_0}\right] > n_0\left(1 - \frac{1}{2}\alpha_g^2 r_0^2\right) \tag{5.50}$$

or

$$1 + l + m < r_0^2 k_0 \alpha_g/2. \tag{5.51}$$

Thus, the core radius r_0 (and the coefficient α_g) determines the number of guided modes in a graded index fiber.

The treatment given here is a very coarse one, since the many approximations used are valid only for weak confinement ($w_0 \gg \lambda_0$). Nonetheless, the main features of gradient index fibers become clear. A more rigorous analysis shows that the mode dispersion of the group velocity is not exactly zero but still much smaller than that of a step index fiber with the same number of modes. For this reason, gradient index fibers are the preferred choice if multimode fibers are to be used. The attractivity of multimode fibers lies in the fact that their core diameter is much larger than that of single mode fibers, so that light insertion and fiber–fiber connection is much less demanding.

5.3   Integrated Optics

Integrated optics comprises optical devices that work without free space propagation and rely on waveguides. The waveguides used can be planar or fibers. Passive components such as splitters, couplers, mirrors and interferometers as well as laser amplifiers and lasers can be realized in an integrated fashion. Optically integrated structures can also be built on a semiconductor substrate, allowing for the integration of light sources and detectors (optoelectronic integrated circuits, OEICs).

5.3.1 Waveguide Couplers

One of the most important components of integrated optics is the waveguide coupler (Fig. 5.21) that allows the controlled exchange of optical energy between waveguides. The operating principle is to use the evanescent field of one waveguide to produce a polarization current in the other. The coupling coefficient that describes the transfer is, as we will see, determined by the overlap integral of the transverse mode profiles of the two modes involved (Fig. 5.26).

The exact solution of the coupling problem requires solving the wave equation under the given geometric conditions. For weak coupling, an elegant approximative solution is provided by the coupled modes formalism (Haus 1984; Yariv 1973) that starts from the modes of the isolated waveguides and treats the interaction between them as small perturbation.

The complex wave function in waveguide (i) is assumed to be

$$E_i = a_i(z)u_i(x, y), \quad i = 1, 2, \tag{5.52}$$

where $u_i(x, y)$ is the transverse mode profile and the amplitude $a_i(z)$ is normalized such that

$$a_i(z)a_i^*(z) \tag{5.53}$$

is the power in waveguide (i) at z. In the framework of perturbation theory, we assume that the presence of a second waveguide in the vicinity of the first one

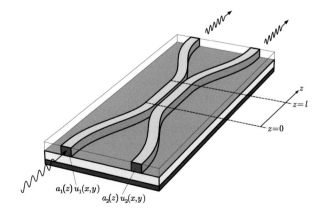

Fig. 5.21 Geometry of a waveguide coupler

leaves the profiles $u_{1,2}(x, y)$ unaltered, influencing only the propagation constants and amplitudes of the modes.

Let the undisturbed modes have the propagation constants $\beta_{1,2}$ so that $a_i(z) = a_i(0)e^{-j\beta_i z}$, and the differential change of a_i is given by

$$\frac{da_i}{dz} = -j\beta_i a_i. \tag{5.54}$$

The perturbation of one mode by the other is taken into account by a cross term $\kappa_{ij}a_j$

$$\frac{da_1}{dz} = -j\beta_1 a_1 + \kappa_{12}a_2$$

$$\frac{da_2}{dz} = -j\beta_2 a_2 + \kappa_{21}a_1, \tag{5.55}$$

where κ_{ij} is the respective coupling coefficient. While the coupling modifies the mode amplitudes, the total power transported in the waveguide system is conserved. With Eqs. (5.53) and (5.55) we can express the differential power change in waveguide (i) as

$$\frac{d(a_i a_i^*)}{dz} = a_i \frac{da_i^*}{dz} + a_i^* \frac{da_i}{dz} = a_i^* \kappa_{ij} a_j + a_i \kappa_{ij}^* a_j^*; \tag{5.56}$$

energy conservation requires

$$\frac{d(a_1 a_1^* + a_2 a_2^*)}{dz} = a_1^* a_2(\kappa_{12} + \kappa_{21}^*) + a_1 a_2^*(\kappa_{12}^* + \kappa_{21}) = 0. \tag{5.57}$$

This must be valid for arbitrary $a_{1,2}$ (for example, $a_1 = a_2 = 1$, or $a_1 = 1, a_2 = j$), so that energy conservation imposes the condition

$$\kappa_{12} = -\kappa_{21}^*. \tag{5.58}$$

We therefore can set $\kappa_{12} =: \kappa$, $\kappa_{21} = -\kappa^*$; in Sect. 5.3.3 we will derive an equivalent relation for counterpropagating modes.

5.3.1.1 Eigenstates of a Waveguide Coupler

At a given point z, the coupled waveguide system can be represented by a vector

$$\psi(z) = \begin{bmatrix} a_1(z) \\ a_2(z) \end{bmatrix}, \tag{5.59}$$

where $a_{1,2}(z)$ are the field amplitudes in the respective waveguide. By definition, eigenmodes are states that conserve the ratio a_1/a_2 during propagation, satisfying the eigenvalue equation $\psi(z) = \lambda\psi(0)$, where $|\lambda| = 1$ in a lossless system. We set $\lambda = e^{-j\beta z}$ (β is the propagation constant of the eigenstate) so that $d\psi(z)/dz = -j\beta\psi(z)$. Substitution in Eq. (5.55) yields

$$\begin{bmatrix} j(\beta - \beta_1) & \kappa \\ -\kappa^* & j(\beta - \beta_2) \end{bmatrix} \begin{bmatrix} a_1 \\ a_2 \end{bmatrix} = \mathbf{0}. \tag{5.60}$$

Existence of non-trivial solutions $a_{i,j} \neq 0$ requires the determinant of the matrix to vanish

$$(\beta - \beta_1)(\beta - \beta_2) - \kappa\kappa^* = 0, \tag{5.61}$$

yielding two propagation constants

$$\beta^\pm = \bar{\beta} \pm K, \tag{5.62}$$

with

$$K = \sqrt{(\Delta\beta)^2 + |\kappa|^2}, \tag{5.63}$$

where

$$\bar{\beta} = \frac{\beta_1 + \beta_2}{2}, \quad \Delta\beta = \frac{\beta_1 - \beta_2}{2}. \tag{5.64}$$

The corresponding eigenvectors follow after substitution of $\beta^\pm$ in Eq. (5.60)

$$\psi^\pm = \begin{bmatrix} \pm j\kappa \\ K \mp \Delta\beta \end{bmatrix} e^{-j\beta^\pm z}. \tag{5.65}$$

Any arbitrary state of the coupled system can be synthesized as a linear combination of these eigenstates

$$\psi(z) = A^+\psi^+ + A^-\psi^-; \tag{5.66}$$

the coefficients $A^\pm$ follow from the boundary conditions $\psi(0)$. The existence of two different propagation constants results in a spatial beating of the amplitudes along the waveguides (Fig. 5.22), very similar to the beating of a superposition of two monochromatic signals in time.

Fig. 5.22 Power transfer between two coupled waveguides for different values of the phase mismatch $\Delta\beta$; only for $\Delta\beta = 0$, complete transfer is possible

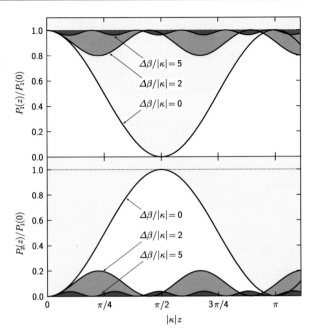

5.3.2 Splitters and Switches

As a simple example, we assume that at $z = 0$, light is launched into waveguide (1) only, so that $a_1(0) = a_0$, $a_2(0) = 0$. The corresponding linear combination turns out to be

$$a_1(z) = a_0 \left(\cos Kz - j\frac{\Delta\beta}{K} \sin Kz \right) e^{-j\bar{\beta}z}$$

$$a_2(z) = -a_0 \frac{\kappa^*}{K} (\sin Kz)\, e^{-j\bar{\beta}z}. \qquad (5.67)$$

In the synchronous case $\beta_1 = \beta_2 = \bar{\beta}$ (coupling of two identical monomode waveguides, for example) this simplifies to

$$a_1(z) = a_0 \cos |\kappa|z\, e^{-j\bar{\beta}z}$$

$$a_2(z) = -a_0 \frac{\kappa^*}{|\kappa|} \sin |\kappa|z\, e^{-j\bar{\beta}z}, \qquad (5.68)$$

Fig. 5.23 Spatial
development of the power
carried in the two branches of
a 3 dB-coupler

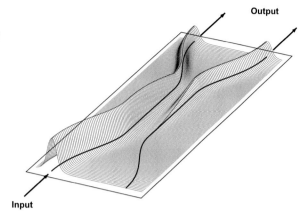

Output

Input

with the guided powers $P_i(z) = a_i a_i^*$

$$P_1(z) = P_1(0) \cos^2 |\kappa| z$$
$$P_2(z) = P_1(0) \sin^2 |\kappa| z. \tag{5.69}$$

The light field, initially confined to waveguide (1), is completely transferred to
waveguide (2) within a distance of $l_0 = \pi/2|\kappa|$ and keeps swinging between the
two waveguides over the entire interaction length (Fig. 5.22); diffusive interaction,
by contrast, would lead to an equilibrium distribution between the two channels
after a sufficiently long interaction distance.

At $z = l_0/2$, 50 % of the power is transferred; a coupler of this length is called a
3 dB-coupler (because $10 \lg 0.5 = -3$ dB) and is a waveguide-implementation of a
1:1 beam splitter (Fig. 5.23).

It is important to note that κ (and in general also $\Delta\beta$) is frequency dependent
because the mode overlap that determines the coupling depends on the wavelength;
for a given interaction length, the splitting ratio therefore may be different for
different frequencies, so that waveguide couplers also have filtering characteristics.
With proper layout, light containing two different frequencies can be split by a
"dichroic" coupler so that the two output branches of the coupler contain only one
of the frequencies each (wavelength selective coupler, WSC).

In the asynchronous case, $\beta_1 \neq \beta_2$, the power transfer is incomplete (Fig. 5.22),
because the phase difference between the fields in the two waveguides changes
during propagation. From Eqs. (5.63) and (5.67) follows

$$\frac{P_2(z)}{P_1(0)} = \frac{|\kappa|^2}{K^2} \sin^2 Kz = \frac{\sin^2\left[\sqrt{1 + (\Delta\beta/|\kappa|)^2}|\kappa|z\right]}{1 + (\Delta\beta/|\kappa|)^2} \leq 1; \tag{5.70}$$

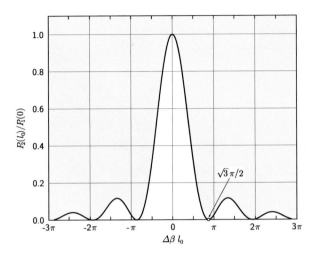

Fig. 5.24 Power transfer ratio of a coupler as a function of the normalized phase mismatch $\Delta\beta/l_0$; $l_0 = \pi/2|\kappa|$

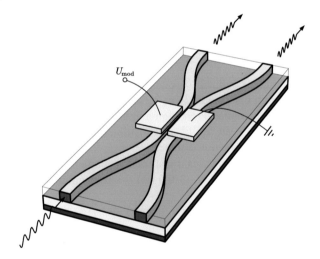

Fig. 5.25 Electro-optically controlled waveguide coupler

this transmission function is shown in Fig. 5.24. Obviously, efficient power transfer is possible only if $(\Delta\beta/|\kappa|)^2 \ll 1$. On the other hand, if one manages to influence $\Delta\beta$ externally, Eq. (5.70) allows controlling the transmission ratio between 0 and 1. Varying $\beta_{1,2}$ is possible via the electro-optic effect (which requires the coupler to be made out of an appropriate medium such as lithium niobate). Such a device is shown in Fig. 5.25; the interaction length is chosen to be $l_0 = \pi/2|\kappa|$, so that for

$\Delta\beta = 0$, the transfer is complete. From Eq. (5.70) we obtain the transfer ratio as a function of $\Delta\beta$

$$\frac{P_2(l_0)}{P_1(0)} = \left(\frac{\pi}{2}\right)^2 \frac{\sin^2 X}{X^2} \tag{5.71}$$

where $X := \sqrt{(\Delta\beta l_0)^2 + (\pi/2)^2}$; for $\Delta\beta l_0 = \sqrt{3}\pi/2$, the transfer is equal to zero (Fig. 5.24). With appropriate waveguide design, the voltage needed to achieve the required detuning $\Delta\beta$ is as low as several Volt; the switching speed is about 10 GHz and the extinction ratio (on/off) is typically 20 dB (10^{-2}).

5.3.2.1 Coupling Coefficient

Physically, the two waveguides are defined by the elevated local susceptibility $[\Delta\chi(x, y)]_{1,2} = [\Delta\varepsilon(x, y)]_{1,2}$ (Fig. 5.26). The evanescent field of waveguide (1) produces a polarization current density within waveguide (2), of which the component $j\omega\varepsilon_0[\Delta\varepsilon(x, y)]_2 E_1(x, y)$ acts as a source of the field in this waveguide; the remaining component, proportional to the substrate susceptibility, belongs to waveguide (1). According to Eq. (1.54), the product of this current density with the field E_2 in waveguide (2) is equal to the temporal change of the local energy density. Using Eq. (1.59), we can calculate the averaged differential power transfer to waveguide (2)

$$\frac{\mathrm{d}(a_2 a_2^*)}{\mathrm{d}z} = -\frac{1}{2}\mathrm{Re}\left[\int E_2^* j\omega\varepsilon_0[\Delta\varepsilon(x, y)]_2 E_1 \, \mathrm{d}A\right]$$

$$= -\frac{1}{4}\left[\int E_2^* j\omega\varepsilon_0[\Delta\varepsilon(x, y)]_2 E_1 \, \mathrm{d}A + c.c.\right]$$

$$= -\frac{1}{4}\left[j\omega a_2^* a_1 \varepsilon_0 \int [\Delta\varepsilon(x, y)]_2 u_2^*(x, y) u_1(x, y) \, \mathrm{d}A + c.c.\right], \tag{5.72}$$

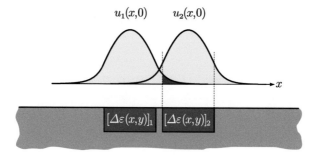

Fig. 5.26 Overlap of the transverse profiles of two coupled waveguides; the *dashed lines* show the integration limits for the calculation of the coupling coefficient κ_{12}

where $\int dA$ denotes the integral over the cross section of waveguide (2). Comparing this result with Eq. (5.56), we obtain

$$\kappa_{21} = -\tfrac{1}{4}j\omega\varepsilon_0 \int [\Delta\varepsilon(x,y)]_2 u_2^*(x,y)u_1(x,y)\, dA. \tag{5.73}$$

Since $[\Delta\varepsilon(x,y)]_2 = 0$ outside the waveguide, the coupling coefficient is determined by the overlap integral of the two transverse mode profiles within the waveguide cross sectional area (Fig. 5.26). The exponential decay of the evanescent field of mode (1) is responsible for the roughly exponential decrease of the coupling coefficient as a function of distance between the waveguides.

5.3.2.2 S-Matrix of a Coupler

For a symmetric waveguide coupler, $\beta_1 = \beta_2 = \bar\beta$ and $\kappa_{12} = \kappa_{21} = \kappa$, which because of Eq. (5.58) implies a purely imaginary value of κ. For the mode profiles shown in Fig. 5.26, Eq. (5.73) yields $\kappa = -j|\kappa|$. The propagation constants are $\beta^\pm = \bar\beta \pm |\kappa|$, corresponding to the eigenmodes

$$\psi^+(z) = \begin{bmatrix} 1 \\ 1 \end{bmatrix} e^{-j(\bar\beta+|\kappa|)z}, \quad \psi^-(z) = \begin{bmatrix} 1 \\ -1 \end{bmatrix} e^{-j(\bar\beta-|\kappa|)z}, \tag{5.74}$$

as illustrated in Fig. 5.26.

An arbitrary state with the inputs $a_1(0)$, $a_2(0)$ can be written as linear combination Eq. (5.66) with the coefficients $A^\pm = [a_1(0) \pm a_2(0)]/2$; propagation over the distance z results in

$$\begin{bmatrix} a_1(z) \\ a_2(z) \end{bmatrix} = \begin{bmatrix} a_1(0)\cos|\kappa|z - ja_2(0)\sin|\kappa|z \\ -ja_1(0)\sin|\kappa|z + a_2(0)\cos|\kappa|z \end{bmatrix} e^{-j\bar\beta z}. \tag{5.75}$$

This can be cast in the form

$$\begin{bmatrix} a_1(z) \\ a_2(z) \end{bmatrix} = S \begin{bmatrix} a_1(0) \\ a_2(0) \end{bmatrix} e^{-j\bar\beta z}, \tag{5.76}$$

where

$$S = \begin{bmatrix} \cos|\kappa|z & -j\sin|\kappa|z \\ -j\sin|\kappa|z & \cos|\kappa|z \end{bmatrix} \tag{5.77}$$

is the scattering matrix (compare Sect. 4.1.3) of the coupler.

5.3.2.3 Coupler as Phase Detector

Of particular interest is the 3 dB-coupler ($|\kappa|z = \pi/4$) already mentioned above, with the scattering matrix

$$S_{3\,\text{dB}} = \frac{\sqrt{2}}{2} \begin{bmatrix} 1 & -j \\ -j & 1 \end{bmatrix} = \frac{\sqrt{2}}{2} \begin{bmatrix} 1 & e^{-j\pi/2} \\ e^{-j\pi/2} & 1 \end{bmatrix}; \tag{5.78}$$

apart from its power splitting capacity, such a coupler serves as phase detector: if signals of equal magnitude but different phase ($a_{1,2}(0) = a_0 e^{\pm j\Delta\phi}$) are launched into its input ports, the output amplitudes according to Eq. (5.76) are

$$\begin{bmatrix} a_{1,\text{out}} \\ a_{2,\text{out}} \end{bmatrix} = \sqrt{2} a_0 e^{-j\pi/4} \begin{bmatrix} \cos(\Delta\phi - \pi/4) \\ -\sin(\Delta\phi - \pi/4) \end{bmatrix}, \tag{5.79}$$

and the respective output powers, with $\cos^2 x = (1 + \cos 2x)/2$, are given by

$$P_{1,\text{out}} = 2|a_0|^2 \cos^2(\Delta\phi - \pi/4) = |a_0|^2 [1 + \cos(2\Delta\phi - \pi/2)]$$
$$P_{2,\text{out}} = 2|a_0|^2 \sin^2(\Delta\phi - \pi/4) = |a_0|^2 [1 - \cos(2\Delta\phi - \pi/2)], \tag{5.80}$$

so that

$$\Delta\phi = \arctan\sqrt{\frac{P_{2,\text{out}}}{P_{1,\text{out}}}} + \frac{\pi}{4}. \tag{5.81}$$

We will return to this result in Sect. 5.3.4 in the context of waveguide interferometers and sensors.

Waveguide couplers can be realized in integrated planar optics, but also in fibers: for this purpose, two fibers are twisted and stretched close to the melting temperature. Stretching reduces the core diameter and increases the extension of the evanescent field so that it can overlap with the core of the second fiber.

5.3.3 Waveguide Gratings

Another important waveguide component is the waveguide grating, that is a periodic waveguide structure that can act as a filter and/or reflector. These components are conceptually similar to dielectric multilayer systems as treated in Sect. 4.2.2. Waveguide gratings are realized by a periodic longitudinal modulation of waveguide parameters such as the core refractive index or the transverse waveguide profile (Fig. 5.27). A waveguide mode travelling in the forward direction is scattered at these inhomogeneities and can, under proper conditions, couple into a backward

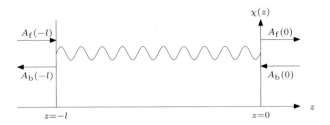

Fig. 5.27 Axial variation of the susceptibility in a waveguide grating

propagating mode. The performance of such structures can be described in the framework of the coupled modes formalism. A treatment analog to that of dielectric multilayer structures is not feasible in general, because the periodic modulation is usually continuous and not step-like.

Consider a waveguide whose core is "perturbed" by a (small) periodic modulation of the propagation index, $n_c(z) = n_{c,0} + \Delta n_c \cos(2\pi z/\Lambda_g)$, where Λ_g is the spatial period of the modulation.[2] Since $n = \sqrt{\chi + 1}$, this corresponds to a modulation of the susceptibility, $\chi(z) = \chi + \Delta\chi(z)\cos(K_g z)$, where $\Delta\chi = 2n_{c,0}\Delta n_c$ and $K_g := 2\pi z/\Lambda_g$. The electric field of a mode propagating in the forward direction with the propagation constant β_f produces a polarization density that contains an alternating component proportional to

$$\cos(K_g z)e^{-j\beta_f z} \propto e^{-j(\beta_f + K_g)z} + e^{-j(\beta_f - K_g)z}. \tag{5.82}$$

These "sidebands" of the unperturbed mode can exchange energy with other modes of the waveguide, provided that their propagation constant is close to $\beta_f \pm K_g$. Here, we are interested in the coupling to the backward propagating mode with the propagation constant $\beta_b = -\beta_f$, which is possible if the so-called Bragg condition

$$\beta_f - K_g = \beta_b \tag{5.83}$$

or, equivalently,

$$K_g = 2\beta_f \tag{5.84}$$

is met; the frequency (wavelength) that corresponds to this condition is called Bragg frequency ω_B (wavelength λ_{0B}). If n_{eff} is the effective propagation index of

---

[2]If the periodic longitudinal modulation is not cosinusoidal, it can be decomposed in a Fourier series, and the following analysis applies to a selected component of this expansion.

the unperturbed mode, so that $\beta = n_{\text{eff}}\omega/c_0 = 2\pi n_{\text{eff}}/\lambda_0$, these parameters are given by

$$\lambda_{0B} = \frac{4\pi n_{\text{eff}}}{K_g} = 2n_{\text{eff}}\Lambda_g \qquad (5.85)$$

$$\omega_B = \frac{c_0}{n_{\text{eff}}}\frac{K_g}{2} = \frac{c_0}{n_{\text{eff}}}\frac{\pi}{\Lambda_g}. \qquad (5.86)$$

In analogy to Eq. (5.55), we can describe the interaction of the two modes by

$$\frac{da_f}{dz} = -j\beta a_f + \kappa_{fb}a_b e^{-jK_g z}$$

$$\frac{da_b}{dz} = j\beta a_b + \kappa_{bf}a_f e^{jK_g z}. \qquad (5.87)$$

The first of the two equations describes the coupling of the reflected mode to the forward propagation mode, the second one relates to the reverse process. Energy conservation for counterpropagating waves demands $d(a_f a_f^*)/dz = d(a_b a_b^*)/dz$, or

$$\frac{d(a_f a_f^* - a_b a_b^*)}{dz} = a_1^* a_2(\kappa_{fb} - \kappa_{bf}^*) + a_1 a_2^*(\kappa_{fb}^* - \kappa_{bf}) = 0. \qquad (5.88)$$

With the same arguments that led to Eq. (5.58), we now obtain the relation

$$\kappa_{fb} = \kappa_{bf}^* =: \kappa. \qquad (5.89)$$

As can be seen from Eq. (5.87), a shift of the axial coordinate $z \rightarrow z + \Delta z$ is equivalent to a change of κ by a factor of $e^{-jK_g\Delta z}$; we therefore can always choose the coordinate system such that κ is real and $\kappa_{fb} = \kappa_{bf} = \kappa$. A shift by half a period, $\Lambda_g/2$, changes κ by a factor of $e^{-j\pi} = -1$.

Near the Bragg wavelength, the propagation constant of the two modes is approximately equal to $\pm K_g/2$; therefore, the modes can be expressed as a product of a slowly varying amplitude $A_{f,b}(z)$ and $e^{\mp j(K_g/2)z}$:

$$a_f = A_f e^{-j(K_g/2)z}$$

$$a_b = A_b e^{j(K_g/2)z}. \qquad (5.90)$$

Substitution in Eq. (5.87) yields the amplitude equations

$$\frac{dA_f}{dz} = -j\delta A_f + \kappa A_b$$

$$\frac{dA_b}{dz} = j\delta A_b + \kappa A_f, \tag{5.91}$$

where

$$\delta := \beta - K_g/2 \tag{5.92}$$

is the deviation from the Bragg condition Eq. (5.84); in terms of frequency, δ is equivalent to a deviation $\Delta\omega$ from the Bragg frequency with

$$\delta \approx \frac{d\beta}{d\omega}\Delta\omega = \frac{\Delta\omega}{v_g}, \tag{5.93}$$

where v_g is the group velocity of the unperturbed mode at ω_B.

Except for a different sign in the second equation, this system is similar to Eq. (5.55) and we can treat it as an eigenvalue problem with eigenstates

$$\psi(z) = \begin{bmatrix} A_f \\ A_b \end{bmatrix} e^{-jBz}; \tag{5.94}$$

note that B is not a propagation constant, but a parameter determining the axial development of the modes. Substitution in Eq. (5.91) yields

$$\begin{bmatrix} -j(\delta - B) & \kappa \\ \kappa & j(\delta + B) \end{bmatrix} \begin{bmatrix} A_f \\ A_b \end{bmatrix} = 0; \tag{5.95}$$

existence of non-zero solutions requires

$$B^2 = \delta^2 - \kappa^2. \tag{5.96}$$

Within the interval $|\delta| < |\kappa|$, B is imaginary

$$B^\pm = \pm jb, \quad b = \sqrt{|\kappa|^2 - \delta^2} \tag{5.97}$$

with corresponding eigenstates

$$\psi^{\pm}(z) = \begin{bmatrix} \pm\kappa \\ b \pm j\delta \end{bmatrix} e^{\pm bz}. \tag{5.98}$$

The coefficients $A^{\pm}$ of the general solution

$$\psi(z) = A^{+}\psi^{+} + A^{-}\psi^{-} \tag{5.99}$$

follow from boundary conditions. In particular, if the amplitudes $A_{f,b}(0)$ at $z = 0$ are given, we obtain

$$A_f(z) = A_f(0)\left(\cosh bz - \frac{j\delta}{b}\sinh bz\right) + A_b(0)\frac{\kappa}{b}\sinh bz$$

$$A_b(z) = A_f(0)\frac{\kappa}{b}\sinh bz + A_b(0)\left(\cosh bz + \frac{j\delta}{b}\sinh bz\right); \tag{5.100}$$

because of the quasi-exponential decay of the forward propagating mode, the interval $|\delta| < |\kappa|$ is called stop band.

5.3.3.1 Reflectance
Relation (5.100) can be cast in matrix form; for a waveguide grating extending between $-l \leq z \leq 0$, as shown in Fig. 5.27, we have

$$\begin{bmatrix} A_f(-l) \\ A_b(-l) \end{bmatrix} = F \begin{bmatrix} A_f(0) \\ A_b(0) \end{bmatrix} \tag{5.101}$$

with the coefficients

$$F_{11} = F_{22}^* = \cosh bl + \frac{j\delta}{b}\sinh bl \tag{5.102}$$

$$F_{12} = F_{21} = -\frac{\kappa}{b}\sinh bl; \tag{5.103}$$

note that as a consequence of energy conservation, $\det F$ must be equal to 1,

$$F_{11}F_{11}^* - F_{12}^2 = 1. \tag{5.104}$$

If the grating is used as a mirror, we can assume that $A_f(-l)$ is known and $A_b(0) = 0$, so that $A_f(-l) = F_{11}A_f(0)$, $A_b(-l) = F_{21}A_f(0)$. Thus, the reflection and transmission coefficients r, t are given by

$$r = \frac{A_b(-l)}{A_f(-l)} = \frac{F_{21}}{F_{11}} = \frac{-\kappa \sinh bl}{b \cosh bl + j\delta \sinh bl}, \tag{5.105}$$

$$t = \frac{A_f(0)}{A_f(-l)} = \frac{1}{F_{11}} = \frac{b}{b \cosh bl + j\delta \sinh bl}. \tag{5.106}$$

If the Bragg condition is exactly met (in the center of the stop band), $\delta = 0$ and $b = |\kappa|$, so that the reflectance rr^* is given by

$$R = |r|^2 = \tanh^2 |\kappa| l. \tag{5.107}$$

For a typical coupling coefficient of $|\kappa| = 3\,\text{cm}^{-1}$ and a grating length of 1 cm, the reflectance amounts to $\approx 99\,\%$.

Outside the stop band, B is real valued and the hyperbolic functions in Eq. (5.100) are replaced by their trigonometric counterparts, resulting in oscillatory solutions (Fig. 5.28). The reflection coefficient is then given by

$$r = \frac{-j\kappa \sin Bl}{jB \cos Bl - \delta \sin Bl} \tag{5.108}$$

and vanishes whenever $Bl = m\pi$. According to Eq. (5.96) this is the case if

$$\delta = \pm\kappa \sqrt{1 + \left(\frac{m\pi}{\kappa l}\right)^2}. \tag{5.109}$$

As Fig. 5.28 shows, the axial power at frequencies outside the stop band can exceed the input power, particularly pronounced at the root $m = 1$ of Eq. (5.109). This resonant enhancement resembles the response of the Fabry–Perot interferometer (Sect. 4.2.3) and is important for the operation of semiconductor lasers (Fig. 7.44).

5.3.3.2 Bandwidth
The range between the two reflectance-minima next to the stop band [$m = \pm1$ in Eq. (5.109)],

$$|\delta| = \left|\frac{\Delta\omega}{v_g}\right| < |\kappa| \sqrt{1 + (\pi/\kappa l)^2}, \tag{5.110}$$

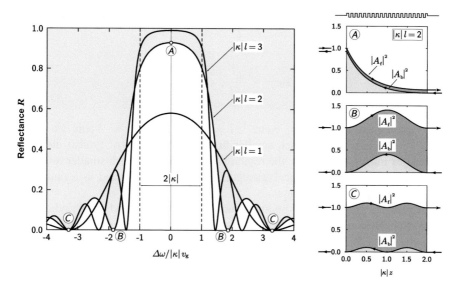

Fig. 5.28 Spectral reflectance of a waveguide grating for different values of $|\kappa|l$; also shown is the the axial intensity of the forward and backward propagation mode, respectively, for selected frequencies (A: center of stop band, B and C: first and second zero of R); note the hyperbolic development inside the stop band and the resonant enhancement at the two zeros of R

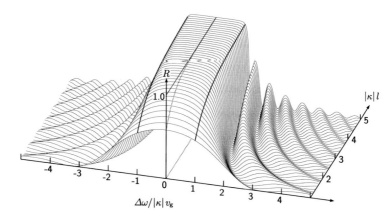

Fig. 5.29 Spectral reflectance as a function of $|\kappa|l$; the stop band is marked by the lines at $\Delta\omega/v_g|\kappa| = 1$

is a measure for the bandwidth of the waveguide-mirror; for short gratings ($l < 1/|\kappa|$), this interval is significantly broader than the stop band, which is determined exclusively by the coupling coefficient (Fig. 5.29). With increasing $l|\kappa|$, the reflectance approaches 1 and the bandwidth reduces to the width of the stop band.

In terms of wavelength, the bandwidth $\Delta\lambda_0$ can be expressed using $d\beta/d\lambda_0 = -2\pi n_{\text{eff}}/\lambda_0^2$

$$\Delta\lambda_0 = \frac{\lambda_{0B}^2}{\pi n_{\text{eff}} l}\sqrt{\pi^2 + (\kappa l)^2}. \tag{5.111}$$

For typical parameters $|\kappa| = 3\,\text{cm}^{-1}$, $l = 1\,\text{cm}$, and $\lambda_{0B} = 1\,\mu\text{m}$, we obtain $\Delta\lambda_0 \approx 0.1\,\text{nm}$, which is much smaller than the bandwidth of typical dielectric multilayer mirrors (Fig. 4.10); the reason is, of course, the much smaller refractive index variation within a typical waveguide grating as compared to a multilayer mirror.

In planar waveguides, the grating structure can be produced by periodic modulation of the core index with ion implantation or by a periodic variation of the core thickness. In (germanium-doped) glass fibers, a refractive index change can be induced by illumination with UV light. A periodic core index modulation can be realized by exposing a fiber to a periodic interference pattern of two UV beams (compare Fig. 4.3) or by employing UV irradiation through periodic transmission masks.

5.3.3.3 Waveguide Gratings with Phase Defect

Waveguide gratings can be combined to produce a variety of devices such as Fabry–Perot resonators; a particularly interesting example is the immediate serial combination of two identical gratings with a phase slip, i.e., with an axial shift of one of the gratings in respect to the other; here, we consider a shift of $\Lambda_g/2 = \lambda_{0B}/4n_{\text{eff}}$ (Fig. 5.30) and assume that the two gratings extend over the ranges $[-l/2, 0]$ and $[0, l/2]$, respectively. For the first grating, Eqs. (5.101)–(5.103) yield

$$\begin{bmatrix} A_{\text{f}}(-l/2) \\ A_{\text{b}}(-l/2) \end{bmatrix} = \begin{bmatrix} F_1 & F_2 \\ F_2 & F_1^* \end{bmatrix} \begin{bmatrix} A_{\text{f}}(0) \\ A_{\text{b}}(0) \end{bmatrix} \tag{5.112}$$

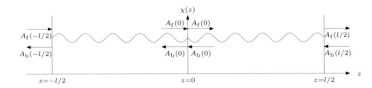

Fig. 5.30 Axial variation of the susceptibility in a waveguide grating with $\lambda/4$-phase defect

with

$$F_1 = \cosh bl/2 + \frac{j\delta}{b} \sinh bl/2 \qquad (5.113)$$

$$F_2 = -\frac{\kappa}{b} \sinh bl/2. \qquad (5.114)$$

The second grating is displaced by $\Lambda_{\rm g}/2$, which is equivalent to the transformation of the coupling coefficient to $-\kappa$, as mentioned above. We therefore find

$$\begin{bmatrix} A_{\rm f}(l/2) \\ A_{\rm b}(l/2) \end{bmatrix} = \begin{bmatrix} F_1' & F_2' \\ F_2' & F_1'^* \end{bmatrix} \begin{bmatrix} A_{\rm f}(0) \\ A_{\rm b}(0) \end{bmatrix} \qquad (5.115)$$

with

$$F_1' = \cosh bl/2 - \frac{j\delta}{b} \sinh bl/2 = F_1^* \qquad (5.116)$$

$$F_2' = -\frac{\kappa}{b} \sinh bl/2 = F_2. \qquad (5.117)$$

We invert Eq. (5.115)

$$\begin{bmatrix} A_{\rm f}(0) \\ A_{\rm b}(0) \end{bmatrix} = \begin{bmatrix} F_1^* & F_2 \\ F_2 & F_1 \end{bmatrix}^{-1} \begin{bmatrix} A_{\rm f}(l/2) \\ A_{\rm b}(l/2) \end{bmatrix} = \begin{bmatrix} F_1 & -F_2 \\ -F_2 & F_1^* \end{bmatrix} \begin{bmatrix} A_{\rm f}(l/2) \\ A_{\rm b}(l/2) \end{bmatrix}, \qquad (5.118)$$

using Eq. (5.104), and substitute the result in Eq. (5.112), yielding

$$\begin{bmatrix} A_{\rm f}(-l/2) \\ A_{\rm b}(-l/2) \end{bmatrix} = F_{\rm s} \begin{bmatrix} A_{\rm f}(l/2) \\ A_{\rm b}(l/2) \end{bmatrix} \qquad (5.119)$$

with

$$F_{\rm s} = \begin{bmatrix} F_1^2 - F_2^2 & -F_1 F_2 + F_1^* F_2 \\ F_1 F_2 - F_1^* F_2 & F_1^{*2} - F_2^2 \end{bmatrix}. \qquad (5.120)$$

The reflection and transmission coefficients of the structure follow from Eqs. (5.105) and (5.106)

$$r = \frac{F_{\rm s21}}{F_{\rm s11}} = \frac{F_2(F_1 - F_1^*)}{F_1^2 - F_2^2} \qquad (5.121)$$

$$t = \frac{1}{F_{\rm s11}} = \frac{1}{F_1^2 - F_2^2}. \qquad (5.122)$$

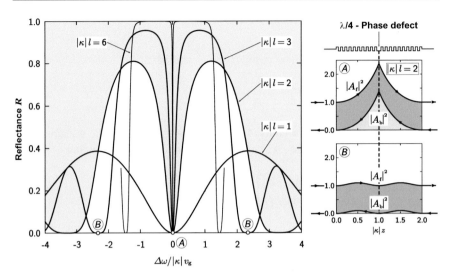

Fig. 5.31 Same as Fig. 5.28 for a waveguide grating with $\lambda/4$-phase defect

Using the identities $2\sinh x \cosh x = \sinh 2x$, $2\sinh^2 x = -1 + \cosh 2x$, $2\cosh^2 x = 1 + \cosh 2x$ and the relation $b^2 = \kappa^2 - \delta^2$ [Eq. (5.97)], we finally obtain

$$r = \frac{-2j\delta\kappa \sinh^2(bl/2)}{\kappa^2 - \delta^2 \cosh bl + j\delta b \sinh bl} \qquad (5.123)$$

$$t = \frac{b^2}{\kappa^2 - \delta^2 \cosh bl + j\delta b \sinh bl}. \qquad (5.124)$$

As shown in Fig. 5.31, the reflectance of this structure has a very narrow dip ($R = 0$) at the center of the stop band that results from a pronounced resonance enhancement within the grating structure. Such gratings can be used as transmission filters or as resonators for semiconductor lasers (Sect. 7.5).

The fact that periodic structures exhibit frequency ranges where waves cannot propagate but decay quasi-exponentially is well known from solid state physics, where a periodic atomic crystal lattice exhibits stop bands (called band gaps) for electronic wave functions. A phase defect as described above also finds its solid state physics analog, since a lattice defect can result in electronic states within the band gap. The analogy between the electronics of periodic lattices and photonics becomes almost complete if the periodic modulation of the optical medium is extended to three dimensions; such structures are called photonic band gap materials (Joannopoulos et al. 2008; Sakoda 2005; Sibilia and Benson 2008).

5.3.4 Waveguide-Interferometers and Modulators

Waveguide integrated interferometers rely on the same operating principles as their conventional counterparts; the advantages of integration are, among others, reduced size, enhanced rigidity, and lower costs.

5.3.4.1 Mach–Zehnder Interferometer

The Mach–Zehnder interferometer has already been discussed briefly in Sect. 4.1; the integrated version is used for sensors, modulators, and switches. It is very convenient to describe the operation of such an interferometer with the S-matrix formalism introduced in Sect. 4.1.3, that relates an input state $[a_1, a_2]$ to the corresponding output state $[b_1, b_2]$

$$\begin{bmatrix} b_1 \\ b_2 \end{bmatrix} = S_{MZ} \begin{bmatrix} a_1 \\ a_2 \end{bmatrix}. \tag{5.125}$$

As an example, we calculate the S-matrix of the electro-optic modulator shown in Fig. 5.32. The two couplers/splitters are represented by the matrix $S_{3\,dB}$ [Eq. (5.78)], while the electro-optic phase shifts of $\pm\Delta\phi/2$ in the two interferometer branches can be accounted for by a diagonal matrix with the components $M_{11} = e^{j\Delta\phi/2}$, $M_{22} = e^{-j\Delta\phi/2}$; the total scattering matrix of the interferometer is thus

$$\begin{aligned} S_{MZ} &= \frac{1}{2} \begin{bmatrix} 1 & -j \\ -j & 1 \end{bmatrix} \begin{bmatrix} e^{j\Delta\phi/2} & 0 \\ 0 & e^{-j\Delta\phi/2} \end{bmatrix} \begin{bmatrix} 1 & -j \\ -j & 1 \end{bmatrix} \\ &= j \begin{bmatrix} \sin\Delta\phi/2 & -\cos\Delta\phi/2 \\ -\cos\Delta\phi/2 & -\sin\Delta\phi/2 \end{bmatrix}. \end{aligned} \tag{5.126}$$

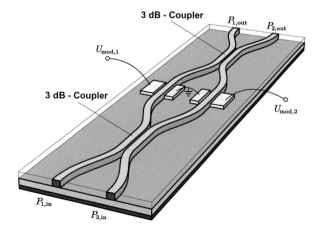

Fig. 5.32 Integrated Mach–Zehnder interferometer with electro-optic phase control

Usually, light is launched into one of the inputs only, so that $a_1 := a_0$, $a_2 = 0$ and

$$\begin{bmatrix} b_1 \\ b_2 \end{bmatrix} = ja_0 \begin{bmatrix} \sin \Delta\phi/2 \\ -\cos \Delta\phi/2 \end{bmatrix}; \qquad (5.127)$$

the output power at the two ports is given by

$$P_{1,\text{out}} = P_{1,\text{in}}(1 - \cos \Delta\phi)/2$$
$$P_{2,\text{out}} = P_{1,\text{in}}(1 + \cos \Delta\phi)/2, \qquad (5.128)$$

where the identity $\sin^2 x = (1 - \cos 2x)/2$ has been used; electrically controlling $\Delta\phi$ allows modulating the output power or switching between the two output ports.

Optical sensors are often realized as fiber integrated interferometers. One of the two interferometer branches is used as reference and isolated from external influences, while the other one is the sensing fiber; in many applications, the phase change in the sensor fiber is brought about by stretching the fiber. Thus, the parameter to be measured has first to be converted into a length change. In this fashion, temperature, pressure, magnetic, or electric fields can be monitored. To linearize the sensor and to stay in the operating point of maximum sensitivity [Eq. (5.80)], the reference fiber can be stretched with a piezoelectric transducer so as to keep the phase difference constant and equal to $\pi/2$. The primary measurement parameter is then the compensation voltage.

5.3.4.2 Fiber Gyroscope

One of the most important waveguide sensors is the fiber gyroscope that is based on the Sagnac interferometer. It allows measuring rotation rates in inertial systems with very high precision. Figure 5.33 shows the basic setup: a 3dB-coupler splits the light coming from a laser and feeds it into the two opposite ports of a fiber loop. The two modes propagating in the loop [clockwise (cw), or counterclockwise (ccw)] are recombined by the same 3-dB coupler, which acts as phase sensitive output coupler according to Eq. (5.81). Since the two modes propagate physically identical paths, the phase difference is expected to be zero for reasons of reciprocity (to

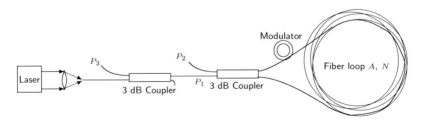

Fig. 5.33 Fiber gyroscope: the output is measured at the reciprocal port P_3, the modulator introduces a dynamic phase shift that allows operation in a linear range (see main text)

rule out phase differences due to polarization dispersion, polarization maintaining fibers and polarized light is used). A second coupler in the input branch of the interferometer allows accession of the output that is reciprocal in the sense that the modes have also taken the same path through the 3-dB coupler (one "reflection" and one "transmission" each).

Because of the reciprocal geometry, a phase difference can only occur if an external parameter affects the two modes in different ways, such as a rotation of the fiber loop. The exact description of this so-called Sagnac effect requires solving the Maxwell equations in an accelerated coordinate system. The resulting phase shift, however, can also be rationalized by a simple comparison of the phase delay times. In a rotating loop, the mode that co-propagates with the rotational movement of the fiber experiences a longer time before it reaches the output coupler, since the fiber (and the coupler) move along during the propagation time, while the reverse applies to the counterpropagating mode. In addition to this geometric effect, the change of the phase velocity in a moving medium (Sect. 2.4.3) has to be taken into account.

We consider a circular fiber coil with radius R and N loops, rotating ccw with angular velocity Ω; then the phase delay time τ_{ccw} and τ_{cw} of the respective modes follow from the equations $c_{\mathrm{ccw}}\tau_{\mathrm{ccw}} = l + R\Omega\tau_{\mathrm{ccw}}$ and $c_{\mathrm{cw}}\tau_{\mathrm{cw}} = l - R\Omega\tau_{\mathrm{cw}}$ to be

$$\tau_{\mathrm{ccw}} = \frac{l}{c_{\mathrm{ccw}} - R\Omega}$$

$$\tau_{\mathrm{cw}} = \frac{l}{c_{\mathrm{cw}} + R\Omega}, \tag{5.129}$$

where the phase velocities $c_{\mathrm{cw,ccw}}$ are given by Eq. (2.200) with $v = R\Omega$:

$$c_{\mathrm{ccw}} = \frac{c_0}{n} + R\Omega - \frac{R\Omega}{n^2}$$

$$c_{\mathrm{cw}} = \frac{c_0}{n} - R\Omega + \frac{R\Omega}{n^2}. \tag{5.130}$$

Thus, the phase delay time difference $\Delta\tau$ is

$$\Delta\tau = \tau_{\mathrm{ccw}} - \tau_{\mathrm{cw}} \approx l\frac{(c_{\mathrm{cw}} - c_{\mathrm{ccw}}) + 2R\Omega}{c_{\mathrm{ccw}}c_{\mathrm{cw}}} \approx \frac{2lR\Omega}{c_0^2}, \tag{5.131}$$

and the Sagnac phase difference is given, with $\omega = 2\pi c_0/\lambda_0$ and $l = 2\pi NR$, by

$$\Delta\phi_{\mathrm{s}} = \omega\Delta\tau = \frac{4\pi lR\Omega}{c_0\lambda_0} = \frac{8\pi AN}{c_0\lambda_0}\Omega; \tag{5.132}$$

it is proportional to the loop area A, the number N of loops, and the angular velocity (more precisely, the component of the angular velocity parallel to the loop axis)

Ω; note that the Sagnac phase shift is independent of the propagation index of the medium.

The output power at the "reciprocal" output port is given by

$$P_{3,\text{out}} = P_{1,\text{out}}/2 = |a_0|^2 \left(1 + \cos \Delta\phi_s\right)/4, \qquad (5.133)$$

where $|a_0|^2$ is the input power at port 1.

For small values of ϕ_s, the sensor characteristic is $\propto 1 + (\Delta\phi_s)^2$ implying that the sensitivity is close to zero

$$\frac{\mathrm{d}P_{3,\text{out}}}{\mathrm{d}\Omega}|_{\Omega\to 0} = 0 \qquad (5.134)$$

and the sign of the rotation cannot be resolved. The introduction of a phase bias of $\pm\pi/2$ would linearize the response

$$P'_{3,\text{out}} = |a_0|^2 \left(1 + \sin \Delta\phi_s\right)/4 \approx |a_0|^2 \left(1 + \Delta\phi_s\right)/4 \qquad (5.135)$$

but cannot be implemented into the interferometer as easily as in the Mach–Zehnder interferometer, since the two modes travel the same path.

A (dynamic) nonreciprocal phase delay, however, can be realized with a time dependent phase modulator that is positioned asymmetrically within the fiber loop; in Fig. 5.33, the modulator is a piezoelectric fiber stretcher located immediately behind the coupler. Let us assume that the modulator changes the length of the fiber (and thus the phase delay) linearly with time, $\phi_m(t) = R_m t$, the resulting phase difference between the cw and ccw mode, respectively, is $\Delta\phi_m = R_m \tau$, where τ is the difference of the arrival time of the respective mode at the modulator. With proper choice of the stretching rate R_m, the desired phase difference of $\pi/2$ can be achieved; in practice, the length modulation of the fiber is not a linear ramp, but an oscillating function giving rise to a phase shift oscillating between $\pm\pi/2$; the operating principle remains the same, however.

With typical design parameters $l = 1$ km, $r = 5$ cm, $\lambda_0 = 600$ nm, the rotational velocity of the earth ($\Omega_E = 7.3 \times 10^{-5}\,\text{s}^{-1}$) produces a phase shift of $\Delta\phi_s = 2.6 \times 10^{-4}$ rad. The sensitivity of commercial fiber gyroscopes can be $<10^{-3}\Omega_E$.

5.3.5 Active Waveguide Components

In Sect. 6.2, we will study the amplification of light by stimulated emission of photons from excited atoms or ions. Such laser-active ions can be implemented in a glass-host and excited by light of a wavelength shorter than the emitted, or amplified light. If the preform of a fiber waveguide is doped with such atoms, a fiber can be employed as amplifier. Popular dopants are rare earth atoms such as erbium, neodymium, or ytterbium, typical doping concentrations are 10^{-4}. Combined with

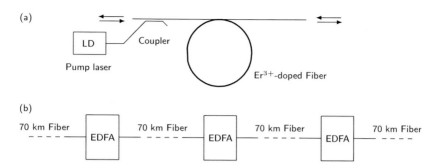

Fig. 5.34 Erbium doped fiber amplifier (EDFA) (**a**); optical long distance link with EDFA-repeaters (**b**)

two waveguide reflectors, an optical fiber amplifier can also be converted into an integrated fiber oscillator (Fig. 7.27).

One of the most important applications of such fiber amplifiers is the signal regeneration in optical communication networks. Even in the transmission optimum of quartz glass, at 1.55 μm, fiber networks require signal amplification in intervals of 70–100 km. Erbium doped fiber amplifiers (EDFAs) provide a broad gain spectrum that allows amplification of many parallel data channels in WDM and are therefore ideally suited for this purpose. Figure 5.34 shows schematically a chain of such repeaters; the radiation required for the excitation (pumping) of the erbium atoms is provided by semiconductor lasers at 1.48 μm and launched into the amplifier fiber with the help of dichroic couplers, that transfer the pump light from the semiconductor laser "pigtail" into the amplifier. The amplifying fiber (with a length of about 10 m) is fusion-spliced into the data-fiber. With a pump power of several mW, a signal gain of 30–40 dB is achieved (Fig. 5.35).

Apart from its simplicity, reliability, and low electric power consumption, the advantage of an EDFA over conventional (electronic) repeaters is that it can handle virtually any signal encoding protocol with a bandwidth of several THz, while electronic repeaters are optimized for a particular format and data rate.

5.3.6   Photonic Band Gap Fibers

Alternative wave guide structures provide guiding not by total reflection but by interference effects: in Sect. 5.3.3, we have encountered high reflecting one-dimensional photonic band gaps resulting from the periodic modulation of the refractive index along the waveguide axis. In photonic band gap fibers, the refractive index is radially modulated instead: a core is surrounded by a periodic structure of high and low refractive index materials, usually glass and air. The (hollow) core constitutes a "defect" in the crystal structure and allows for a propagating mode within the band gap, comparable to the narrow band transmission feature of a waveguide grating with phase defect in Fig. 5.31.

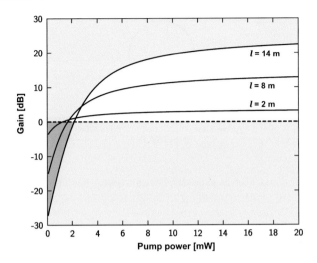

Fig. 5.35 Typical power amplification of an EDFA as a function of pump power

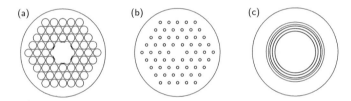

Fig. 5.36 Photonic crystal fibers (PCFs): (**a**) hollow core photonic band gap fiber, (**b**) index-guided solid core PCF, (**c**) hollow core Bragg fiber; the small circles indicate hollow channels in the glass matrix of the fiber; see, e.g., Bjarklev et al. (2003) and Russell (2006)

   Photonic band gap fibers belong to the wider class of photonic crystal fibers (PCFs, Fig. 5.36); they can also be realized with a solid core surrounded by air holes. These fibers, however, rely on conventional guiding by total reflection, and the holes only serve to reduce the refractive index of the medium surrounding the core. Because of the large refractive index contrast, light can be confined to a very small mode area, which makes such fibers very suitable for nonlinear optical applications.

5.4 Summary

Many of the characteristic properties of dielectric waveguides become clear in the analysis of simple (symmmetric) planar waveguides. First of all, guiding requires that the propagation constant of the guided wave is larger than that in the medium surrounding the guiding core. An additional self-consistency condition reduces the number of possible guided waves to a finite number of modes. This number depends on the normalized frequency, a dimensionless parameter that includes information on the frequency of the mode, the thickness of the core and the refractive index step between the core and the surrounding medium. Symmetric waveguides always support at least one guided mode, which can be polarized in two orthogonal states.

The propagation constant of a given mode depends on its frequency, since the width of the core, measured in units of the wavelength, changes with frequency. Just like any other dispersion mechanism, this geometric waveguide dispersion leads to the broadening of transmitted light pulses, but can be compensated by the material dispersion of the waveguide medium.

Cylindrical waveguides (fibers) share these fundamental properties with planar waveguides and are of utmost importance for optical communications. For this purpose, silica single mode step index fibers are used at the wavelength of 1.55 μm, where absorption is minimal.

Graded index fibers are frequently used for low cost local area network applications. For the lack of boundary conditions, they require a mathematical treatment different from step index waveguides: our approach is to treat the fiber as extended graded index lens and to use the ABCD formalism introduced in Chap. 3 to find self-consistent solutions.

Part of this Chapter is devoted to waveguide couplers, gratings, filters, and interferometers. The physics underlying fiber gratings is similar to that of the dielectric multilayer structures treated in Sect. 4.2, but the mathematical treatment is quite different: for lack of boundary conditions, we analyze these devices in terms of mode coupling mediated by a cross talk between the coupled modes. Efficient energy transfer requires phase matching between the modes; waveguide dispersion, i.e., the frequency dependence of the propagation constant, can thus be used to realize dichroic filters and couplers. We also employ the S-matrix formalism of Chap. 4 to describe waveguide interferometers and modulators in a concise manner.

5.5 Problems

1. Assume a step index fiber with a core refractive index of 1.5 and a 1 % smaller cladding index. What is the NA of this fiber? What is the maximum core diameter for the waveguide to be a single mode fiber at 632 nm?

2. Derive the self-consistency condition for an asymmetric planar waveguide and find numerically the propagation constants of the eigenmodes for $n_g = 1.5$, $n_s = 1.4$, $n_c = 1$ and various ratios a/λ_0. Show that there is an absolute cutoff wavelength, above which no guided modes exist.

3. Plot the transverse mode profiles of the lowest mode of the asymmetric planar waveguide of problem 2 and observe what happens near the cutoff wavelength.

4. A semiconductor laser is used for an optical monomode fiber link with 10 Gbit/s transmission rate. The laser operates at 890 nm and oscillates at two adjacent longitudinal modes; the cavity length is 100 μm, the cavity refractive index is 3.5. Calculate the maximum permissible length of the link if the dispersion coefficient is $20 \, \text{ps nm}^{-1} \, \text{km}^{-1}$ [use Eq. (4.82) to calculate the mode spacing]. Neglecting

losses, what is the maximum length of the link if operated with a single mode laser? What is the maximum link length if a single mode laser is used with a step index fiber supporting two waveguide modes (assume $V = 3.5$, NA $= 0.1$ and neglect material dispersion).

5. Reproduce Fig. 5.28, including the insets.

References and Suggested Reading

Abramowitz, M., & Stegun, I. A., (2014). *Handbook of mathematical functions*. New York: Martino Publishing.

Agrawal, G. P. (2012). *Nonlinear fiber optics*. New York: Academic Press.

Bass, M., & van Stryland, E.W. (2001). *Fiber optics handbook*. New York: McGraw-Hill.

Bjarklev, A., Broeng, J., Bjarklev, A. S. (2003). *Photonic crystal fibres*. New York: Springer.

Bottacchi, S. (2014). *Theory and design of terabit optical fiber transmission systems*. New York: Cambridge University Press.

Desurvire, E. (2001). *Erbium doped fiber amplifiers*. New York: Wiley.

Gao, J. (2010). *Optoelectronic integrated circuit design and device modeling*. New York: Wiley.

Hasegawa, A. (2003). *Optical solitons in fibers*. New York: Springer.

Haus, H. A. (1984). *Waves and fields in optoelectronics*. Englewood Cliffs, NJ: Prentice Hall.

Joannopoulos, J. D., Johnson, S. G., Winn, J. N., Meade, R. D. (2008). *Photonic crystals*. Princeton: Princeton University Press.

Lifante, G. (2003). *Integrated photonics*. New York: John Wiley.

Marcuse, D. (1991). *Theory of dielectric optical waveguides*. New York: Academic Press.

Mitschke, F. (2010). *Fiber optics: Physics and technology*. New York: Springer.

Nishihara, H., Haruna, M., Suhara, T. (1989). *Optical integrated circuits*. New York: McGraw-Hill.

Pollock, C., & Lipson, M. (2003). *Integrated photonics*. New York: Springer.

Reed, G. T., & Knights, A. P. (2004). *Silicon photonics*. New York: John Wiley.

Russell, P. S. J. (2006). Photonic-crystal fibers. *Journal of Lightwave Technology, 24*(12), 4729–4749. http://jlt.osa.org/abstract.cfm?URI=jlt-24-12-4729

Sakoda, K. (2005). *Optical properties of photonic crystals*. New York: Springer.

Saleh, B. E., & Teich, M. C. (2007). *Fundamentals of photonics*. New York: Wiley.

Sibilia, C., & Benson, T. (2008). *Photonic crystals*. New York: Springer.

Snyder, A. W. (2010). *Optical waveguide theory*. New York: Springer.

Tamir, Th. (Ed.). (1995). *Guided wave optoelectronics*. New York: Springer.

Venghaus, H. (2006). *Wavelength filters in fibre optics*. New York: Springer.

Wartak, M. S. (2012). *Computational photonics*. New York: Cambridge University Press.

Yariv, A. (1973). Coupled-mode theory for guided-wave optics. *IEEE Journal of Quantum Electronics, 9*(9), 919–933.

Light–Matter Interaction

6

The classical linear oscillator model of Sect. 2.2.1 provides useful qualitative insights into the interaction of light with matter. In particular, it yields essentially correct results about the *complex* character of the electric susceptibility and its resonant behavior, and the frequency dependence (dispersion) of the refractive index is explained in a simple and intuitive way. For a more quantitative and detailed treatment of light-matter interaction, however, a quantum mechanical treatment is required. Since the optical response of matter is dominated by the electrons, the following discussion refers to the interaction of light with electrons, bound in atoms, molecules, or semicondutors.

6.1 Optical Interactions with Two Level Systems

The fundamental quantum mechanical equation is the Schrödinger equation

$$\left[-\frac{\hbar^2}{2m}\nabla^2 + V(\mathbf{x})\right]\Psi(\mathbf{x},t) = -j\hbar\frac{\partial\Psi(\mathbf{x},t)}{\partial t}, \tag{6.1}$$

where $\hbar = h/2\pi$ and $h = 6.63 \times 10^{-34}\,\mathrm{J\,s}$ is Planck's constant, $V(\mathbf{x})$ is the potential of the electron, and the term $[-(\hbar^2/2m)\nabla^2 + V(\mathbf{x})]$ is the Hamilton or energy operator $\mathcal{H}_0$; the wave function $\Psi(\mathbf{x},t)$ comprises the complete information on the particle.

A formal solution of Eq. (6.1) is

$$\Psi(\mathbf{x},t) = \psi(\mathbf{x})\mathrm{e}^{\mathrm{j}(E/\hbar)t}, \tag{6.2}$$

© Springer International Publishing Switzerland 2016
G.A. Reider, *Photonics*, DOI 10.1007/978-3-319-26076-1_6

provided that ψ fulfills the time independent Schrödinger equation

$$\mathcal{H}_0\psi(\mathbf{x}) = E\psi(\mathbf{x}). \tag{6.3}$$

The set of solutions ψ of this equation depends on the potential V and constitutes the "spectrum" (discrete or continuous) of eigenstates of the Hamilton operator with corresponding eigenvalues E that denote the energy of the state. Due to the linearity of the Schrödinger equation, any linear combination of solutions is also a solution. Moreover, the set of eigenfunctions is complete in the sense that any possible solution of Eq. (6.1) can be "synthesized" as a linear combination of eigenfunctions,

$$\Psi = \sum_n c_n \psi_n e^{j(E_n/\hbar)t}. \tag{6.4}$$

In the absence of a potential ($V = 0$), the solutions of Eq. (6.1) are plane waves $\Psi(\mathbf{x},t) = e^{-j\mathbf{k}\cdot\mathbf{x}}e^{j(E/\hbar)t}$ (DeBroglie waves); the relation between the $\mathbf{k}$-vector and E is the E–k-dispersion relation for free electrons,

$$E = \frac{\hbar^2|\mathbf{k}|^2}{2m}. \tag{6.5}$$

For attractive potentials such as the Coulomb potential of the atomic core, the spectrum consists of a set of discrete eigenfunctions ψ_n with eigenvalues E_n (representing the bound states), and a continuum of plane waves.

Similar to electrodynamics, where the absolute square of the complex wave function is a measure of the local energy density of the light wave, the absolute square $|\Psi(\mathbf{x},t)|^2 = \Psi(\mathbf{x},t)\Psi^*(\mathbf{x},t)$ is a measure of the probability density of the particle, that is the probability to find it at the point $\mathbf{x}$. Accordingly, the wave function must be normalized such that the volume integral of $\Psi(\mathbf{x},t)\Psi^*(\mathbf{x},t)$ is equal to 1:

$$\int \psi_m^* \psi_m \, dV = 1. \tag{6.6}$$

Furthermore, eigenfunctions are mutually orthogonal in the sense that

$$\int \psi_m^* \psi_n \, dV = \begin{cases} 1 & \text{for} \quad m = n \\ 0 & \text{for} \quad m \neq n. \end{cases} \tag{6.7}$$

Note that if the particle is in an eigenstate of the Hamilton operator (i.e., its wave function is an eigenfunction), the probability density $|\Psi(\mathbf{x}, t)|^2$ is time independent, i.e., the probability density of an eigenstate is stationary

$$|\Psi(\mathbf{x}, t)|^2 = \left[\psi_n(\mathbf{x})e^{j(E_n/\hbar)t}\right]\left[\psi_n^*(\mathbf{x})e^{-j(E_n/\hbar)t}\right] = |\psi_n(\mathbf{x})|^2; \tag{6.8}$$

in other words, the electron density of an eigenstate is stationary and the electron consequently does not emit any electromagnetic radiation.

In contrast, the probability density of a superposition of eigenstates oscillates at frequencies that are determined by the energy differences of the involved eigenstates; the superposition of two states

$$|\Psi(\mathbf{x}, t)|^2 = \left|c_1\psi_1 e^{j(E_i/\hbar)t} + c_2\psi_2 e^{j(E_2/\hbar)t}\right|^2$$

$$= |c_1\psi_1|^2 + |c_2\psi_2|^2 + 2\mathrm{Re}\left[c_1 c_2^* \psi_1 \psi_2^* e^{j[(E_1-E_2)/\hbar]t}\right], \tag{6.9}$$

for example, oscillates at the frequency $|E_1 - E_2|/\hbar$ (Fig. 6.1).

6.1.1 Perturbations

Let us now study the effect of a time varying "perturbation," such as an electromagnetic field, on a quantum mechanical system. Any such perturbation can be expressed as a time dependent contribution to the potential V; denoting the stationary "back ground" potential as V_0 and the external perturbation as $V'(t)$, the Hamilton operator is

$$\mathcal{H} = \mathcal{H}_0 + \mathcal{H}'(t) \tag{6.10}$$

with $\mathcal{H}' = V'$. While it is possible, in principle, to solve the Schrödinger equation with a time dependent Hamilton operator, an approximative solution can be obtained in form of a (time dependent) linear combination of the unperturbed solutions ψ_n, provided that the perturbation is small in comparison to $\mathcal{H}_0$:

$$\Psi = \sum_n c_n(t)\psi_n e^{j(E_n/\hbar)t}; \tag{6.11}$$

$c_n(t)$ are the time dependent "mixing" coefficients. The absolute square $|c_n(t)|^2$ of these coefficients can be interpreted as probability to find the system in state ψ_n, if a measurement of the energy of the system is taken at time t (in quantum mechanics, measuring a certain observable always returns an eigenvalue of the respective operator); since the set of eigenfunctions is complete, $\sum_n |c_n(t)|^2 = 1$ to warrant $\int |\psi|^2 \, dV = 1$.

To determine the mixing coefficients $c_n(t)$, we substitute Eq. (6.11) into the time dependent Schrödinger equation Eq. (6.1)

$$\sum_n (\mathcal{H}_0 + \mathcal{H}')c_n\psi_n e^{j(E_n/\hbar)t} = -j\hbar \sum_n \left[c_n\psi_n \frac{jE_n}{\hbar} + \dot{c}_n\psi_n \right] e^{j(E_n/\hbar)t}. \qquad (6.12)$$

According to Eq. (6.3), $\sum_n c_n \mathcal{H}_0 \psi_n = \sum_n c_n E_n \psi_n$, so that

$$-j\hbar \sum_n \dot{c}_n\psi_n e^{j(E_n/\hbar)t} = \sum_n \mathcal{H}' c_n\psi_n e^{j(E_n/\hbar)t}. \qquad (6.13)$$

Multiplication of both sides with ψ_m^* and applying the orthonormality relations Eq. (6.7), we obtain

$$-j\hbar\dot{c}_m e^{j(E_m/\hbar)t} = \sum_n c_n \int \psi_m^* \mathcal{H}' \psi_n e^{j(E_n/\hbar)t} \, dV. \qquad (6.14)$$

The integral

$$H'_{mn} := \int \psi_m^* \mathcal{H}' \psi_n \, dV \qquad (6.15)$$

represents the impact of the perturbation $\mathcal{H}'$ on the set ψ_m, ψ_n of states and is called the (m, n)-th element of the perturbation matrix. With this definition, Eq. (6.14) can be written as

$$\dot{c}_m(t) = \frac{j}{\hbar} \sum_n c_n(t)H'_{mn} e^{j[(E_n - E_m)/\hbar]t}, \qquad (6.16)$$

which is a set of coupled differential equations for the mixing coefficients.

We now restrict the discussion to a system of two eigenstates $n = i, f$; this allows us to derive simple, yet very important results and also describes many situations in optics quite well, as we shall see. Equation (6.16) then simplifies to

$$\dot{c}_i(t) = \frac{j}{\hbar} \left[c_i(t)H'_{ii} + c_f(t)H'_{if} e^{j\omega_0 t} \right]$$

$$\dot{c}_f(t) = \frac{j}{\hbar} \left[c_i(t)H'_{fi} e^{-j\omega_0 t} + c_f(t)H'_{ff} \right], \qquad (6.17)$$

where $|E_f - E_i|/\hbar =: \omega_0$. We further assume that the perturbations starts at $t = 0$, with the system being in the initial (eigen)state ψ_i,[1] $c_i(0) = 1$, $c_f(0) = 0$.

Perturbation theory is an iterative, approximative technique to solve equations such as Eq. (6.17): in lowest (zero) order, one simply neglects the perturbation, so that, $c_i(t) = 1$, $c_f(t) = 0$. Substituting this "solution" into Eq. (6.17) yields the first order approximation

$$\dot{c}_i(t) = \frac{j}{\hbar} H'_{ii} \tag{6.18}$$

$$\dot{c}_f(t) = \frac{j}{\hbar} H'_{fi} e^{-j\omega_0 t}. \tag{6.19}$$

We further assume a periodic, harmonic time dependence of the perturbation (which is equivalent to picking a certain Fourier component of it),

$$H'_{mn}(t) = H'^{0}_{mn} \cos \omega t = \frac{1}{2} H'^{0}_{mn} \left[e^{j\omega t} + e^{-j\omega t} \right], \tag{6.20}$$

so that integration of Eq. (6.19) from 0 to t yields

$$c_f(t) = \frac{H'^{0}_{fi}}{2\hbar} \left[\frac{e^{j(\omega - \omega_0)t} - 1}{\omega - \omega_0} - \frac{e^{-j(\omega + \omega_0)t} - 1}{\omega + \omega_0} \right]. \tag{6.21}$$

If the frequency of the perturbation is comparable to ω_0, the second term in parenthesis can be neglected because of the much larger denominator. Introducing the "detuning" $\Delta \omega := \omega - \omega_0$, we obtain

$$|c_f(t)|^2 = \frac{|H'^{0}_{fi}|^2}{\hbar^2} \left[\frac{\sin \Delta \omega t/2}{\Delta \omega} \right]^2, \tag{6.22}$$

where the identity $1 - \cos x = 2 \sin^2(x/2)$ was used.

6.1.1.1   Fermi's Golden Rule

Of particular interest is the *rate* of change of the probability $|c_f(t)|^2$ to find the system in the "final" state, the so-called transition rate $W_{if} = \frac{|c_f(t)|^2}{t}$. Using the approximation (valid for $t \to \infty$)

$$\left[\frac{\sin \Delta \omega t/2}{\Delta \omega} \right]^2 \to \frac{\pi}{2} \delta(\Delta \omega) t, \tag{6.23}$$

---

[1] The subscripts i and f refer to *initial* and *final*; note that the energy E_i of the initial state is not necessarily lower than E_f.

where δ is the Dirac distribution with the properties $\delta(x \neq 0) = 0$ and $\int \delta(x)\, dx = 1$, we obtain the approximative result

$$W_{if} = \frac{|c_f(t)|^2}{t} = \frac{\pi |H_{fi}^{'0}|^2}{2\hbar^2} \delta(\Delta\omega),\qquad (6.24)$$

known as Fermi's golden rule; it essentially states that

- a transition $\Psi_i \to \Psi_f$ requires the frequency ω of the perturbation to coincide with $\omega_0 = (E_f - E_i)/\hbar$;
- the transition rate W_{if} is proportional to the square of the perturbation matrix element $|H_{fi}^{'0}|^2$;
- since $\mathcal{H}'$ is a Hermitian operator with the property $H_{if}' = H_{fi}'^*$, the transition rate for $\Psi_f \to \Psi_i$ is equal to that of the reverse process $\Psi_i \to \Psi_f$.

In practice, the perturbative interaction does not last for an infinite time. For this and other reasons, the frequency dependence of the transition rate (the so-called line shape) is not an infinitely narrow Dirac delta function, but a line function $g(\Delta\omega)$ peaking at $\omega = \omega_0$ with $\int g(\Delta\omega)\, d\omega = 1$; Eq. (6.24) then assumes the form

$$W_{if} = \frac{\pi |H_{fi}^{'0}|^2}{2\hbar^2} g(\Delta\omega);\qquad (6.25)$$

some of the reasons for line broadening will be discussed in Sect. 6.1.4.

6.1.1.2 Dipole Interaction

The most important optical interaction is the dipole interaction, that is the interaction between the electric field and the electric dipole constituted by the electron and atomic core. The corresponding potential is

$$\mathcal{H}' = -e\mathbf{E} \cdot \mathbf{x},\qquad (6.26)$$

where $\mathbf{E}(t) = \mathbf{E}_0 \cos \omega t$ is the electric field, $-e$ the electron charge, and $\mathbf{x}$ the displacement of the electron in respect to the core. The perturbation matrix element hence is

$$H_{mn}' = -e \int \psi_m^* \mathbf{E} \cdot \mathbf{x} \psi_n\, dV.\qquad (6.27)$$

At optical wavelengths, the electric field is practically constant over the extension of an atom, so that with Eq. (6.20)

$$H_{mn}^{'0} = -e\mathbf{E}_0 \cdot \int \psi_m^* \mathbf{x} \psi_n \, dV = \mathbf{E}_0 \cdot \boldsymbol{\mu}. \tag{6.28}$$

The so-called dipole matrix element

$$\boldsymbol{\mu} := \boldsymbol{\mu}_{mn} = \boldsymbol{\mu}_{nm} = -e \int \psi_m^* \mathbf{x} \psi_n \, dV \tag{6.29}$$

is a measure for the dipole moment that is associated with the superposition of the states ψ_m and ψ_n.

The vectors $\boldsymbol{\mu}$ and $\mathbf{E}$ are not necessarily parallel; therefore

$$|H_{fi}^{'0}|^2 = |\mathbf{E}_0 \cdot \boldsymbol{\mu}|^2 = E_0^2 |\boldsymbol{\mu}|^2 \cos^2 \theta, \tag{6.30}$$

where θ is the angle between the two vectors. If the orientation of $\boldsymbol{\mu}$ is equally distributed over the spatial angle Ω, the average value of the factor $\cos^2 \theta$ is given by

$$\langle \cos^2 \theta \rangle = \frac{\int \cos^2 \theta \, d\Omega}{\int d\Omega} = \frac{1}{4\pi} \int_0^{2\pi} \int_0^\pi \cos^2 \theta \sin \theta \, d\theta \, d\varphi = \frac{1}{3}, \tag{6.31}$$

and

$$\langle |H_{fi}^{'0}|^2 \rangle = \langle \cos^2 \theta \rangle E_0^2 |\boldsymbol{\mu}|^2 = \frac{1}{3} E_0^2 |\boldsymbol{\mu}|^2. \tag{6.32}$$

For dipole interaction, Eq. (6.25) thus can be expressed as

$$W = W_{if} = W_{fi} = \frac{\pi}{6\hbar^2} E_0^2 |\boldsymbol{\mu}|^2 g(\Delta\omega). \tag{6.33}$$

6.1.1.3 Interaction Cross Section

We can express the electric field in Eq. (6.33) in terms of its intensity $I = \sqrt{\varepsilon\varepsilon_0/\mu_0} E_0^2 / 2$ [Eq. (1.71)] to obtain

$$W = \frac{\pi}{3n\varepsilon_0 c_0 \hbar^2} |\boldsymbol{\mu}|^2 g(\Delta\omega) I, \tag{6.34}$$

where $n = \sqrt{\varepsilon}$ and $c_0 = 1/\sqrt{\varepsilon_0\mu_0}$. As we have seen, the energy exchange between an atom (electron) and the field (or any other perturbation) is quantized in the sense that the energy difference of the atom before and after a transition from ψ_i to ψ_f is equal to $\hbar\omega_0$; the atom cannot exchange fractions of that energy with the field. Conversely, the electromagnetic field can exchange energy only in integer multiples of this energy because of the atomic structure of matter. The concept of photons comprises much more than this "granular" currency of energy exchange, but it is very convenient even at this level to express the electromagnetic energy flow density I as a flow density F of energy quanta, or photons

$$F = \frac{I}{\hbar\omega}, \qquad (6.35)$$

where ω is the frequency of the field.[2] With this relation, Eq. (6.34) can be written as

$$W = \frac{\pi\omega}{3n\varepsilon_0 c_0\hbar}|\boldsymbol{\mu}|^2 g(\Delta\omega)F =: \sigma F = \sigma\frac{I}{\hbar\omega}, \qquad (6.36)$$

where the interaction or transition cross section σ

$$\sigma(\Delta\omega) = \frac{\pi\omega}{3n\varepsilon_0 c_0\hbar}|\boldsymbol{\mu}|^2 g(\Delta\omega) \qquad (6.37)$$

has been introduced. The interaction cross section has the dimension of an area (usually given in cm^2) and has a very intuitive meaning: just as a target disk of area σ in a stream of point-like bullets is hit at a rate that is equal to the flux density of the bullets times the target area, an atom in a stream of photons undergoes transitions with the rate σF. To induce one transition in the time interval τ, an intensity of $\hbar\omega/\sigma\tau$ is required. With an exemplary peak value for $\sigma = 10^{-19}\,cm^2$ and a transition energy of $1\,eV = 1.6\times10^{-19}\,J$, it takes an intensity of $\approx 1.6\,W/cm^2$ to statistically hit every atom of an ensemble once per second. Note that off resonance, the cross section vanishes and the atom becomes "invisible."

Finally, a relation between the electromagnetic energy density ρ_{em} and the transition rate is useful if the interaction happens in a resonator cavity; we assume that the cavity of volume V contains q photons; then the photon density $\rho_{ph} = q/V$ is related to the energy density by $\rho_{em} = \hbar\omega q/V$; ρ_{ph} is related to F by $F = c\rho_{ph}$,

---

[2] See Table 1.1 for different units of $\hbar\omega$.

since the photons travel at the speed $c = c_0/n$. Thus we can express Eq. (6.36) in the form

$$W = \sigma c \rho_{ph} = \frac{\sigma c \rho_{em}}{\hbar \omega}. \tag{6.38}$$

6.1.1.4 Selection Rules

For a transition to happen, three conditions must be fulfilled according to Fermi's rule: the frequency of the light field must meet the resonance condition $|E_f - E_i|/\hbar = \omega$, the intensity must be non-vanishing, and the matrix element μ_{if} [Eq. (6.29)] must be non-zero. The latter condition is a so-called selection rule, which limits the possible choice of states participating in transitions. A hydrogen atom, for example, with its spherically symmetric potential, has eigenstates that are either symmetric $\psi(-\mathbf{x}) = \psi(\mathbf{x})$ or anti-symmetric $\psi(-\mathbf{x}) = -\psi(\mathbf{x})$—they are of even or odd parity, respectively. It is evident that the dipole matrix element Eq. (6.29) vanishes

$$\mu_{if} = -e \int \psi_f^* \mathbf{x} \psi_i \, dV = 0, \tag{6.39}$$

if ψ_{if} are of same parity, and a dipole-transition between such states is therefore "forbidden"[3]; the dipole selection rule requires states of different parity.

It is quite instructive to visualize these important implications graphically: in Fig. 6.1, the probability density of a superposition of two pairs of eigenstates of a hydrogen atom is shown as a function of time; panel (a) shows the superposition of the 1s and the 2s states, both of even parity; (b) shows the mixing of an 1s (even) with a 2p-state (odd parity). While both superpositions oscillate, only (b) develops a dipole moment; in (a), the center of gravity of the oscillating electron wave remains in the positive core of the atom (breathing sphere).

6.1.2 Absorption and Stimulated Emission

The introduction of the photon as a "currency" for energy exchange Eq. (6.36) allows us to set up balance equations for the number of atoms in a particular eigenstate on the one hand, and the number of photons on the other: the transition of an atom from a lower to a higher state consumes one photon (so-called absorption), while the reverse process is equivalent to the generation (emission) of one photon; it is important to note that this additional photon is indistinguishable from the

---

[3]A transition might still be possible because of higher order interactions such as quadrupole interactions, but the cross section is smaller by several orders of magnitude in this case.

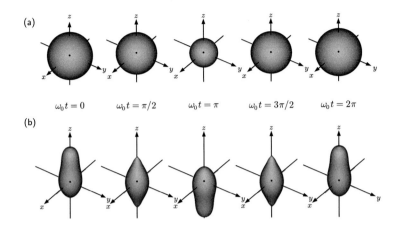

Fig. 6.1 Probability density of an electron in a hydrogen atom: (**a**) superposition of 1s and 2s states, (**b**) superposition of 1s and 2p-states; the *dark dot* in the center represents the positive core

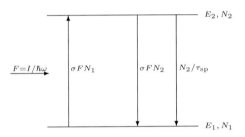

Fig. 6.2 Transition rates in a two-level system

field that has perturbed the atom, it has the same phase, frequency, polarization, and wave vector—in short, it belongs to the same electromagnetic mode that has stimulated the transition. This coherent emission process is called "stimulated." The probability for the respective process is given by W_{if}, as defined in Eq. (6.24). In the following, we discuss the resulting balances for an ensemble of two-level atoms (Fig. 6.2) that can be either in a "ground state" 1 or in an "excited state" 2. We assume the density of atoms (number of atoms per unit volume) to be N; the density of atoms in states 1 and 2 are denoted as N_1 and N_2, respectively; $N_{1,2}$ is also called population density of the respective state. Since there are no other states, the population densities are related by $N_1 + N_2 = N$. According to Eq. (6.24) (which refers to *one* atom), the interaction with the light field changes the population densities with the rates $dN_2/dt|_{abs} = -dN_1/dt|_{abs} = N_1 W_{12}$ (absorption), and $dN_1/dt|_{se} = -dN_2/dt|_{se} = N_2 W_{21}$ (stimulated emission). In addition to the stimulated emission, there is also a certain probability that an excited atom returns to the ground state without stimulation, i.e., in the absence of a light field. This "spontaneous" emission, which is not predicted by Fermi's rule, happens at a rate that is proportional to the cross section for stimulated emission and to the number of possible modes of the electromagnetic field at the transition frequency. The spontaneously emitted photon is stochastic in terms of phase, polarization, and wave vector; the frequency is distributed within the bandwidth of the cross section. The transition rate due to this process can be expressed by the average life time τ_{sp} of the excited state (see Sect. 6.1.3).

Taking these three processes into account, we obtain, for the population density N_2, the rate equation

$$\frac{dN_2}{dt} = N_1 W_{12} - N_2 W_{21} - \frac{N_2}{\tau_{sp}}. \qquad (6.40)$$

Because of the one-to-one correspondence of atomic transitions and photon annihilation or creation, respectively, we obtain

$$\frac{d\rho_{ph}}{dt} = -N_1 W_{12} + N_2 W_{21} = W(N_2 - N_1), \qquad (6.41)$$

where ρ_{ph} refers exclusively to the photons of the interacting light mode, while spontaneous photon emission is not taken into account.

Let us now look at a light field with photon flux density $F = c\rho_{ph}$, propagating in z-direction through a volume filled with (excited) atoms of density N_1 and N_2. In a slice of thickness dz within the medium, the total temporal derivative of the photon density is

$$\frac{d\rho_{ph}}{dt} = \frac{\partial\rho_{ph}}{\partial t} + \frac{\partial\rho_{ph}}{\partial z}\frac{dz}{dt}; \qquad (6.42)$$

the first term on the right side describes an explicit temporal change of the local photon density, the second one represents the difference between the photons flowing in and out of the volume element. In combination with Eqs. (6.41) and (6.38), and using $dz/dt = c$, we obtain

$$\frac{\partial\rho_{ph}}{\partial t} + c\frac{\partial\rho_{ph}}{\partial z} = c\sigma\rho_{ph}(N_2 - N_1). \qquad (6.43)$$

If we assume $N_{1,2}$ and $F = c\rho_{ph}$ to be stationary, then $\partial\rho_{ph}/\partial t = 0$ and

$$\frac{dF}{F} = \frac{dI}{I} = (N_2 - N_1)\sigma\, dz. \qquad (6.44)$$

If we further assume that $N_{1,2}$ is constant over the interaction length l, then integration yields

$$\frac{F(z)}{F(0)} = \frac{I(z)}{I(0)} = e^{(N_2-N_1)\sigma z}. \tag{6.45}$$

Comparison with the classical result Eq. (2.70)

$$\frac{I(z)}{I(0)} = e^{-2\kappa k_0 z} = e^{-\alpha z} \tag{6.46}$$

allows us to identify

$$\alpha = -(N_2 - N_1)\sigma = (N_1 - N_2)\sigma. \tag{6.47}$$

With Eqs. (2.71) and (2.76) we obtain the following relations between the cross section σ and the imaginary parts of the refractive index and the susceptibility, respectively:

$$\kappa = (N_1 - N_2)\sigma/2k_0 \tag{6.48}$$

$$\chi''_{\text{dot}} = 2(N_2 - N_1)\sigma n_w/k_0. \tag{6.49}$$

Of particular interest is the fact that Eq. (6.45) implies an (exponential) growth of the intensity if $N_2 > N_1$, an effect known as Light Amplification by Stimulated Emission of Radiation. This process is of utmost importance for the field of photonics and will be discussed in detail in Sect. 6.2.

6.1.3 Spontaneous Emission

The semiclassical treatment of light–matter interaction as outlined above treats the electromagnetic field classically, with the result that in the absence of perturbations such as electromagnetic radiation, eigenstates of the energy operator are stable; spontaneous emission is not possible in this framework. By the same token, the stationary populations of a two-level system in the presence of thermal (or any other) radiation are predicted by Eq. (6.40) to be $N_2 = N_1 = N/2$, in contradiction to the thermodynamic population ratio $N_2/N_1 = e^{-\hbar\omega/k_B T}$, where $k_B = 1.38 \times 10^{-23}$ J K^{-1} is Boltzmann's constant.

Spontaneous emission can only be explained satisfactorily in the framework of a quantum theory of electromagnetism. In this theory, an electromagnetic mode behaves like a quantum mechanical oscillator whose energy is represented by the photons in the mode. Just as its quantum mechanical counterpart, the oscillator in its ground state (corresponding to zero photons) is not at rest but fluctuates. These so-called vacuum fluctuations of the electromagnetic field stimulate the "spontaneous" emission and are responsible for the instability of excited states.

Planck postulated the quantization of the electromagnetic field energy to explain the spectral features of thermal radiation, at a time when quantum mechanics was not yet known. He assumed that the energy of an electromagnetic mode at frequency ω is not continuous but an integer multiple of $\hbar\omega$; these energy "quanta" were later denoted as photons. The electromagnetic energy spectrum in thermodynamic equilibrium is then the product of the density of modes Eq. (4.107)

$$N(\omega) = \frac{\omega^2 n^3}{\pi^2 c_0^3}, \tag{6.50}$$

the average number of photons per mode, and the photon energy $\hbar\omega$. The average number of photons per mode will be derived below [Eq. (9.20)] and is equal to

$$\bar{n}_{\text{ph}} = \frac{1}{e^{\hbar\omega/k_B T} - 1}. \tag{6.51}$$

Thus, one obtains the spectral energy density

$$\rho_{\text{em}}(\omega) = \hbar\omega N(\omega)\bar{n}_{\text{ph}} = \frac{\hbar\omega^3 n^3}{\pi^2 c_0^3} \frac{1}{e^{\hbar\omega/k_B T} - 1}. \tag{6.52}$$

Einstein derived a structurally equivalent expression by postulating three fundamental processes constituting the interaction of atoms with electromagnetic radiation: absorption, stimulated emission, and spontaneous emission; an ensemble of $N = N_1 + N_2$ two-level atoms, exposed to thermal radiation undergoes transitions between the two levels with the rates $A(\omega)N_2$ (spontaneous emission), $B_{21}(\omega)\rho_{\text{em}}(\omega)N_2$ (stimulated emission), and $B_{12}(\omega)\rho_{\text{em}}(\omega)N_1$ (absorption). In equilibrium,

$$B_{21}(\omega)\rho_{\text{em}}(\omega)N_2 + A(\omega)N_2 = B_{12}(\omega)\rho_{\text{em}}(\omega)N_1. \tag{6.53}$$

Substituting the above mentioned Boltzmann distribution N_2/N_1 into Eq. (6.53), we obtain

$$\rho_{\text{em}}(\omega) = \frac{A(\omega)}{B_{12}(\omega)} \frac{1}{e^{\hbar\omega/k_B T} - B_{21}(\omega)/B_{12}(\omega)}. \tag{6.54}$$

A comparison with Eq. (6.52) shows, in agreement with Eq. (6.33), that $B_{21}(\omega)/B_{12}(\omega) = 1$; moreover,

$$A(\omega) = \frac{\hbar\omega^3 n^3}{\pi^2 c_0^3} B_{21}(\omega). \tag{6.55}$$

Since $\omega^2 n^3/\pi^2 c_0^3$ is the density of modes at ω, the factor $\hbar\omega^3 n^3/\pi^2 c_0^3$ corresponds to one photon per mode. Spontaneous emission is therefore equivalent to an emission stimulated by one photon per mode; in regard to emission, a mode containing a number m of photons ($m = 0, 1, 2, \ldots$) acts as if there were $m + 1$ photons. Because of the quadratic frequency dependence of the density of modes, spontaneous emission becomes more and more prevalent with growing frequency.

The coefficient $B_{21}(\omega)$ can be calculated from Eqs. (6.37) and (6.38)

$$B_{21}(\omega) = \frac{\pi}{3n^2\varepsilon_0\hbar^2}|\boldsymbol{\mu}|^2 g(\Delta\omega), \tag{6.56}$$

so that Eq. (6.55) assumes the form

$$A(\omega) = \frac{\omega^3 n}{3\pi\varepsilon_0\hbar c_0^3}|\boldsymbol{\mu}|^2 g(\Delta\omega). \tag{6.57}$$

For narrow lines, the spectral distribution of spontaneous emission is therefore essentially given by the line function of the transition cross section $\sigma(\omega)$. Note that the spontaneous emission rate grows with the third power of ω.

To obtain the spontaneous life time [Eq. (6.40)], we have to integrate $A(\omega)$ over all frequencies

$$\frac{1}{\tau_{\text{sp}}} = \int A(\omega)\,d\omega; \tag{6.58}$$

assuming a narrow line function, the variable ω in the integrand Eq. (6.57) can be replaced by the resonance frequency ω_0, so that we obtain, using $\int g(\Delta\omega)\,d\omega = 1$,

$$\frac{1}{\tau_{\text{sp}}} = \frac{\omega_0^3 n|\boldsymbol{\mu}|^2}{3\pi\varepsilon_0\hbar c_0^3}. \tag{6.59}$$

6.1.3.1 The Füchtbauer–Ladenburg Equation

Equation (6.59) allows casting the interaction cross section Eq. (6.37) in the form

$$\sigma(\Delta\omega) = \frac{1}{\tau_{sp}} \frac{\pi^2 c_0^2}{n^2 \omega_0^2} g(\Delta\omega). \tag{6.60}$$

The spectral distribution of the spontaneously emitted light can be determined experimentally, usually as a function $I(\lambda_0)$ of the wavelength. For narrow line widths, one can therefore assume the (normalized) line function $g(\lambda_0)$ to be

$$g(\lambda_0) = \frac{I(\lambda_0)}{\int I(\lambda_0)\, d\lambda_0}. \tag{6.61}$$

From $\int g(\Delta\omega)\, d\omega = \int g(\lambda_0)\, d\lambda_0$ follows $g(\Delta\omega) = g(\lambda_0)\frac{\lambda_0^2}{2\pi c_0}$, so that Eq. (6.60) can be expressed as a function of λ_0,

$$\sigma(\lambda_0) = \frac{1}{\tau_{sp}} \frac{\lambda_0^4}{8\pi c_0 n^2} \frac{I(\lambda_0)}{\int I(\lambda_0)\, d\lambda_0}. \tag{6.62}$$

This important relation is known as Füchtbauer–Ladenburg equation (see Fowler and Dexter 1962).

6.1.4 Line Broadening

6.1.4.1 Homogeneous Line Broadening

In Eq. (6.25) we have replaced the δ-line function by a broadened line function $g(\Delta\omega)$. A fundamental broadening mechanism, also called natural broadening, is due to the finite, spontaneous life time of an excited state. As an ensemble of excited atoms decays exponentially in time, one can, in a semiclassical picture, view the spontaneous emission of a single atom as an exponentially decaying field at the carrier frequency $\omega_0 = (E_2 - E_1)/\hbar$ (Fig. 6.3). Assuming the power to decay as $e^{-t/\tau_{sp}}$, the field decays as $e^{-t/2\tau_{sp}}$. The normalized power spectrum (absolute square of the Fourier transform) of $e^{-t/2\tau_{sp}} e^{j\omega_0 t}$ is the line function

$$g(\Delta\omega) = \frac{1}{\pi} \frac{2\tau_{sp}}{1 + (2\tau_{sp}\Delta\omega)^2}. \tag{6.63}$$

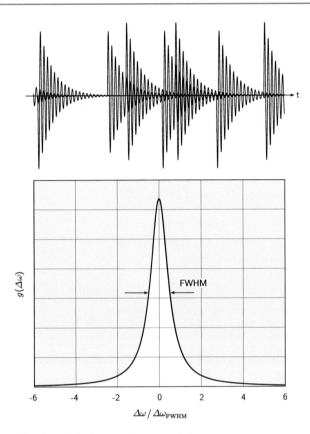

Fig. 6.3 Natural line broadening by spontaneous emission

This function drops to half of the peak value at $\Delta\omega = 1/2\tau_{\mathrm{sp}}$ so that we obtain the FWHM line width

$$\Delta\omega_{\mathrm{FWHM}} = \frac{1}{\tau_{\mathrm{sp}}}. \qquad (6.64)$$

The line function Eq. (6.63) is of the same Lorentzian shape $1/(1 + (\Delta x)^2)$ already encountered as the frequency dependence of absorption by a linear oscillator [Eq. (2.61)].

A similar impact on the line shape results from dephasing (Fig. 6.4), for example, by statistical collisions of atoms with others. The time between collisions in a gas is distributed exponentially with the decay time T_2 (dephasing time). In the frequency

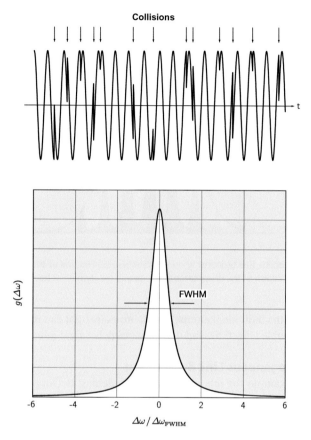

Fig. 6.4 Dephasing and line broadening induced by collisions

domain, this results again in a Lorentz line shape[4]

$$g(\Delta\omega) = \frac{1}{\pi} \frac{T_2}{1 + (\Delta\omega T_2)^2}. \tag{6.65}$$

These broadening mechanisms affect each individual atom, and thus an ensemble of atoms in the same way; they represent what is known as homogeneous line broadening. By contrast, statistically distributed shifts of the resonance frequency of individual atoms result in the broadening of the line function of an ensemble, without affecting the individual line width. If the range of frequency shifts exceeds

---

[4]For a derivation see, e.g., Svelto (2010).

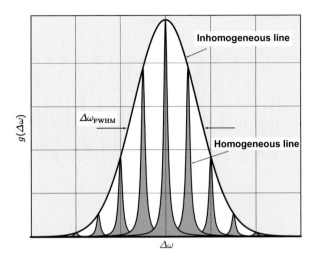

Fig. 6.5 Inhomogeneous line broadening due to statistical distribution of line shifts; the resulting line shape is a convolution of the distribution function and the individual line width

the line width of the individual atoms, monochromatic light can interact only with a certain sub-ensemble (Fig. 6.5), and the broadening is called inhomogeneous.

6.1.4.2 Inhomogeneous Line Broadening

An instructive example for inhomogeneous broadening is Doppler broadening in gases. The thermal velocity of the atoms in a gas gives rise to a Doppler shift of the apparent transition frequency, whenever the atom emits or absorbs light. According to Eq. (2.197), the resonance frequency ω_0 of an atom moving at velocity v along a certain direction is shifted to

$$\omega_0' = \omega_0 \sqrt{\frac{1 + v/c_0}{1 - v/c_0}} \approx \omega_0(1 + v/c_0), \tag{6.66}$$

when observed along this direction in a coordinate system at rest. For small velocities $|v|/c_0 \ll 1$, the Doppler shift is therefore

$$\omega_0' - \omega_0 \approx \omega_0 \frac{v}{c}. \tag{6.67}$$

The velocities (and their component along a given direction) of the atoms in thermal equilibrium are distributed according to Boltzmann's distribution $\propto e^{-E/k_B T}$, where E is the kinetic energy of the atoms and T is the temperature of the gas

$$p_v \, dv = \left(\frac{M}{2\pi k_B T}\right)^{1/2} e^{-Mv^2/2k_B T} \, dv, \tag{6.68}$$

where M is the mass of the atom. The prefactor is chosen so that $\int p_v \, dv = 1$. Because of the linear relation Eq. (6.67) between Doppler shift and velocity, the distribution of frequency shifts is given by the velocity distribution

$$p(\omega_0' - \omega_0) \, d\omega_0' = p_v \, dv;$$ (6.69)

from Eq. (6.67) follows $d\omega_0' = (\omega_0/c) \, dv$, so that

$$p(\omega_0' - \omega_0) = \frac{c}{\omega_0} \left(\frac{M}{2\pi k_B T} \right)^{1/2} e^{-\frac{Mc^2}{2k_B T} \frac{(\omega_0' - \omega_0)^2}{\omega_0^2}}.$$ (6.70)

$p(\omega_0' - \omega_0) \, d\omega_0'$ is the probability to find the apparent transition frequency of the moving atom in the interval $[\omega_0', \omega_0' + d\omega_0']$. The FWHM-width of this Gaussian distribution is

$$\Delta\omega_{FWHM} = 2\omega_0 \sqrt{2 \ln 2} \sqrt{k_B T / Mc^2}.$$ (6.71)

The line function $g_{ih}(\Delta\omega)$ of the ensemble of gas atoms is the convolution of this distribution function with the individual (homogeneous) line function of the atom

$$g_{ih}(\omega - \omega_0) = \int_0^\infty p(\omega_0' - \omega_0) g_h((\omega - \omega_0) - (\omega_0' - \omega_0)) \, d\omega_0'$$

$$= \int_0^\infty p(\omega_0' - \omega_0) g_h(\omega - \omega_0') \, d\omega_0'.$$ (6.72)

If the homogeneous line width is negligible in comparison to the width of the inhomogeneous distribution, it can be replaced by $\delta(\omega - \omega_0')$ and $g_{ih}(\omega - \omega_0) = p(\omega - \omega_0)$.

Another inhomogeneous broadening mechanism that can be very significant is crystal field broadening: it affects atoms and ions in a (transparent) solid state host material. The transition frequency of atoms is determined not only by the atomic field but also by the electric field in its microscopic environment. If this environment varies for different atoms, the optical response of the ensemble is inhomogeneously broadened. This effect is particularly pronounced in amorphous host materials such as glasses.

Very broad line functions can be observed in large (organic) molecules and certain crystal hosts (e.g., sapphire) doped with transition metals (e.g., titanium). In these materials, the electronic states of the electrons are split up in a wide manifold of closely spaced vibrational and rotational levels that overlap at room temperature and form broad absorption and emission bands. Semiconductors, on the other hand, display quasi-continuous bands of electronic states, also resulting in very broad emission and absorption lines (Sect. 6.3).

6.1.5   Saturation of Absorption

In thermodynamic equilibrium, the levels E_1 and E_2 of an atom are populated according to Boltzmann's distribution, $N_2/N_1 = e^{-(E_2-E_1)/k_B T}$. For optical transition energies and at room temperature, $E_2 - E_1 \gg k_B T$, so that $N_2 = 0$ and $N_1 = N$; practically all atoms are in the ground state and available for absorption. According to Eq. (6.47), the absorption coefficient then is

$$\alpha_0 := N\sigma. \tag{6.73}$$

We now want to discuss the absorption process Eqs. (6.40)–(6.47) in more detail by taking the population changes due to the irradiation into account. We write Eq. (6.40) in the form

$$\frac{dN_2}{dt} = -W(N_2 - N_1) - \frac{N_2}{\tau_{sp}} \tag{6.74}$$

and introduce the so-called inversion density ΔN

$$\Delta N := N_2 - N_1, \tag{6.75}$$

so that Eq. (6.47) becomes

$$\alpha = -\sigma \Delta N. \tag{6.76}$$

The total density of atoms is $N = N_1 + N_2$, so that

$$N_1 = \frac{N - \Delta N}{2}, \quad N_2 = \frac{N + \Delta N}{2}. \tag{6.77}$$

Eq. (6.74) can now be written as

$$\frac{d\Delta N}{dt} = -\Delta N \left(2W + \frac{1}{\tau_{sp}} \right) - \frac{N}{\tau_{sp}}. \tag{6.78}$$

Under stationary conditions $d/dt = 0$, we obtain

$$-\frac{\Delta N}{N} = \frac{1}{1 + 2W\tau_{sp}}. \tag{6.79}$$

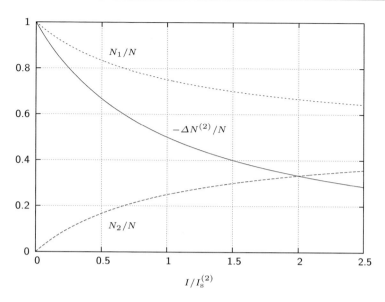

Fig. 6.6 Population densities in a two-level system as function of the normalized signal intensity

With $W = \sigma F = \sigma I/\hbar\omega$ [Eq. (6.36)], we can cast Eq. (6.79) in the form

$$-\frac{\Delta N}{N} = \frac{1}{1 + I(2\sigma\tau_{sp}/\hbar\omega)} := \frac{1}{1 + I/I_s^{(2)}}, \tag{6.80}$$

where

$$I_s^{(2)} := \frac{\hbar\omega}{2\sigma\tau_{sp}} \tag{6.81}$$

is the so-called saturation intensity of a two-level system. In Fig. 6.6, the populations $N_{1,2}$ and $\Delta N/N$ are shown as functions of the normalized intensity $I/I_s^{(2)}$. With Eqs. (6.73) and (6.80), we obtain the intensity dependence of the absorption coefficient

$$\alpha(I) = \alpha_0 \frac{1}{1 + I/I_s^{(2)}}; \tag{6.82}$$

the reduction of the absorption coefficient by the incident light is called saturation of absorption (Fig. 6.7).

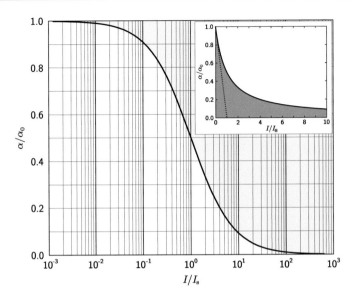

Fig. 6.7 Saturation of absorption (bleaching) at high signal intensity; note the logarithmic scale—the *inset* shows a linear scale)

As long as the signal intensity is very small compared to the saturation intensity, the absorption coefficient is α_0; at higher intensity it is reduced (because there are less atoms available for absorption, and stimulated emission from excited atoms partially compensates the photon losses); at $I_s^{(2)}$ the absorption coefficient is only half of the "small signal" value α_0 (this situation corresponds to a ground state population of 75 % and an excited state population of 25 %). At very high intensity, the absorber becomes transparent (it is bleached), since the two populations approach 50 % each.

6.1.5.1 Saturation and Line Function

The absorption coefficient reflects the frequency dependence of the transition cross section, $\alpha_0(\omega) = N\sigma(\omega)$. When irradiated by a strong monochromatic light field (Fig. 6.8), a homogeneously broadened medium reacts according to Eq. (6.82); note that $I_s^{(2)}(\omega)$ is a function of frequency having a minimum value at the peak of $\sigma(\omega)$.

In an *inhomogeneously* broadened absorber, however, the light interacts with (and saturates) only the sub-ensemble of atoms that is in resonance with the light field. This selective saturation can be experimentally observed by measuring, with a weak, tunable "probe" beam, the complete absorption spectrum of the absorber bleached by a strong, monochromatic light (Fig. 6.9). The (transient) creation of a dip in the absorption spectrum is known as spectral hole burning. The width of the "hole" in the absorption spectrum equals the homogeneous bandwidth of the individual atoms in the sub-ensemble.

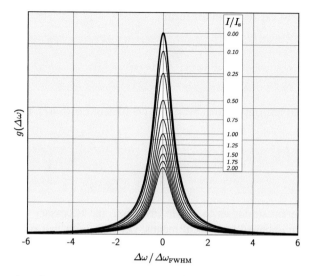

Fig. 6.8 Saturation of an homogeneously broadened absorber

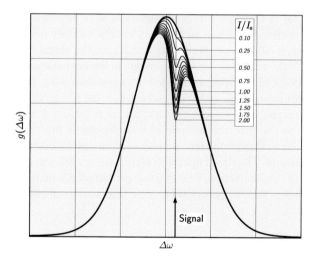

Fig. 6.9 Saturation of an inhomogeneously broadened absorber

6.2 Light Amplification by Stimulated Emission

As remarked in the discussion of Eq. (6.45), $N_2 - N_1 > 0$ results in optical amplification. The situation $\Delta N = N_2 - N_1 > 0$ is called population inversion because it is opposite to the thermodynamic equilibrium situation $N_2 < N_1$. It is convenient to cast Eq. (6.44) in the form

$$\frac{dI}{dz} = I\sigma\Delta N \tag{6.83}$$

and to define, in analogy to the absorption coefficient Eq. (6.47), a gain coefficient $\gamma = -\alpha$

$$\gamma = \sigma\Delta N. \tag{6.84}$$

There are different strategies to obtain inversion: the necessary excitation of the E_2 level can be obtained by optical means, i.e. with light, but other mechanisms such as collisions with energetic free electrons in gases or electron injection in semiconductors are also possible and technologically important. According to Eq. (6.82), a two-level system cannot be inverted by optical radiation, no matter how intense. The light that is used to induce inversion (usually called the pump light) must differ in frequency from the light that is to be amplified (the signal), which can be achieved by using an auxiliary energy level $E_3 > E_2$ to absorb the pump light (Fig. 6.10). Under appropriate conditions, the atoms excited to E_3 can relax to state E_2 (by releasing the excess energy to the host material in the form of heat, for example) and accumulate there to form a population N_2. The pump process requires a photon energy $\hbar\omega_p = E_3 - E_0$, where E_0 is the energy of the ground state. Ideally, the transition from E_3 to E_2 is very fast, so that $N_3 \approx 0$ and stimulated or spontaneous emission from the pump level is negligible. In principle, the resulting population N_2 can be close to the total number of atoms, provided that the pump light is sufficiently intense. The signal light with photon energy $\hbar\omega_s$ can now interact with these atoms. For this interaction, there are two prototypical schemes, as depicted in Fig. 6.10. The lower state E_1 participating in the interaction can either be the ground

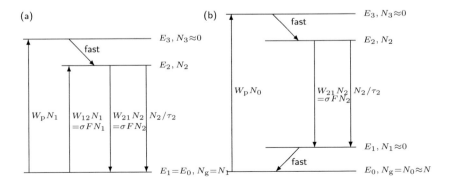

Fig. 6.10 Energy levels and transitions (**a**) in a three-level system, (**b**) in a four-level system

state, $E_1 = E_0$ (three-level system), or another auxiliary state (four-level system). It is immediately clear that in a three-level system, at least 50 % of the atoms need to be excited from the ground state to make inversion possible. In an ideal four-level system, the population of the E_1 is negligible ($N_1 \approx 0$) and the inversion

$$\Delta N = N_2; \tag{6.85}$$

can be obtained with very weak pump.

A requirement for the operation of the four-level scheme is that $E_1 - E_0 \gg k_B T$ ($k_B T$ is about 26 meV at room temperature), so that the thermal population of level E_1 according to Boltzmann's distribution is negligible. Moreover, the life time of the atoms in state E_1 should be very short to prevent a congestion by transitions from E_2 to E_1. Again, the transition from E_1 to the ground state is usually mediated by thermal (nonradiative) interaction with the environment.

Obviously, an ideal atomic amplifier system has to meet a manifold of spectroscopic requirements, and in fact the number of atomic elements (or their ions) that have proven useful as optical amplifiers is rather limited: neodymium, titanium, helium, argon, chromium, copper, ytterbium, and several others. Some of them are four-level systems, others are three-level systems, such as erbium; at the technologically very important wavelength of 1.5 μm, the erbium amplifier is by far the most attractive despite its three-level structure.

6.2.1 Four-Level Amplifier

To describe the operation of an atomic four-level amplifier system, we adapt Eq. (6.40) by adding a pump rate that transports atoms from the ground state via E_3 to level E_2. We denote with W_p the pump transition probability of an atom, and the density of atoms in the ground state with N_g; we further assume N_1 and N_3 to be negligible, so that $N_g = N - N_2$. Then we obtain for N_2 the rate equation

$$\frac{dN_2}{dt} = W_p N_g - W_{21} N_2 - \frac{N_2}{\tau_2}. \tag{6.86}$$

To take possible nonradiative de-excitation processes from E_2 into account, we add to the spontaneous emission rate $1/\tau_{sp}$ a nonradiative rate $1/\tau_{nr}$ so that the total decay rate $1/\tau_2$ is

$$\frac{1}{\tau_2} = \frac{1}{\tau_{sp}} + \frac{1}{\tau_{nr}}; \tag{6.87}$$

for the sake of simplicity we will refer to this combined decay rate as spontaneous, however.

If the signal is very small or zero, $W_{21} \approx 0$ and the stationary inversion is

$$N_{2,0}(W_p) = W_p N_g \tau_2 = \frac{W_p N \tau_2}{1 + W_p \tau_2} \approx W_p N \tau_2; \tag{6.88}$$

the linear approximation is valid as long as $N_2 \ll N$, so that the depletion of the ground state by the pump process can be neglected. Eq. (6.88) describes the equilibrium between the pump and the spontaneous relaxation. The small signal gain coefficient is then

$$\gamma_0 = N_{2,0} \sigma. \tag{6.89}$$

Similar to what we have seen in Sect. 6.1.5 regarding absorption, a sufficiently strong signal modifies the population densities and thus the gain coefficient. The signal stimulates additional decay with a rate $\propto W_{21} = \sigma I/\hbar\omega$ and reduces the upper state population to

$$N_2(W_p, I) = \frac{W_p N \tau_2}{1 + W_p \tau_2 + I/I_s^{(4)}} \approx N_{2,0} \frac{1}{1 + I/I_s^{(4)}}, \tag{6.90}$$

where the saturation intensity is now defined as

$$I_s^{(4)} := \frac{\hbar\omega}{\sigma \tau_2}. \tag{6.91}$$

Note that $I_s^{(4)} = 2 I_s^{(2)}$; while each transition in the four-level system changes the inversion by 1, in a two- (or three-) level system, ΔN changes by 2 per transition.

The signal dependent gain coefficient is then

$$\gamma(I) = \gamma_0 \frac{1}{1 + I/I_s^{(4)}}. \tag{6.92}$$

Equation (6.92) describes the gain saturation by the signal: compared to the small signal gain coefficient γ_0, the gain coefficient drops to one half when the emission rate stimulated by the signal equals the decay rate $1/\tau_2$. The signal intensity required for this is one photon per cross section within the time interval τ_2 [Eq. (6.91)]. Except for the different value of the saturation intensity, saturation of gain and absorption show very similar saturation effects; in particular, Figs. 6.7 and 6.8 apply to both processes.

Since the signal modifies the local gain coefficient, Eq. (6.83) is a nonlinear differential equation, which in general can only be integrated numerically. Two limiting cases, however, yield quite simple solutions. If the input signal is so small that even the output signal is well below the saturation intensity, then saturation is negligible and

$$\frac{I(l)}{I(0)} = e^{\gamma_0 l};\qquad (6.93)$$

$e^{\gamma_0 l}$ is the small signal gain factor. If, on the other hand, $I(0)/I_s^{(4)} \gg 1$, then

$$\frac{\mathrm{d}I}{\mathrm{d}z} = \gamma_0 \frac{I(z)}{I(z)/I_s^{(4)}} = \gamma_0 I_s^{(4)}\qquad (6.94)$$

and accordingly

$$I(l) - I(0) = \gamma_0 I_s^{(4)} l = W_p N_g l \hbar \omega.\qquad (6.95)$$

The increase of the signal photon flux density is thus equal to $W_p N_g l A = W_p N_g V$, where A and V are the cross section and volume of the amplifier, respectively: in the case of very high saturation, every pump photon is converted into a signal photon and added to the signal flux density; note that the temporal and spatial shape of a signal is generally not conserved in that way.

6.2.2   Three-Level Amplifier

While a four-level system Fig. 6.10 without pump is transparent at the signal frequency and can be inverted by an arbitrarily weak pump, three-level systems require a minimum pump rate to become transparent or inverted. The rate equation for this system is

$$\frac{\mathrm{d}N_2}{\mathrm{d}t} = W_p N_g + \sigma F N_1 - \sigma F N_2 - \frac{N_2}{\tau_2}.\qquad (6.96)$$

Ideally, $N_3 = 0$ (the transition E_3–E_2 is very fast), so that $N_g = N_1 = (N - \Delta N)/2$ and $N_2 = (N + \Delta N)/2$; the stationary small signal gain coefficient (at $F \approx 0$) is then

$$\gamma_0 = \sigma \Delta N_0 = \sigma N \frac{W_p \tau_2 - 1}{W_p \tau_2 + 1},\qquad (6.97)$$

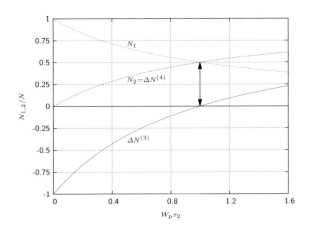

Fig. 6.11 Populations and inversion of a three-level system as a function of the normalized pump rate; for comparison, the inversion of a four-level system is also shown

and the pump rate required for transparency ($\gamma = 0$) is given by $W_p\tau_2 = 1$. Figure 6.11 shows the populations $N_{1,2}$ as well as the inversion as a function of the normalized pump rate.

6.2.3   Pulse Amplification and Absorption

The discussion above applies to signals that are not explicitly time dependent; if light *pulses* are amplified, saturation also modifies the *shape* of the pulse envelope: while the leading edge of the light pulse experiences the undepleted small signal gain, the later sections of the pulse encounter only the inversion that is left over by the preceding part of the pulse. While this nonlinearity of the amplification process may be undesirable (it is actually often desirable, as we shall see), amplification under saturation conditions provides high energy extraction from the gain medium. In the linear, small signal regime, most of the inversion is not utilized for amplification and ultimately decays by spontaneous emission, whereas in the highly saturated regime the energy increase of the signal may be close to the energy stored in the medium.

With $F = c\rho_{ph}$, the photon transport equation Eq. (6.43) for a four-level system can be written as

$$\frac{\partial F(z,t)}{c\partial t} + \frac{\partial F(z,t)}{\partial z} = \sigma F(z,t)N_2(z,t); \qquad (6.98)$$

if the pulse duration τ_p is so short that pump and spontaneous emission can be neglected during the pulse, the inversion density $N_2(z,t)$ develops according to

$$\frac{dN_2}{N_2} = -\sigma F(z,t)\,dt. \qquad (6.99)$$

In the context of pulse amplification, it is convenient to introduce the energy fluence of the pulse,

$$\Phi := \int \hbar\omega F(t)\, dt \tag{6.100}$$

in units of $[\mathrm{J\,m^{-2}}]$. Integration of Eq. (6.99) over the pulse duration yields

$$N_2(z) = N_2(z,0) \exp\left(-\Phi(z)/\Phi_s^{(4)}\right) \tag{6.101}$$

for the inversion left over after the pulse has passed. The material specific fluence

$$\Phi_s^{(4)} := \hbar\omega/\sigma \tag{6.102}$$

is called saturation fluence; it is the fluence that reduces the initial inversion to a fraction $1/e$. For a typical interaction cross section σ of $10^{-19}\,\mathrm{cm^2}$ and a photon energy of $1\,\mathrm{eV}$, $\Phi_s^{(4)}$ is about $1.6\,\mathrm{J\,cm^{-2}}$.

The system Eqs. (6.98)–(6.99) of coupled, nonlinear differential equations can be solved numerically; for selected pulse profiles, analytical solutions have been derived, known as Frantz–Nodvik equations (Frantz and Nodvik 1963). A particularly instructive (if also unrealistic) case is that of a rectangular input pulse with an input flux density F_0 for $0 < t < \tau_p$ and zero otherwise. The amplifier is assumed to have a length of l and an initial inversion density of $N_{2,i}$; the output flux density then turns out to be

$$F(l,t') = \frac{F_0}{1 - [1 - \exp(-\sigma N_{2,i}l)]\exp(-\sigma F_0 t')}, \quad \text{for} \quad 0 < t' < \tau_p, \tag{6.103}$$

and zero otherwise, where $t' = t - l/c$ is a time coordinate retarded by the transit time l/c of the pulse; the second exponential function in the denominator represents the gain depletion [see Eq. (6.101)]. We can express the above result in terms of the input fluence $\Phi_0 = \hbar\omega F_0 \tau_p$, the saturation fluence Eq. (6.102), and the energy stored in the amplifier per unit cross sectional area, $\Phi_{sto} = \hbar\omega N_{2,i}l$; note that $\exp(\Phi_{sto}/\Phi_s^{(4)}) = \exp(\sigma N_{2,i}l)$ is the small signal gain. With these definitions, Eq. (6.103) is

$$F(l,t') = \frac{F_0}{1 - [1 - \exp(-\Phi_{sto}/\Phi_s^{(4)})]\exp[-(\Phi_0/\Phi_s^{(4)})t'/\tau_p]}. \tag{6.104}$$

Figure 6.12 shows the temporal pulse profile at the output of the amplifier for different values of input fluence. In all cases, the pulse front is amplified by the

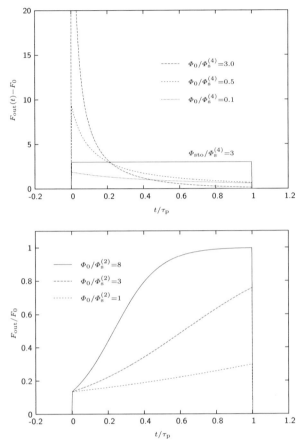

Fig. 6.12 Effect of gain saturation on a rectangular input pulse of duration τ_p for different ratios of the input signal fluence Φ_0 to the saturation fluence Φ_s. Comparison of the area under the pulse with a rectangle that contains the entire energy stored in the gain medium (*solid line*) allows us to estimate the energy extraction efficiency; the output flux density is normalized to $\Phi_s^{(4)}/\tau_p$

Fig. 6.13 Effect of absorption saturation on a rectangular input pulse; the small signal absorption is chosen to be equal to e^{-2}

undepleted gain $\exp(\sigma N_{2,i}l)$; depending on the magnitude of the input fluence, the pulse envelope then drops more or less precipitously and the later sections of the pulse gain little, if any energy; the pulse duration is considerably shortened by the process. Note that even input pulses of relatively low fluence show saturation effects, because their fluence increases during propagation.

Equation (6.103) also describes the saturation of a (two-level) absorber, if the depletion factor in the denominator is replaced by $\exp(-2\sigma F_0 t')$ to account for the fact that one transition changes the population difference by two (accordingly, the saturation fluence for a two-level system is defined as $\Phi_s^{(2)} := \hbar\omega/2\sigma$). Figure 6.13 shows output pulses for different values of input fluence. In contrast to the case of gain saturation, the input fluence needs to be comparable to the saturation fluence or higher to induce significant saturation effects, since the fluence gets lower during propagation. Also in contrast to gain saturation, the leading section of the pulse is now distorted. Once the leading part of the pulse has saturated the absorber, the absorber becomes transparent and transmits the rest of the pulse almost without loss.

Saturable absorbers can therefore be employed for optical switching (Sect. 7.3.1) or formation of ultrashort pulses (Sect. 7.3.2). Note that gain saturation shortens the tail of a pulse, while absorption saturation chops off its head.

6.3 Optical Interactions with Semiconductors

6.3.1 Electronic States in Semiconductors

Semiconductors do not behave like an ensemble of independent atoms, but rather like a huge, covalently bound molecule. The overlap of sp^3-hybrid orbitals of adjoining atoms constitutes a set of bonding and (energetically higher lying) anti-bonding molecular orbitals (Fig. 6.14). The electrons in these orbitals are delocalized over the entire crystal and shared by all atoms. Because of the directionality of the sp^3-orbitals, the resulting molecule displays crystalline order, usually of diamond or zinc blende structure. In an ideal semiconductor, the number of bonding states is exactly equal to the total number of sp^3-electrons. Since every state can be occupied by not more than one electron (Pauli exclusion principle), all bonding states are occupied if the semiconductor is in its ground state.

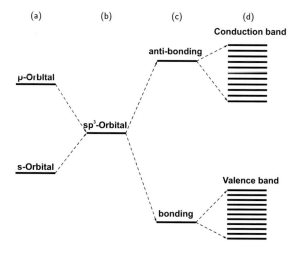

Fig. 6.14 Formation of energy bands in a semiconductor (silicon): (**a**) s- and p-orbitals of the individual atom form sp^3-hybrid orbitals (**b**); the overlap with an adjoining sp^3-orbital results in the splitting of the orbitals into a low lying bonding orbital and a higher lying anti-bonding orbital (**c**); addition of more and more atoms results in the splitting of these states into a manifold of closely lying states, so-called energy bands (**d**); the bonding and the anti-bonding bands are separated by a gap where no states exist

6.3.1.1 Electronic States, Density of States

According to Bloch's theorem (see, e.g., Burns 1985), the energy eigenstates of the electrons in a semiconductor crystal can be written as a product

$$\psi(\mathbf{x}) = u_{B,\mathbf{k}}(\mathbf{x})e^{-j\mathbf{k}\cdot\mathbf{x}} \qquad (6.105)$$

of a function $u_{B,\mathbf{k}}(\mathbf{x})$ that exhibits the periodicity of the crystal lattice, and a plane carrier wave with the wave vector $\mathbf{k}$; the subscript B refers to the band (v indicating the valence and c the conduction band, respectively). For a semiconductor crystal of macroscopic dimensions $d_{x,y,z}$, it is convenient to apply so-called periodic boundary conditions to find (approximate) values of the electronic wave vector, by assuming that the wave function "repeats" itself after the distance $d_{x,y,z}$; consequently, the components of the wave vector assume the discrete values

$$k_i = m_i \frac{2\pi}{d_i}, \quad m_i = \ldots, -2, -1, 0, 1, 2, \ldots \qquad (6.106)$$

While m_i has no upper limit in principle, the periodicity of the crystal lattice implies that the wave vectors within the first Brillouin zone are sufficient to identify all distinct wave functions in a unique way; all wave functions with a wave vector outside this zone are equivalent to a wave function within the first zone. If we assume, without going into details, that the borders of the first Brillouin zone are given by $\pm\pi/a$, where a is the lattice constant of the semiconductor, then $-\pi/a < k_i < \pi/a$, and the index m_i of unique wave vectors is limited by $|m_i| < d_{x,y,z}/2a$. In a macroscopic crystal, $d_{x,y,z}/a$ is a very large number, so that the wave vectors are very closely spaced within the Brillouin zone. The density $\rho_B(k)$ of states, i.e., the number of states per unit volume in the interval $[k, k + dk]$ can then be calculated in a way analogous to Eq. (4.106) and is given by

$$\rho_B(k) = \frac{k^2}{\pi^2}; \qquad (6.107)$$

the two possible polarization states per wave vector of an electromagnetic mode correspond to two different spin states of the electrons. Note that because of different boundary conditions, Eq. (6.106) includes positive as well as negative values of k_i; the restriction to positive k_i-values in Eq. (4.105) is compensated by the smaller mode spacing; the main motivation for applying the more general boundary conditions Eq. (6.106) is that wave vectors of opposite sign are needed to describe electron transport, while the electromagnetic modes in a cavity are standing waves.

Fig. 6.15 Schematic band structure of a semiconductor in the vicinity of the valence band maximum

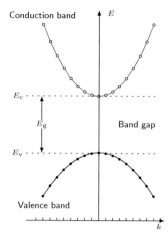

The momentum of an electron is given by $\hbar\mathbf{k}$, and its kinetic energy is accordingly equal to $\hbar^2 k^2/2m$; the energy of a quasi-free electron in the conduction band is therefore

$$E(k) = E_c + \frac{\hbar^2 k^2}{2m_c}, \qquad (6.108)$$

where E_c is the conduction band minimum or edge (Fig. 6.15). In comparison to expression Eq. (6.5) for a free electron, the mass is replaced by an effective mass m_c to account for the interaction of the electron with the crystal lattice. A similar relation holds for the valence band, with the mass $-m_v$ and the band edge E_v

$$E(k) = E_v - \frac{\hbar^2 k^2}{2m_v}; \qquad (6.109)$$

the effective masses of gallium arsenide (GaAs) are $m_c \approx 0.068\, m_e$ and $m_v \approx 0.5\, m_e$.

Equation (6.108) allows us to express the density of states as a function of energy: with $\rho(E)\,dE = \rho(k)\,dk$ and $dk = (m_c/\hbar^2 k)\,dE$, we obtain for the conduction band

$$\rho_c(E - E_c) = \rho_c(k)\frac{dk}{dE} = \frac{1}{2\pi^2}\left(2m_c/\hbar^2\right)^{3/2}\sqrt{E - E_c}; \qquad (6.110)$$

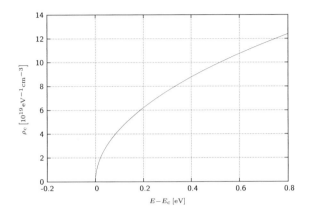

Fig. 6.16 Conduction band density of states $\rho_c(E - E_c)$ for GaAs; the valence band density of states $\rho_v(E_v - E)$ is larger by a factor of $(m_v/m_c)^{3/2} \approx 20$

for the valence band we use $dk = -(m_v/\hbar^2 k)\,dE$ and $-\rho(-E)\,dE = \rho(k)\,dk$ to obtain

$$\rho_v(E_v - E) = \frac{1}{2\pi^2}\left(2m_v/\hbar^2\right)^{3/2}\sqrt{E_v - E}. \tag{6.111}$$

Within the band gap $E_v < E < E_c$, the density of states is zero, above and below it increases with the square root of the energy (Fig. 6.16).

6.3.1.2 (Quasi)-Fermi Distribution

Within a band, electrons can rapidly alter their state by exchanging energy and momentum with the crystal lattice vibrations (phonons); such transitions are called intraband transitions in contrast to interband transition between two different bands (Fig. 6.17). In thermodynamic equilibrium with the lattice, the states are statistically occupied. Other than the atoms in a classical gas, however, the electrons in a semiconductor are (a) indistinguishable and (b) subject to the Pauli exclusion principle that implies that an individual state cannot be occupied by more than one electron (because of its half-integer spin). For these reasons, the occupation probability does not follow Boltzmann's distribution, but an equilibrium distribution known as Fermi–Dirac distribution (Fig. 6.18)

$$f(E - E_F) = \frac{1}{e^{(E-E_F)/k_B T} + 1}. \tag{6.112}$$

In an intrinsic (i.e., undoped) semiconductor, the total number of states available in the valence band equals exactly the total number of valence electrons. Each electron in the conduction band must therefore correspond to a vacancy (hole) in

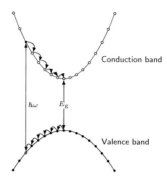

Fig. 6.17 An optical interband transition followed by intraband transitions (thermalization)

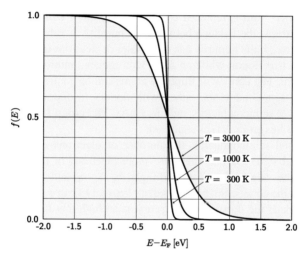

Fig. 6.18 Fermi–Dirac distribution at different temperatures; the energy range of partial occupation is centered around the Fermi energy E_F and has a width of several $k_B T$

the valence band so that the respective densities are equal, $n_e = n_h$. In thermal equilibrium, n_e is given by the integral over the conduction band density of states times the respective occupation probability

$$n_e = \int_{E_c}^{\infty} \rho_c(E) f(E - E_F) \, dE. \tag{6.113}$$

The density of holes in the valence band, on the other hand, is given by

$$n_h = \int_{-\infty}^{E_v} \rho_v(E) [1 - f(E - E_F)] \, dE, \tag{6.114}$$

since $1-f(E)$ is the probability that a state is empty. The Fermi energy E_F establishes itself such that

$$\int_{E_c}^{\infty} \rho_c(E)f(E - E_F)\,dE = \int_{-\infty}^{E_v} \rho_v(E)[1 - f(E - E_F)]\,dE. \qquad (6.115)$$

Because the conduction band density of states differs from that of the valence band by a factor of $(m_c/m_v)^{3/2}$, the Fermi energy is not precisely in the center of the band gap, but shifted slightly towards the band with smaller effective mass (usually the conduction band). Once the value of E_F is established, the equilibrium n_e follows from Eq. (6.113).

6.3.1.3 Doping of Semiconductors

If a regular lattice atom is replaced by an impurity atom with a different number of valence electrons, the balance between the number of valence band states and electrons is altered; if the impurity has more valence electrons than the regular atom, it serves as electron donor, while an atom with reduced number of electrons is an electron acceptor. The density of impurity atoms can be controlled technologically by doping the semiconductors, either with donors (n-doping) or acceptors (p-doping). In III–V semiconductors such as GaAs, Zn (group II) is frequently used as p-dopant and Se (group VI) as n-dopant.

In an n-doped semiconductor, the number of electrons exceeds the number of valence band states so that there are more electrons in the conduction band than holes in the valence band. Accordingly, $n_e > n_h$ and the Fermi energy is shifted towards the conduction band to satisfy Eq. (6.113); in a p-doped material the opposite applies. If the dopant concentration is so high that the Fermi energy lies

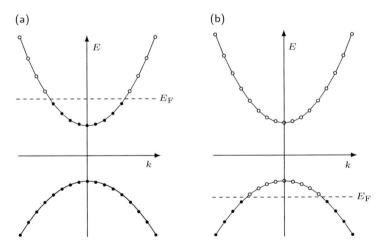

Fig. 6.19 Position of the Fermi level in a highly doped (degenerate) semiconductor: (**a**) n-doped, (**b**) p-doped semiconductor

within one of the bands, the semiconductor is called degenerate and actually behaves like a metal (Fig. 6.19). While doping is of central importance in electronics, it plays a rather minor role in photonics. Throughout the following, undoped ("intrinsic") semiconductors are assumed unless explicitly stated otherwise.

6.3.1.4 Excitation of Semiconductors

Excitation of electrons from the valence into the conduction band increases n_e and n_h, but the balance $n_e = n_h$ is, of course, conserved. Excited electrons stay in the conduction band until they recombine with the holes by radiative or nonradiative transitions at a rate n_e/τ_{rec}, where τ_{rec} is the so-called recombination time, typically of the order of nanoseconds. During this time, the electrons undergo transitions between the closely lying states of the band (intraband transitions) on a time scale of picoseconds. Energy and momentum is conserved by exchange with lattice vibrations (phonons). Because of the much higher rate of intraband transitions, a thermal quasi-equilibrium distribution of the electrons is established within the conduction band, that is again a Fermi distribution

$$f_c(E) = \frac{1}{e^{(E-E_{F,c})/k_B T} + 1} \tag{6.116}$$

but with a quasi-Fermi energy $E_{F,c}$ that is such that

$$n_e = \int_{E_c}^{\infty} \rho_c(E) f(E - E_{F,c}) \, dE. \tag{6.117}$$

In the same fashion, holes occupy the valence band states according to the distribution

$$f_v(E) = \frac{1}{e^{(E-E_{F,v})/k_B T} + 1} \tag{6.118}$$

characterized by a quasi-Fermi energy $E_{F,v}$ that satisfies

$$n_h = \int_{-\infty}^{E_v} \rho_v(E)[1 - f(E - E_{F,v})] \, dE; \tag{6.119}$$

Fig. 6.20 shows the position of the two quasi-Fermi levels of a highly excited semiconductor.

The difference $E_{F,c} - E_{F,v}$, which is zero in thermal equilibrium, grows with increasing excitation; Fig. 6.21 shows the dependence of $E_{F,c}$ and $E_{F,v}$ on n_e for two different temperatures. For $E_{F,c} - E_{F,v} < E_g$, the Fermi levels vary, in good approximation, logarithmically with the carrier density; near the band edge, the dependence becomes more pronounced. At a certain carrier density, the difference

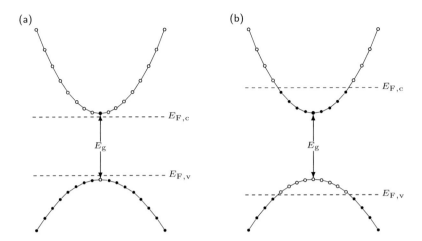

Fig. 6.20 Quasi-Fermi levels of an excited semiconductor at two different degrees of excitation: (**a**) below inversion, (**b**) inverted; see also Fig. 6.21

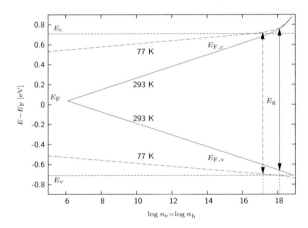

Fig. 6.21 Position of the quasi-Fermi levels in GaAs (Table 6.1) as a function of the carrier density for two different temperatures. The transparency carrier density, defined by $E_{F,c} - E_{F,v} = E_g$, is lower by approximately one order of magnitude at 77 K as compared to room temperature (*vertical arrows*)

between the quasi-Fermi levels is equal to the gap energy; for reasons that will become clear in the following, this carrier density is called transparency carrier density. For temperatures approaching 0 K, the integral Eq. (6.117) can be easily evaluated because the Fermi distribution converges into simple step function, and we obtain

$$E_{F,c} = E_c + \frac{\hbar^2}{2m_c}\left(3\pi^2 n_e\right)^{2/3} \qquad (6.120)$$

and

$$E_{F,v} = E_v - \frac{\hbar^2}{2m_v} \left(3\pi^2 n_h\right)^{2/3} . \tag{6.121}$$

6.3.2 Optical Transitions in Semiconductors

As in atoms, transitions between the states in the valence and conduction bands can be driven by electromagnetic radiation and are accompanied by the creation or annihilation of photons. Since the states are distributed over practically continuous bands, the resonance condition required by Fermi's golden rule Eq. (6.24) can be met by photons of a wide frequency range, provided that $\hbar\omega \geq E_g$. The most important selection rule for an optical interband transition in a semiconductor follows from momentum conservation, $\hbar k_f - \hbar k_i = \pm \hbar k_{ph}$, where $k_{i,f}$ are the initial and final electron wave vectors and k_{ph} is the optical wave vector; since the wave vector of an 1 eV-photon ($\approx 10^5$ cm^{-1}) can be neglected in comparison with typical electronic wave vectors (10^7 cm^{-1}), the selection rule simplifies to $k_f - k_i = 0$. In the E–k band diagram, optical transitions must therefore be "vertical" (Fig. 6.22). Transitions between states of different wave vector (called indirect transitions) are possible in principle, but require the simultaneous interaction with a phonon (lattice vibration) to conserve momentum; accordingly, the probability of indirect transitions is very small in comparison with direct transitions.

The actual band structure of semiconductors is usually quite complicated. From the viewpoint of light–matter interaction, one of the most important features of the band diagram is the position of the conduction band minimum in respect to the maximum of the valence band. If the two points have the same k-vector, the band gap is denoted as direct. Since electrons accumulate in the conduction band minimum, and holes in the valence band maximum, such semiconductors, when

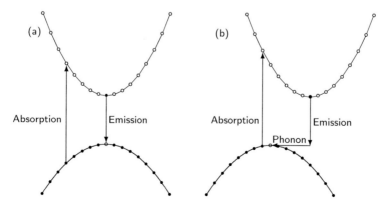

Fig. 6.22 Energy bands and optical transitions: (**a**) direct semiconductor, (**b**) indirect semiconductor

excited, can recombine radiatively. Indirect semiconductors (such as Si) cannot efficiently emit light because the electrons in the conduction band cannot reach empty states in the valence band via direct transitions (Fig. 6.22). In regard to absorption, the difference between direct and indirect band gap semiconductors is irrelevant, and both types are suitable as photodetectors.

6.3.2.1 Joint Density of States

To analyze the interaction of light with semiconductors, we first have to find, for a given of photon energy $\hbar\omega$, the states E_a, E_b in the valence and conduction band, respectively, that fulfill the requirements $E_b - E_a = \hbar\omega$ and $\mathbf{k}_a = \mathbf{k}_b$ (Fig. 6.23a). With Eqs. (6.108) and (6.109), these conditions can be combined in the equation

$$\hbar\omega = \frac{\hbar^2 k^2}{2m_r} + E_g, \tag{6.122}$$

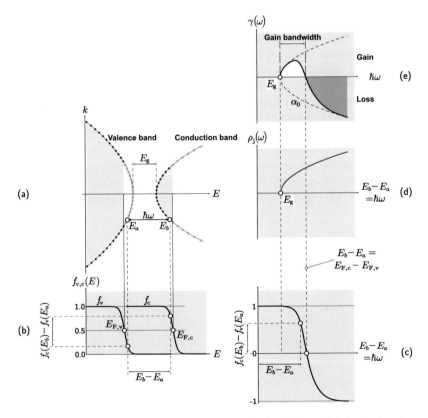

Fig. 6.23 Band structure (**a**), quasi-Fermi distributions (**b**), Fermi factor (**c**), joint density of states (**d**), and gain coefficient $\gamma(\omega) = \alpha_0(\omega)[f_c(E_b) - f_v(E_a)]$ (**e**), for $E_{F,c} - E_{F,v} > E_g$

where $k^2 = k_a^2 = k_b^2$, $E_c - E_v = E_g$ and the reduced electron mass m_r is defined by

$$\frac{1}{m_r} := \frac{1}{m_c} + \frac{1}{m_v}. \tag{6.123}$$

The solution of Eq. (6.122),

$$k^2(\omega) = \frac{2m_r}{\hbar^2}(\hbar\omega - E_g) \tag{6.124}$$

provides the wave number of the states that are coupled by direct optical transitions at ω. The corresponding energy levels in the two bands follow from Eqs. (6.108) and (6.109)

$$E_a(\omega) = E_v - \frac{m_r}{m_v}(\hbar\omega - E_g),$$

$$E_b(\omega) = E_c + \frac{m_r}{m_c}(\hbar\omega - E_g). \tag{6.125}$$

With Eqs. (6.107) and (6.124), we can calculate the so-called joint density $\rho_j(k)$ of such pairs of states in the interval $[k, k + dk]$

$$\rho_j(k) = \frac{k^2(\omega)}{\pi^2} = \frac{2m_r}{\pi^2\hbar^2}(\hbar\omega - E_g); \tag{6.126}$$

the (more relevant) joint density of states (Fig. 6.23d) in the interval $[\omega, \omega + d\omega]$ can be obtained from Eq. (6.126) using the identity $\rho_j(\omega)\, d\omega = \rho(k)\, dk$; with $dk/d\omega = m_r/\hbar k$ following from Eq. (6.124), we obtain

$$\rho_j(\omega) = \frac{1}{2\pi^2}\frac{(2m_r)^{3/2}}{\hbar^2}(\hbar\omega - E_g)^{1/2}. \tag{6.127}$$

6.3.2.2 Fermi Factor

If a (two-level) atom is in one of the states participating in an optical transition, then the other state is unoccupied and one does not have to care about the blocking of a transition by the Pauli exclusion principle. In semiconductors, states are statistically occupied according to the Fermi distribution Eq. (6.112), and the final state of a transition may be occupied, i.e., the transition is blocked with a certain probability. The (absorptive) transition between a state E_a in the valence band

and the corresponding conduction band state E_b requires, for example, that E_a is occupied and E_b is unoccupied, which happens with the probability

$$f_{\text{abs}} = f_{\text{v}}(E_a)[1 - f_{\text{c}}(E_b)]; \qquad (6.128)$$

note that the Fermi levels in the two bands are generally different, as indicated by the subscripts v and c in the distribution function. Analog considerations for the emissive transition from E_b to E_a yield

$$f_{\text{em}} = f_{\text{c}}(E_b)[1 - f_{\text{v}}(E_a)]; \qquad (6.129)$$

the net-transition rate is thus proportional to the so-called Fermi factor (Fig. 6.23c)

$$f_{\text{em}} - f_{\text{abs}} = f_{\text{c}}(E_b) - f_{\text{v}}(E_a) \begin{Bmatrix} \leq \\ = \\ > \end{Bmatrix} 0 \quad \text{if } E_b - E_a \begin{Bmatrix} \geq \\ = \\ < \end{Bmatrix} E_{\text{F,c}} - E_{\text{F,v}}; \qquad (6.130)$$

the inequalities follow after substitution of the distribution functions Eqs. (6.116) and (6.118). With $E_b - E_a = \hbar\omega$, we find that the Fermi factor is zero if $\hbar\omega = E_{\text{F,c}} - E_{\text{F,v}}$, and positive (negative) for lower (higher) values of $\hbar\omega$; note that a zero Fermi factor implies that the semiconductor is transparent at the respective frequency, independent of the joint density of states.

While the quasi-Fermi levels $E_{\text{F,c}}(n_e)$ and $E_{\text{F,v}}(n_h)$ are determined by the carrier density $n_e = n_h$ (Fig. 6.21), the energies E_a and E_b are functions of ω [Eq. (6.125)]. Thus, the Fermi factor is completely determined by n_e and ω

$$f_{\text{c}}(E_b) - f_{\text{v}}(E_a) = f_{\text{c}}\left(E_{\text{c}} + \frac{m_{\text{r}}}{m_{\text{c}}}(\hbar\omega - E_{\text{g}})\right) - f_{\text{v}}\left(E_{\text{v}} - \frac{m_{\text{r}}}{m_{\text{v}}}(\hbar\omega - E_{\text{g}})\right). \qquad (6.131)$$

6.3.2.3 Absorption Coefficient

For the calculation of the absorption coefficient, we can follow the derivation of Eq. (6.47), replacing the population difference $N_2 - N_1$ by the product $\rho_j(\omega)(f_{\text{em}} - f_{\text{abs}})$. The transition probability for a given pair of states E_a, E_b is determined by Fermi's golden rule Eq. (6.24) and is characterized by an interaction cross section that is non-zero only within a small bandwidth centered at $(E_b - E_a)/\hbar$. According to Eq. (6.60), this cross section can be expressed as $\sigma = (\pi^2 c^2/\omega^2 \tau_{\text{r}})g(\Delta\omega)$, where τ_{r} is the *radiative* recombination time.

Table 6.1 Selected material properties of intrinsic GaAs

E_g [eV]	n	(m_c/m_e)	(m_v/m_e)	(m_r/m_e)	τ_r [ns]
1.42	3.55	0.068	0.5	0.06	2

Under assumption of a narrow line $g(\Delta\omega) \approx \delta(\Delta\omega)$ of the individual transition, the absorption coefficient thus is given by

$$\alpha(\omega) = -\frac{\pi^2 c^2}{\omega^2 \tau_r} \rho_j(\omega) [f_c(E_b) - f_v(E_a)]. \tag{6.132}$$

At sufficiently low temperatures ($k_B T \ll E_g$) and in the absence of any other excitation, $f_v(E_a) = 1$ and $f_c(E_b) = 0$ so that the Fermi factor is equal to -1 and we obtain, with Eq. (6.127)

$$\alpha_0(\omega) = \frac{c^2}{(\hbar\omega)^2} \frac{(2m_r)^{3/2}}{2\tau_r} (\hbar\omega - E_g)^{1/2}. \tag{6.133}$$

For GaAs (Table 6.1) and near the band gap, the prefactor in Eq. (6.133) amounts to $c^2(2m_r)^{3/2}/2(\hbar\omega)^2\tau_r \approx 5 \times 10^3 \text{ cm}^{-1} \text{ eV}^{-1/2}$; only slightly above the band gap, at $\hbar\omega - E_g = 0.01$ eV, the absorption coefficient is already as large as $5 \times 10^2 \text{ cm}^{-1}$.

6.3.3 Optical Gain Condition

In an excited semiconductor, the absorption coefficient Eq. (6.132) turns negative provided that $\hbar\omega > E_g$ and the Fermi factor assumes a positive value, which requires $\hbar\omega < E_{F,c} - E_{F,v}$ (Fig. 6.23e). The semiconductor then acts as optical amplifier with the gain coefficient $\gamma = -\alpha$ following from Eq. (6.132)

$$\gamma(\omega) = -\alpha(\omega) = \alpha_0(\omega) [f_c(E_b) - f_v(E_a)], \tag{6.134}$$

with α_0 given by Eq. (6.133). Gain is provided within the interval

$$E_g < \hbar\omega < E_{F,c} - E_{F,v}. \tag{6.135}$$

The gain condition, equivalent to the inversion condition, therefore requires that the difference between the quasi-Fermi levels is larger than the band gap. As already mentioned, the carrier density needed to reach $E_{F,c} - E_{F,v} = E_g$ is called the transparency carrier density n_{tr}: with this carrier density, the semiconductor is transparent for a photon energy just equal to the band gap.

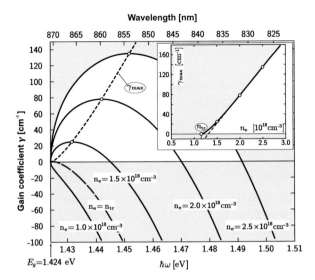

Fig. 6.24 Spectral dependence of the gain coefficient γ of GaAs (calculated) at room temperature for different values of n_e (cm^{-3}); the *inset* shows the peak gain coefficient as a function of n_e

Figure 6.24 shows the gain coefficient as a function of n_e. Because of the increasing density of states, the peak of the gain spectrum moves to higher frequencies with increasing carrier density; the peak gain coefficient is approximately given by the semi-empirical formula

$$\gamma(n_e) \simeq \alpha_0 \left(\frac{n_e}{n_{tr}} - 1 \right). \tag{6.136}$$

6.3.4 Low Dimensional Semiconductors

The position of the quasi-Fermi level depends on the carrier density, the temperature and the density of states: a smaller density of states, for example, implies that states of higher energy must be filled to accommodate a given number of electrons in the conduction band. As we will see in the following, the density of states can be modified (lowered) by spatial confinement of the carriers in one or more dimension. By the same token, the carrier density necessary to reach a certain difference of the quasi-Fermi level is reduced in such structures.

6.3.4.1 Quantum Wells

In a macroscopic crystal, the points in k-space representing the electronic wave functions are so closely packed that they form practically continuous bands (Sect. 6.3.1.1); accordingly, the density of states [Eq. (6.108)], given in $[\text{cm}^{-3}\,\text{eV}^{-1}]$, is independent of the size and shape of the crystal. If, however, the dimension of the crystal in one or more directions approaches the lattice constant of the crystal, the spacing between the points in k-space becomes significant and the density of states is not a continuous function of the wave number or the energy, respectively, but starts to display discontinuities. Let us consider a semiconductor slab with very small thickness d_z (but macroscopic extensions $d_{x,y}$); if we model the confinement of the carriers in the z-direction by a one-dimensional rectangular potential well with infinite barriers, the electron wave function must vanish at the walls, forming a standing wave in the confinement direction, just as the electromagnetic modes in a perfectly conducting cavity. We therefore can use the result Eq. (4.105) to find

$$k_z = n_z \frac{\pi}{d_z}, \quad n_z = 1,2,3\ldots, \tag{6.137}$$

while $k_{x,y}$ is given by Eq. (6.106). Because d_z is assumed to be very small, this means that the tips of the wave vectors are arranged in distinct planes normal to the k_z direction and separated by π/d_z; to each value of n_z corresponds a manifold of densely packed states $k_{x,y}$ representing a two-dimensional sub-k-space (Fig. 6.25).

To estimate the number of states in such a sub-space in the interval $[k, k + dk]$ (where $k = \sqrt{k_x^2 + k_y^2}$), we divide the volume $2\pi^2 k^2/d_z\,dk$ of a cylindrical ring of radius k, radial thickness dk, and height π/d_z by the volume $4\pi^3/d_x d_y d_z$ that a state consumes and multiply by 2 to account for the two possible spins; the density of states is obtained after dividing the result by the volume $d_x d_y d_z$ of the crystal

$$\rho_{n_z,\text{c}}(k) = \frac{k}{\pi d_z}, \tag{6.138}$$

in contrast to the bulk-expression $\rho_\text{c}(k) = k^2/\pi^2$ [Eq. (6.107)].

Fig. 6.25 Location of electronic states of a two-dimensional semiconductor in k-space, forming closely packed planar sub-spaces; also shown is the volume element that a state occupies

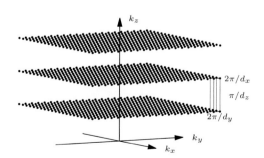

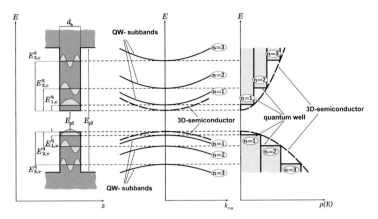

Fig. 6.26 Band structure and density of states of a quantum well-structure; $E_{g1,2}$ are the bulk band gaps of the two materials forming the structure

The energy of the conduction band states is [see Eq. (6.108)]

$$E(k) = E_c + E_{n_z,c}^q + \frac{\hbar^2 k^2}{2m_c} \tag{6.139}$$

with

$$E_{n_z,c}^q := \frac{\hbar^2 (n_z \pi / d_z)^2}{2m_c}; \tag{6.140}$$

for every value of n_z, there exists a separate parabolic sub-band (Fig. 6.26).

To find the density of conduction band states as a function of energy, we express k by E,

$$k = \sqrt{(2m_c/\hbar^2)(E - E_c - E_{n_z,c}^q)}. \tag{6.141}$$

From $\rho_{n_z,c}(k)\, dk = \rho_{n_z,c}(E)\, dE$ and

$$dk = \frac{\sqrt{2m_c/\hbar^2}}{2\sqrt{E - E_c - E_{n_z,c}^q}}\, dE \tag{6.142}$$

we find for $E \geq E_c + E_{n_z,c}^q$

$$\rho_{n_z,c}(E) = \frac{m_c}{\pi \hbar^2 d_z}, \tag{6.143}$$

which is obviously constant. The total density of states of the manifold of sub-bands is obtained by summation over n_z; it is a step function enclosed by the bulk density of states (Fig. 6.26).

The valence band density of states is obtained by replacing the effective mass of the conduction band by that of the valence band:

$$\rho_{n_z,\mathrm{v}}(E) = \frac{m_\mathrm{v}}{\pi \hbar^2 d_z}. \tag{6.144}$$

Replacing the effective mass by the reduced mass and applying the selection rules $\Delta k = 0$, $\Delta n_z = 0$, we obtain for $n_z = 1$ the optical joint density of states at $\omega = E/\hbar$

$$\rho_{1,\mathrm{j}}(\omega) = \frac{m_\mathrm{r}}{\pi \hbar^2 d_z}, \quad \text{for} \quad \omega > E_\mathrm{g} + E_{1,\mathrm{c}}^\mathrm{q} + E_{1,\mathrm{v}}^\mathrm{q}. \tag{6.145}$$

Technologically, quantum wells are produced by growing few atomic layers of a semiconductor material such as GaAs, sandwiched between two layers of a wide band gap material such as AlAs.

The modified density of states of quantum well-structures has important consequences:

- the transparency carrier density is significantly reduced
- the constant joint density of states provides almost constant gain within one subband
- increasing the carrier density increases the gain bandwidth without affecting the peak gain.

For these reasons, quantum well semiconductor amplifiers and -lasers are widely applied in photonics.

6.3.4.2 Quantum Wires and Quantum Dots

Carrier confinement can be extended to two or three dimensions; a needle shaped semiconductor crystal of microscopic measures in two directions (say y and z) is called a quantum wire, while microscopic semiconductor grains are referred to as quantum dots. In a quantum wire, two components of the k-vector become discrete

$$k_{y,z} = n_{y,z} \frac{\pi}{d_{y,z}}, \quad n_{y,z} = 1, 2, 3 \dots \tag{6.146}$$

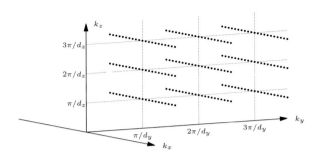

Fig. 6.27 Location of electronic states of a quantum wire in k-space, forming closely packed linear sub-spaces

while $k_x = k$ is quasi-continuous and given by Eq. (6.106). The resulting sub-spaces are parallel lines of densely packed states k_x in k-space (Fig. 6.27). The conduction band energy of sub-band n_y, n_z is given by

$$E(k) = E_c + E_{n_y,c}^q + E_{n_z,c}^q + \frac{\hbar^2 k^2}{2m_c} \qquad (6.147)$$

with

$$E_{n_{y,z},c}^q := \frac{\hbar^2 (n_{y,z}\pi/d_{y,z})^2}{2m_c}. \qquad (6.148)$$

The density of states as a function of k is constant and equal to $1/(\pi d_y d_z)$; expressed as a function of energy, the density of states of a given sub-band is

$$\rho_{n_y,n_z,c}(E) = \frac{1}{\pi d_y d_z} \frac{\sqrt{m_c}/\hbar}{\sqrt{2(E - E_c - E_{n_y,c}^q - E_{n_z,c}^q)}} \qquad (6.149)$$

for $E > E_c + E_{n_y,c}^q + E_{n_z,c}^q$.

Quantum confinement in three dimensions results in quantum dots. Since all components of the k-vector are discrete, no bands are formed. The energy levels are discrete

$$E = E_c + \sum_{i=x,y,z} \frac{\hbar^2 (n_i\pi/d_i)^2}{2m_c}, \qquad (6.150)$$

similar to that of an atom. Consequently, the term density of states is not appropriate any more and is replaced by the density of quantum dots, if there is more than one. Since it is technologically difficult to produce many quantum dots of exactly the same dimensions, an ensemble of quantum dots is usually inhomogeneously broadened. An important application of quantum dots is the conversion of UV light from luminescence diodes (UV-LED) into visible light, exploiting the manifold of radiative transitions within a quantum dot.

6.3.4.3 Inter-Sub-Band Transitions

Optical transitions are also possible between different sub-bands of a quantum well, wire, or dot. The energy spacing is generally much smaller than the band gap and can be designed by appropriate choice of the geometric dimensions and the material constituents. Such inter-sub-band transitions are the base for IR emitting so-called quantum cascade lasers (see, e.g., Faist 2011).

6.3.5 Carrier Induced Refractive Index Change

An important side effect of increased carrier density is the change of the refractive index of the semiconductor upon excitation. The electrons in the conduction band (and holes in the valence band) act as a free electron gas (Sect. 2.2.3), which, for frequencies above the plasma-frequency Eq. (2.87) provides a negative contribution to the total permittivity. For typical carrier densities in semiconductor amplifiers, the plasma frequency is well below the band gap. To estimate the change of the permittivity, we treat the carriers as having the reduced electron mass m_r and obtain, according to Eq. (2.88), the estimate

$$\Delta\varepsilon = -\frac{n_e e^2}{\varepsilon_0 m_r \omega^2}. \tag{6.151}$$

In GaAs, at a carrier density of $n_e = 2 \times 10^{18}\,\text{cm}^{-3}$ we get $\Delta\varepsilon \approx -0.016$ at a wavelength of $\lambda_0 = 830\,\text{nm}$ (1.5 eV). With $n = \varepsilon^{1/2}$, this corresponds to a refractive index change of $\Delta n \approx \Delta\varepsilon/2n \approx -0.005$.

A second contribution to the refractive index of an excited semiconductor follows from the Kramers–Kronig relations Eq. (2.103) that relate the real and imaginary parts, n and κ, of the refractive index. With Eq. (2.71) and $k_0 = \omega/c_0$, the absorption coefficient α of the semiconductor in the absence of excitation is related to κ by $\kappa(\omega) = \alpha(\omega)c_0/2\omega$, while in the inverted, amplifying medium we have $\kappa(\omega) = -\gamma(\omega)c_0/2\omega$. Equation (2.103) then yields

$$\Delta n(\omega) = \frac{2}{\pi}\int_0^\infty \frac{\omega'\Delta\kappa(\omega')}{\omega'^2 - \omega^2}\,d\omega' = -\frac{c_0}{\pi}\int_0^\infty \frac{\Delta\gamma(\omega')}{\omega'^2 - \omega^2}\,d\omega', \tag{6.152}$$

where $\Delta\gamma(\omega') = \gamma(\omega') + \alpha(\omega')$. This contribution to the refractive index is also negative at the peak of the gain spectrum and usually exceeds the plasma-contribution. The dependence of the refractive index on the carrier density has important implications for the operation of semiconductor lasers (Sect. 7.5.3). It also can be used to realize semiconductor based interferometric modulators, since the refractive index change extends to light frequencies within the band gap, where the semiconductor is transparent.

6.4   Summary

Schrödinger's equation is introduced as the wave equation of electrons. For a given atomic potential, the solutions of this equation are the electronic energy eigenstates of the atom with corresponding eigenvalues. In the absence of a perturbation by, for example, an electromagnetic field, an atom in any of these eigenstates is stable. The effect of a (periodic) perturbation is that the exposed atom assumes a state that is a superposition of its eigenstates; such a superposition is not stationary anymore but oscillates at a frequency equal to the difference of the respective energy eingenvalues, divided by Planck's constant. The perturbation is therefore only efficient if its frequency is equal or close to the oscillation frequency of the atom.

After the perturbation, the atom is found, with a certain probability, in a different state—it has undergone a transition. The energy difference between the initial and the final is balanced by the electromagnetic field in the process of absorption or stimulated emission. The probability of such a transition can be expressed as a product of an interaction cross section and the fluence of light quanta the atom is exposed to. We derive Fermi's golden rule which provides, in good approximation, the transition probability and the interaction cross section. We discuss various mechanisms that lead to line broadening, i.e., to the broadening of the frequency range where transitions are possible.

Stimulated emission is the base of light amplification in a laser. We introduce the concept of rate equations to describe the interaction of a laser mode with the laser medium. A gain condition is formulated and techniques to reach the necessary population inversion are described. Amplification relies on the transition of atoms from a higher energy state to a lower, and the amplification of intense signals reduces the number of excited atoms and thus the gain. This saturation process is discussed for continuous waves as well as light pulses.

We extend the discussion of light–matter interaction to semiconductors, where the electrons are not allocated to an individual atom, but are delocalized over the entire crystal. The gain condition therefore includes statistical components, taking the Fermi distribution of electrons within the energy bands of the semiconductor into account. The number of pairs of semiconductor states that are eligible for a transition at a given frequency is expressed in the optical joint density of states. This important quantity also depends on the geometric dimensions of the semiconductor:

quantum wells, wires, and dots have a reduced density of states which generally is advantageous for laser applications.

6.5 Problems

1. The power spectrum of a black body at temperature T is given by Eq. (6.52): (a) calculate the frequency ω_{max} of spectral maximum; (b) use the identity $\rho(\omega)\,d\omega = \rho(\lambda)\,d\lambda$ to express the power spectrum as a function of the vacuum wavelength; calculate the wavelength $\lambda_{0,max}$ of spectral maximum. Explain why $\omega_{max}\lambda_{0,max} \neq 2\pi c_0$, although $\omega\lambda_0 = 2\pi c_0$. What does a grating spectrometer measure, $\rho(\omega)$ or $\rho(\lambda)$?

2. Assume a 1 mmol/l solution of Rhodamine 6G in ethanol (the molecular weight and absorption spectrum of Rhodamine 6G can be found in many data bases); calculate σ_{max} of the dye-molecule and the loss of monochromatic light at 530 nm upon transmission through a 1 (5) cm thick layer of this solution. Further assume "white light" with a constant spectral power $I(\omega) = dP/d\omega = $ const. between 300 and 700 nm (and zero outside this range) transmitted through this solution. What is the approximate loss (in percent) (approximate the absorption band by a rectangle); note that $I(\lambda_0)$ of the input light is *not* a constant. The spontaneous life time of the molecule is 10 ns; calculate the fluence of an ultrashort 530 nm pulse that reduces the absorption of a thin layer of this solution to 50 %. How long does it take the dye solution to recover after the light is switched off?

3. Atomic sodium vapor in a gas discharge lamp emits a strong line at 589 nm; assuming that the vapor has a temperature of 2000°, what is the range of Doppler frequency shifts due to thermal motion, if we assume that the atoms have a kinetic energy of $k_B T$?

4. Assume a four-level amplifier with spatially homogeneous inversion density and length l. A very short (but energetic) pulse with an energy fluence $\Phi_0 = \hbar\omega F_0\tau_p$ is transmitted through the medium. Calculate the inversion distribution immediately after the transmission for different values of Φ_0. Neglect spontaneous emission and pump processes.

5. Same as before, but with a pulse of finite duration and initially rectangular pulse shape: calculate the pulse shape at the end of the amplifier by numerical integration for different values of Φ_0; for this purpose, slice the pulse into a train of very short pulses of appropriate fluence. Compare the results with that of the Frantz–Nodvik equations.

6. Calculate numerically the quasi-Fermi levels in intrinsic GaAs as a function of the carrier density n_e at room temperature. Evaluate the transparency carrier concentration as a function of T. See Table 6.1 for the properties of GaAs.

7. Consider a GaAs quantum well with a d_a of 10 nm. Calculate numerically the quasi Fermi levels as a function of the carrier density and the temperature T. Evaluate the transparency carrier density as a function of T. Compare these results with the bulk results.

References and Suggested Reading

Allen, L., & Eberly, J. H. (1987). *Optical resonance and two-level atoms*. New York: Dover.

Bass, M., (Ed.). (2010). *Handbook of optics*. New York: McGraw-Hill.

Burns, G. (1985). *Solid state physics*. Orlando: Academic Press.

Capasso, F. (1990). *Physics of quantum electron devices*. New York: Springer.

Cerullo, G., Longhi, S., Nisoli, M., Stagira, S., Svelto, O. (2001) *Problems in laser physics*. New York: Springer.

Chow, W. W., & Koch, S. W. (1999). *Semiconductor-laser fundamentals*. New York: Springer.

Connelly, M. J. (2002). *Semiconductor optical amplifiers*. New York: Springer.

Desurvire, E. (2001). *Erbium doped fiber amplifiers*. New York: Wiley.

Faist, J. (2011). *Quantum cascade lasers*. London: Oxford University Press.

Fowler, W., & Dexter, D. L. (1962). Relation between absorption and emission probabilities in luminescent centers in ionic solids. *Physical Review, 128*(5), 2154.

Fowles, G. F. (1989). *Introduction to modern optics*. New York: Dover.

Frantz, L. M., & Nodvik, J. S. (1963). Theory of pulse propagation in a laser amplifier. *Journal of Applied Physics, 34*(8), 2346–2349.

Ghafouri-Shiraz, H. (1995). *Fundamentals of laser diode amplifiers*. New York: Wiley.

Graham-Smith, F. (2007). *Optics and photonics*. New York: Wiley.

Herzberg, G. (2010). *Atomic spectra and atomic structure*. New York: Dover.

Hodgson, J. N. (1970). *Optical absorption and dispersion in solids*. New York: Springer.

Loudon, R. (2000). *The quantum theory of light*. Oxford: Oxford University Press.

Miller, D. A. B. (2008). *Quantum mechanics for scientists and engineers*. New York: Cambridge University Press.

Numai, T. (2004). *Fundamentals of semiconductor lasers*. Berlin: Springer.

Omar, M. A. (1993). *Elementary solid state physics*. New York: Pearson Education.

Paschotta, R. (2009). *Encyclopedia of laser physics and technology*. New York: John Wiley.

Saleh, B. E., & Teich, M. C. (2007). *Fundamentals of photonics*. New York: Wiley.

Siegman, A. E. (1986). *Lasers*. Mill Valley, CA: University Science Books.

Svelto, O. (2010). *Principles of lasers*. New York: Plenum Press.

Wooten, F. (1972). *Optical properties of solids*. New York: Elsevier.

Yariv, A. (1982). *An introduction to theory and applications of quantum mechanics*. New York: Dover.

Yariv, A., & Yeh, P. (2006). *Photonics*. Oxford: Oxford University Press.

Optical Oscillators

7

Positive feedback converts an amplifier into an oscillator. In optics, feedback can be provided by mirrors at the two ports of an amplifier (Fig. 7.1). While the acronym "laser" stands for light *amplification* by stimulated emission of radiation, it usually refers to an optical *oscillator* that relies on amplification by stimulated emission; we shall see in Sect. 7.6 and Chap. 8 that there are alternative amplification mechanisms that can be used to build optical oscillators.

The optical properties of a Fabry–Perot resonator, in particular its modes, have been described in Sect. 4.3. Because of the coherent nature of the amplification process, a laser oscillator can produce light consisting of a huge number of photons that belong to one single resonator mode; such a light field is practically monochromatic and exhibits the spatial properties of a coherent light beam such as a Gaussian beam (Sect. 3.1.2). By contrast, thermal light inside a cavity kept at a temperature of, say, 3000 K ($k_B T = 0.26$ eV) contains an average of about 0.0004 photons per mode in the visible (≈ 2 eV), as we will see in Sect. 9.3, Eq. (9.20). Light from a thermal source is therefore an incoherent superposition of a huge number of hardly occupied different spatial and temporal modes.

There exists a wide variety of laser materials, from atomic gases to molecules in liquid solution, from ions in dielectric host materials to semiconductors. Accordingly, the technical implementation, and the pump process in particular, varies greatly. The following discussion of some important aspects of laser oscillators refer to optically pumped atomic lasers; the fundamental results derived in this section apply to all lasers, however.

7.1 Stationary Performance

7.1.1 Rate Equations, Four-Level System

We assume an ideal four-level atomic system (Sect. 6.2), where transitions $E_3 \rightarrow E_2$ and $E_1 \rightarrow E_0$ are so fast that one can set $N_{1,3} = 0$ (Fig. 7.2); thus, the population

© Springer International Publishing Switzerland 2016
G.A. Reider, *Photonics*, DOI 10.1007/978-3-319-26076-1_7

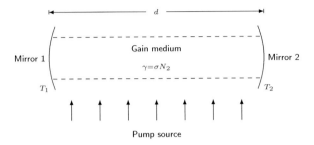

Fig. 7.1 Schematic of a laser oscillator: a Fabry–Perot resonator formed by two mirrors with transmittance $T_{1,2}$ contains a gain medium serving as amplifier

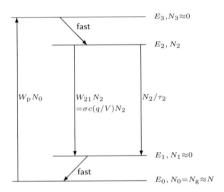

Fig. 7.2 Four-level laser

inversion is given by $\Delta N = N_2$. Assuming $N_2 \ll N$, we can neglect the depletion of the ground state by the pump process and replace N_g by the total density of atoms N. The light field within the resonator is characterized by the number q of photons, related to the photon density ρ_{ph} by $\rho_{ph} = q/V$ with $V := Ad$. Since the resonator modes are not spatially homogeneous (having a Gaussian transverse profile and axial nodes), V is an effective mode volume that can be expressed as a product of an effective cross sectional area A and the length of d the resonator. The stimulated transition probability is then, according to Eq. (6.38), given by $W = \sigma c q/V$.

The rate equation for N_2 follows from Eq. (6.86)

$$\frac{dN_2}{dt} = W_p N - \sigma c N_2 \frac{q}{V} - \frac{N_2}{\tau_2}. \qquad (7.1)$$

The corresponding equation for the photon number q contains the stimulated emission term of Eq. (7.1), multiplied with the mode volume (we assume that the inversion density is constant over the mode volume), and a linear loss term q/τ_{res},

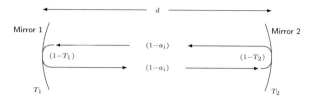

Fig. 7.3 Resonator losses; a_i is the fractional internal loss

where τ_{res} is the resonator life time, i.e., the average time that a photon spends within the resonator:

$$\frac{dq}{dt} = q\left(\sigma c N_2 - \frac{1}{\tau_{res}}\right). \qquad (7.2)$$

The resonator losses comprise the (intended) transmission losses of photons through the output coupler(s) and various internal losses (Fig. 7.3). The loss factor per round trip can be expressed as the product $(1-T_1)(1-T_2)(1-a_i)^2$, where $(1-a_i)$ represents the internal losses. It is convenient to introduce "distributed" loss coefficients

$$\alpha_{1,2} := -\frac{\ln(1-T_{1,2})}{2d} \qquad (7.3)$$

$$\alpha_i := -\frac{\ln(1-a_i)}{d} \qquad (7.4)$$

$$\alpha_{res} := \alpha_1 + \alpha_2 + \alpha_i, \qquad (7.5)$$

so that the loss per round trip can be expressed by $e^{-\alpha_{res}2d}$ and $dF/dz = -F\alpha_{res}$, where F is the photon flux circulating between the resonator mirrors. Since $dF/F = dq/q$ and $dz = c\,dt$,

$$\left.\frac{dq}{q\,dt}\right|_{loss} = c\left.\frac{dF}{F\,dz}\right|_{loss} = -c\alpha_{res}. \qquad (7.6)$$

Comparison with Eq. (7.2) allows us to relate the photon life time and the loss coefficient

$$\alpha_{res} = \frac{1}{c\tau_{res}}; \qquad (7.7)$$

most lasers have only one output coupler (e.g., mirror 2), so that $T_1 = 0$, and $\alpha_2 = \alpha_{res} - \alpha_i$ represents the output coupling coefficient

$$\alpha_2 = -\frac{1}{c}\frac{dq}{q\,dt}\bigg|_{loss,out}. \tag{7.8}$$

With the gain coefficient $\gamma = \sigma N_2$ [Eq. (6.84)], the round trip gain equals $e^{\sigma N_2 2d}$ (provided that the length of the resonator is equal to the length of the amplifying medium). To exactly compensate the round trip losses, the gain and loss coefficients must be equal,

$$\gamma_c = \alpha_{res}; \tag{7.9}$$

$\gamma_c = \sigma N_{2,c}$ is called the critical gain coefficient and

$$N_{2,c} = \frac{\alpha_{res}}{\sigma} = \frac{1}{c\sigma\tau_{res}} \tag{7.10}$$

is the critical inversion density.

7.1.2   Laser Output Characteristic

We now want to study the stationary behavior of the laser as a function of the pump rate W_pN (Fig. 7.4). With increasing pump rate, the inversion increases, but as long as $N_2 < N_{2,c}$, the resonator losses exceed the gain and the stationary solution of Eq. (7.2) is $q = 0$. In this operating regime, the balance between pump and

Fig. 7.4   Inversion N_2 and photon number $q_0 \propto P_{out}$ of a four-level laser as a function of the pump rate

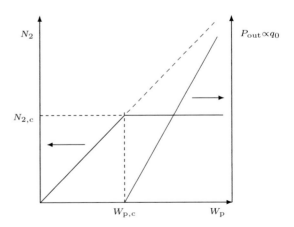

spontaneous emission provides the stationary inversion population $N_2 = W_p N \tau_2$; this applies to pump rates $W_p < W_{p,c}$, where

$$W_{p,c} := \frac{N_{2,c}}{N\tau_2} = \frac{1}{\sigma c N \tau_{res} \tau_2} \tag{7.11}$$

is the critical pump rate, required to reach $N_{2,c}$. Above the critical pump rate, the small signal gain initially exceeds the resonator losses and the photon number, starting from few spontaneously emitted photons, starts to grow exponentially. The details of the transient behavior will be discussed in Sect. 7.2.2 and Sect. 7.2.4, but it is clear that the stationary situation, which requires the balance of loss and gain and consequently $N_2 = N_{2,c}$, can only be reached if the stimulated emission reduces (saturates) the small signal inversion to the critical value $N_{2,c}$. The photon number required for this saturation follows from Eq. (7.1) after the substitution of N_2 by $N_{2,c}$ and represents the stationary photon number q_0

$$q_0(W_p) = (W_p - W_{p,c})NV\tau_{res} = \left(\frac{W_p}{W_{p,c}} - 1\right)NVW_{p,c}\tau_{res}. \tag{7.12}$$

This result can also be understood in the sense that every atom excited into the upper laser level by the excess pump rate $W_p - W_{p,c}$ is de-excited by stimulated emission.
    Introducing the normalized pump rate

$$p := \frac{W_p}{W_{p,c}}, \tag{7.13}$$

we can cast Eq. (7.12) in the form

$$q_0 = (p-1)\frac{V}{\sigma c \tau_2}. \tag{7.14}$$

According to Eq. (7.8) the fraction $q_0 c \alpha_2$ is transmitted through the output mirror, and the output power P_{out} is thus

$$P_{out} = (p-1)\frac{\alpha_2 \hbar \omega V}{\sigma \tau_2}. \tag{7.15}$$

With $V = Ad$, $\alpha_2 = -\ln(1 - T_2)/2d$, and $\ln(1 - x) \approx -x$ this can also be written as

$$P_{\text{out}} = (p - 1)\,\alpha_2 I_s V$$
$$\approx (p - 1)\,\frac{T_2 A I_s}{2}, \qquad (7.16)$$

where $I_s = \hbar\omega/\sigma\tau_2$ is the saturation intensity [Eq. (6.91)]. The appearance of the saturation intensity in this important result can be understood in a very intuitive way: at $p = 2$, the inversion in the absence of stimulated emission would be twice the equilibrium inversion. The saturation intensity I_s is, by definition, the signal intensity that reduces the inversion by a factor of 2—i.e., in our case, to the stationary inversion. Since the atoms are exposed to two counterpropagating photon streams of approximately equal magnitude, the intensity impinging on the output mirror is $I_s/2$, and the output power is $T_2 A I_s/2$, in agreement with Eq. (7.15).

7.1.2.1 Slope Efficiency

The stationary performance of the laser, as shown in Fig. 7.4, is characterized by the laser threshold at $W_{p,c}$ and a strictly linear increase of the output power with the pump rate above the threshold. Assuming that the pump rate is proportional to the power P_p consumed by the pump source, we can set $p \approx P_p/P_{p,c}$, where $P_{p,c}$ is the pump power required to reach the threshold, and obtain from Eq. (7.15)

$$\eta_1 := \frac{\mathrm{d}P_{\text{out}}}{\mathrm{d}P_p} \approx \frac{\alpha_2 \hbar\omega V}{\sigma\tau_2 P_{p,c}}. \qquad (7.17)$$

Defining the pump efficiency

$$\eta_p := \frac{W_p N V_p \hbar\omega_p}{P_p}. \qquad (7.18)$$

as the ratio between the optical power $W_p N V_p \hbar\omega_p$ transferred to the pump volume V_p by absorbing pump photons of energy $\hbar\omega_p$, and the primary pump power P_p, we can use Eqs. (7.11) and (7.7) to obtain the critical pump power $P_{p,c} = V_p \hbar\omega_p \alpha_{\text{res}}/\eta_p \sigma\tau_2$ and to express the slope efficiency by

$$\eta_1 = \frac{\omega}{\omega_p}\,\frac{\alpha_2}{\alpha_i + \alpha_2}\,\frac{V}{V_p}\,\eta_p. \qquad (7.19)$$

The first factor is the so-called quantum efficiency, equal to the fraction of the pump photon energy that is converted to signal photon energy. The second term represents

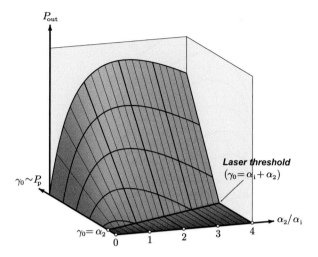

Fig. 7.5 Output power as a function of pump power for different values of α_2; both, threshold and slope efficiency increase with the output coupling coefficient. The optimum output coupling depends on the operating pump power employed

the ratio between useful and total losses, and the third term is the ratio between the inverted volume of the gain medium and the mode volume.

While it is obvious that optimum laser design requires to maximize ω/ω_p and V/V_p, and to minimize the internal losses α_i, maximizing the *total* efficiency requires to optimize the output coupling coefficient α_2 (i e , T_2): larger values of α_2 increase the slope efficiency, but also push the threshold to higher values (Fig. 7.5), so that, at a given pump power, the output power may decrease or even vanish with increasing T_2. To find the output coupling coefficient that maximizes the output power at a given pump power, we express P_{out} as a function of α_2, using $p = \gamma_0/\gamma_c$ and $\gamma_c = \alpha_{\text{res}} = \alpha_i + \alpha_2$

$$P_{\text{out}} = \left(\frac{\gamma_0}{\alpha_i + \alpha_2} - 1 \right) \alpha_2 I_s V; \tag{7.20}$$

$dP_{\text{out}}/d\alpha_2 = 0$ then yields the optimum output coupling coefficient

$$\alpha_{2\text{opt}} = \sqrt{\gamma_0 \alpha_i} - \alpha_i, \tag{7.21}$$

which increases with the operating pump power (Fig. 7.6).

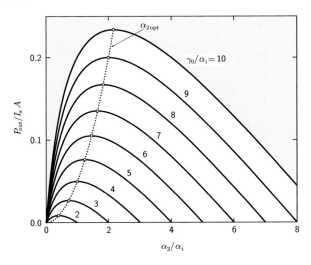

Fig. 7.6 Output power as a function of the output coupling coefficient α_2, for different values of $\gamma_0 \propto P_p$

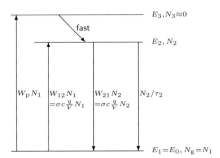

Fig. 7.7 Atomic three-level laser

7.1.3 Three-Level Laser

In a three-level laser, the lower level of the laser transition is also the ground state (Fig. 7.7). It therefore takes a very significant pump rate to establish inversion (to exceed absorption by the laser medium), and additional pump power to reach laser threshold (to compensate resonator losses). The relevant rate equations are

$$\frac{dN_2}{dt} = W_p N_1 - \sigma c(N_2 - N_1)\frac{q}{V} - \frac{N_2}{\tau_2} \qquad (7.22)$$

$$\frac{dq}{dt} = q\left[\sigma c(N_2 - N_1) - \frac{1}{\tau_{\text{res}}}\right]. \qquad (7.23)$$

With $\Delta N = N_2 - N_1$ and $N = N_1 + N_2$ we can write

$$\frac{\mathrm{d}\Delta N}{\mathrm{d}t} = W_{\mathrm{p}}(N - \Delta N) - 2\sigma c \Delta N \frac{q}{V} - \frac{N + \Delta N}{\tau_2}$$

$$\frac{\mathrm{d}q}{\mathrm{d}t} = q\left[\sigma c \Delta N - \frac{1}{\tau_{\mathrm{res}}}\right]. \tag{7.24}$$

The factor 2 in the stimulated emission term in the first equation takes into regard that every transition affects both, the stimulated emission as well as the absorption. The photon rate equation, expressed in terms of the *inversion* remains unaltered in comparison to the four-level system Eq. (7.10); the critical pump rate, however, is now given by

$$W_{\mathrm{p,c}} = \frac{1}{\tau_2} \frac{N + \Delta N_{\mathrm{c}}}{N - \Delta N_{\mathrm{c}}}. \tag{7.25}$$

The increased laser threshold is the most important difference between the two laser schemes; the ratio of the respective threshold pump rates is

$$\frac{W_{\mathrm{p,c}}^{(4)}}{W_{\mathrm{p,c}}^{(3)}} \approx \frac{\Delta N_{\mathrm{c}}}{N} \ll 1. \tag{7.26}$$

7.2   Frequency and Time Behavior of Lasers

7.2.1   Multi-Line vs. Single Line Operation

Many laser media have more than just one pair of levels that allow amplification by stimulated emission. Ionized Argon (Ar^+), e.g., provides a manifold of laser transitions between sub-levels of the 4p- and 4s-states (Fig. 7.8). These transitions allow not only laser operation at different wavelengths but also multi-line operation under proper conditions. Multi-line operation can be avoided by introducing a frequency dependent loss mechanism, for example, a dielectric output mirror with high transmittance at the wavelengths that are to be suppressed. Such mirrors increase the laser threshold for the respective wavelengths so that the oscillator does not produce this radiation. Alternative wavelength selecting devices are dispersive glass prisms (Fig. 7.9) in the cavity, or diffraction gratings under oblique angle of incidence acting as reflectors (Fig. 7.10). According to Eq. (4.38), a plane wave of

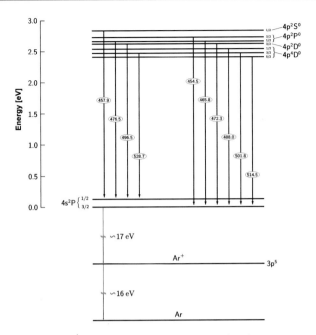

Fig. 7.8 Energy levels of Ar$^+$; the transition wavelengths are given in nm

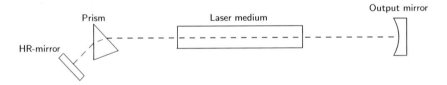

file: prismlas.pic

Fig. 7.9 Line selection using an intracavity prism

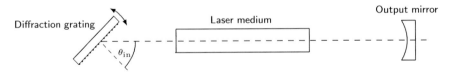

Fig. 7.10 Optical diffraction grating serving as frequency selective retroreflector in the resonator

wavelength λ_0 incident on a line grating is retroreflected ($\theta_{in} = -\theta_{out}$), if

$$2\Lambda \sin \theta_{in} = m\lambda_0, \qquad (7.27)$$

where Λ is the grating period and m is an integer. With such a setup, the laser can be tuned to a desired transition wavelength by adjusting θ_{in}.

7.2.2 Mode Selection

As we have seen in Sect. 4.3, a Fabry Perot resonator supports modes with discrete frequencies spaced by $\Delta\omega_r = c\pi/d$ [Eq. (4.82)]. On the other hand, a laser medium can provide gain over a certain bandwidth determined by the line shape of the stimulated emission cross section $\sigma(\omega)$ (see Sect. 6.1.4). In principle, any mode that experiences sufficient gain to compensate its specific losses can oscillate in such a resonator. There is, however, the very interesting mechanism of mode competition that may reduce the number of oscillating modes in a self-organized way. Right after the start of the pump process, for example, several modes may experience net gain and start to oscillate with increasing photon numbers (Fig. 7.11). As described above, the photon number increases at the expense of the inversion density (gain saturation). In the case of a homogeneously broadened gain medium, the saturation affects the entire gain profile to the same extent. Due to saturation, modes that are situated off the peak of the gain profile still experience gain, but not sufficient to compensate their losses—and perish. The modes that lie close to the gain peak keep growing—at the expense of other modes, until the saturated gain exactly balances the losses of a single, optimally situated mode that survives, and the laser oscillates practically monochromatically.

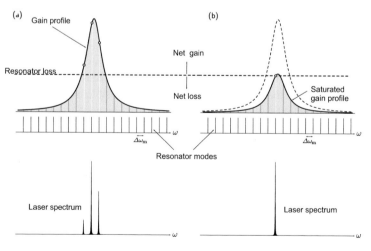

Fig. 7.11 Mode competition in a homogeneously broadened gain medium: (**a**) transient behavior after turn-on, (**b**) stationary situation

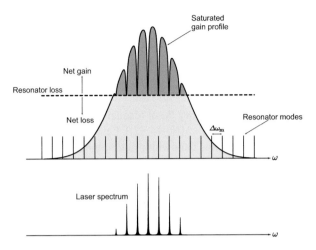

Fig. 7.12 Mode competition in an inhomogeneously broadened laser

The situation is quite different for inhomogeneously broadened gain media: different modes interact—and saturate—only the sub-ensemble of the gain medium that coincides with the mode frequency (Fig. 6.9). Each mode "burns" a spectral hole into the gain profile by saturation, and adjacent modes do not compete for the same gain, provided the homogeneous line width of the sub-ensemble is smaller than the mode spacing (Fig. 7.12).

The statement about the self-organized single mode operation of homogeneously broadened lasers requires some qualification: as we have seen in Sect. 4.3, a given resonator mode forms a standing wave (Fig. 4.15) with a photon density $\propto \cos^2[z(\omega_m/c)]$. In the nodes of this distribution, little or no inversion is consumed, and saturation occurs only in the volume between the nodes (spatial hole burning). Since each mode m has a different axial pattern, it can exploit the gain left over by other modes, making multimode operation possible despite spectral mode competition. Spatial hole burning can be avoided by using ring-shaped (triangular) resonators that include a Faraday isolator (Sect. 2.4.2.1), so that only one mode can propagate and the counterpropagating mode is suppressed.

Independent of the nature of gain broadening, single frequency operation can be obtained by introducing a Fabry–Perot interferometer (etalon, Sect. 4.2.3) into the resonator that transmits the mode frequency of interest and has a free spectral range that is larger than the gain bandwidth (Fig. 7.13).

Spherical mirror resonators support not only different axial, but also *transverse* modes (Sect. 4.3.1). Many applications require transverse single mode operation, i.e., a Gaussian (TEM$_{00}$) mode. Since the transverse mode diameter grows with the mode order, higher order modes can be suppressed by inserting circular apertures into the resonator, with a diameter that matches the TEM$_{00}$ mode (spatial filtering). Like all other mode selection processes mentioned, the principle of operation is not

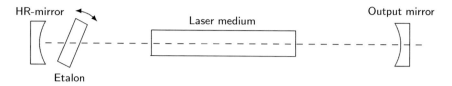

Fig. 7.13 Single mode operation by implementation of an etalon into the resonator; the oscillating mode can be selected by tilting the etalon [see Eq. (4.73)]

to attenuate an oscillating mode, but to prevent it from oscillation by increasing the specific loss above the available gain.

7.2.3   Laser Line Width

Because of the coherent nature of the stimulated emission process, one might assume laser radiation to be completely monochromatic under stationary (single mode) conditions [see the discussion following Eq. (4.80)]; perfect coherence, however, is disturbed by photons emitted spontaneously into the laser mode. The resulting photon number deviation is quickly attenuated by a process described in Sect. 7.2.4, broadening the laser line only very slightly. By contrast, there is no mechanism pinning the *phase* of the laser field to certain value, so that the stochastic phase of the spontaneous photons results in a random walk of the phase of the laser field, reducing the coherence time and imposing a fundamental lower limit on the bandwidth of the laser spectrum. This so-called Schawlow–Townes limit (see, e.g., Yariv and Yeh 2006)

$$\Delta\omega_{min} = \frac{\hbar\omega(\Delta\omega_{res})^2}{P_{out}} \tag{7.28}$$

depends on the bandwidth of the resonator modes $\Delta\omega_{res}$, [Eq. (4.84)] and is inversely proportional to the output power. The inverse dependence of the Schawlow–Townes limit on the laser power reflects the fact that the spontaneous emission rate of a given laser is constant (because N_2 is constant), while the total number of photons is proportional to the output power. For a typical Helium–Neon Laser, $P_{out} = 1\,\text{mW}$ and $\Delta\omega_r = 10^8\,\text{s}^{-1}$, the Schawlow–Townes limit amounts to about 1 Hz. The actual bandwidth of lasers is usually larger by several orders of magnitude and results from technical noise contributions.

7.2.4   Relaxation Oscillations and Gain Modulation

We now want to study the transient behavior of a laser in some detail, i.e., the temporal variation of the photon number following a deviation from the stationary

Fig. 7.14 Laser relaxation
oscillations following startup

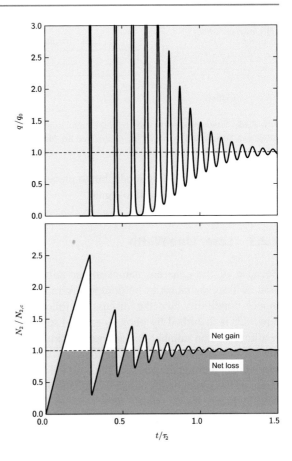

equilibrium. Using Eq. (7.10), (7.2) can be cast in the form

$$\frac{dq}{dt} = \frac{q}{\tau_{\text{res}}}\left(\frac{N_2}{N_{2,\text{c}}} - 1\right).$$

(7.29)

A positive excursion of the inversion from the equilibrium value $N_{2,\text{c}}$, for example,
entails a growth $dq/dt > 0$ of the photon number, which in turn reduces the
inversion by increased stimulated emission, so that $dN_2/dt < 0$. The inversion
actually drops below the critical value $N_{2,\text{c}}$, which implies $dq/dt < 0$ and
consequently $dN_2/dt > 0$. This regulatory cycle is repeated, resulting in (damped)
relaxation oscillations until the equilibrium is established (Figs. 7.14 and 7.15).

Fig. 7.15 Relaxation
oscillations following startup,
shown in the $q - N_2$ phase
space; *arrows* indicate the
temporal evolution towards
the stationary state

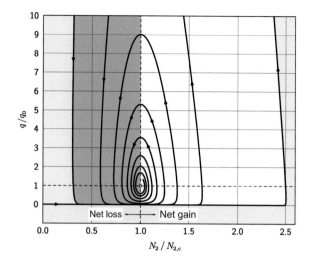

For a simplified analysis, we start with rate equations (7.1) and (7.2), neglecting
pump depletion ($N_g = N$) and assuming a constant pump rate $W_{p,0}$; $N'(t)$ and $q'(t)$
are the deviations (assumed to be small) from the equilibrium values

$$N_2(t) = N_{2,c} + N'(t) \tag{7.30}$$

$$q(t) = q_0 + q'(t). \tag{7.31}$$

Later, we will also assume a time dependent pump rate

$$W_p = W_{p,0} + W_p'(t). \tag{7.32}$$

Substitution of Eq. (7.31) into Eqs. (7.1) and (7.2) yields

$$\frac{dN'}{dt} = W_p' N - \frac{\sigma c}{V} \left(q' N_{2,c} + N' q_0 \right) - \frac{N'}{\tau_2}$$

$$\frac{dq'}{dt} = \sigma c q_0 N', \tag{7.33}$$

where quadratic terms such as $N'q'$ have been neglected. From the second equation
follows

$$N' = \frac{dq'}{dt} \frac{1}{\sigma c q_0}, \tag{7.34}$$

which allows us, using Eqs. (7.10) and (7.14), to write the first equation as

$$\frac{d^2q'}{dt^2} + \frac{p}{\tau_2}\frac{dq'}{dt} + \frac{p-1}{\tau_{res}\tau_2}q' = W_p'(t)NV\frac{p-1}{\tau_2}. \tag{7.35}$$

For $W_p' = 0$, this is the equation of a damped linear oscillator, which can be solved using the ansatz $q' \propto e^{st}$, where s is a complex number; substitution in Eq. (7.35) yields $s = -(1/\tau_{rel}) \pm j\omega_{rel}$ with the relaxation frequency ω_{rel} and the decay time τ_{rel}

$$\omega_{rel} = \sqrt{\frac{p-1}{\tau_{res}\tau_2} - \left(\frac{p}{2\tau_2}\right)^2} =: \sqrt{\omega_0^2 - \left(\frac{1}{\tau_{rel}}\right)^2} \tag{7.36}$$

$$\omega_0^2 = \frac{p-1}{\tau_{res}\tau_2}, \tag{7.37}$$

$$\tau_{rel} = \frac{2\tau_2}{p}. \tag{7.38}$$

Both, the relaxation oscillations and the damping get faster with increasing pump rate p. Of the two characteristic time constants τ_2 and τ_{res} of the system, the upper state life time τ_2 determines the decay time, while the geometric mean value $\sqrt{\tau_2\tau_{res}}$ determines the relaxation oscillation period.

7.2.4.1 Gain Modulation
An important way to modulate the output power of a laser is to control the gain via the pump rate. To get some insight into this process, we assume a harmonic modulation $W_p'(t) = W_p'\cos\omega t$ of the pump, superimposed on a constant background $W_{p,0}$. The photon number will then also oscillate around the stationary value q_0 with the modulation frequency ω and the amplitude $q'(\omega)$. We can solve Eq. (7.35) in the same way as Eq. (2.51) and obtain

$$\left|\frac{q'}{q_0'}\right| = \frac{\omega_0^2}{\left|\omega_0^2 - \omega^2 + 2j\omega/\tau_{rel}\right|} = \frac{\omega_0^2}{\left|(\omega_{rel} - \omega + j/\tau_{rel})(\omega_{rel} + \omega - j/\tau_{rel})\right|}, \tag{7.39}$$

where

$$q_0' = N\tau_{res}VW_p' \tag{7.40}$$

Fig. 7.16 Simulated modulation response of a gain modulated laser for different pump values p; the parameters used are typical for semiconductor lasers ($\tau_{res} = 1$ ps, $\tau_2 = 1$ ns)

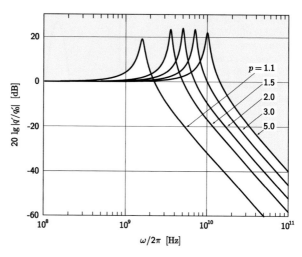

is the modulation amplitude at very low modulation frequency, which follows from Eq. (7.35) for $d/dt \rightarrow 0$. Figure 7.16 shows the frequency response $|q'/q'_0|$: there is a significant resonance enhancement at ω_{rel}; above the resonance, the response rolls off quickly (40 dB/decade). Well below the resonance, the response is independent on the modulation frequency. For a given laser, the resonance frequency increases with the pump ratio p, but more importantly it is determined by $\sqrt{\tau_2 \tau_{res}}$. Semiconductor lasers, with their very short resonator length and high coupling losses [see Eq. (7.7)] have resonator life times of picoseconds and allows for gain modulation up to several 10 GHz.

7.3   Pulsed Lasers

Lasers can also be operated in a pulsed mode; the pulse duration is limited by the gain bandwidth of the laser medium and can be as short as one oscillation cycle of the radiation in principle. At a given average optical power P_{avg}, a laser emitting pulses of duration τ_p at a repetition rate of R_{rep} produces a peak power of $P_{avg}/(R_{rep}\tau_p)$; the optical peak power of a table-top laser with a few Watt average power, a pulse duration of some 10 fs and a repetition rate of $100\,\mathrm{s}^{-1}$, for example, is in the 10^{12}-W range. The high peak output power is one of the reasons for the development of pulsed lasers, another being the very high temporal resolution, if the laser light is used to take snapshots of processes.

A direct way to produce pulses is gain switching, i.e., the rapid switching of the pump source. Semiconductor lasers are well suited for this, since they are pumped directly by electric current. Figure 7.17 shows the output of such a laser.

The full potential of lasers to produce pulses, however, is exploited by two techniques: Q-switching and mode locking. Q-switching takes advantage of the energy storage capability of a laser medium and is used to generate nanosecond

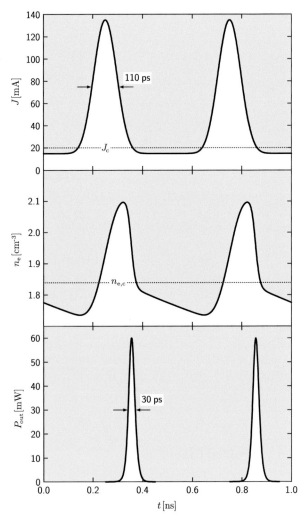

Fig. 7.17 Temporal evolution of inversion and photon number of a gain switched semiconductor laser. The pump rate is proportional to the operating current J; n_e and $n_{e,c}$ are the (critical) carrier densities

pulses at low repetition rate but with high energy, while mode locking refers to the production of trains of pulses that are potentially as short as allowed by the gain bandwidth, separated by the resonator round trip time; this requires control over the phase of the frequency components that build up the pulse. In the following, we will briefly discuss both techniques.

7.3.1   Q-Switching

The idea of Q-switching is to pump a laser, usually with a pulsed pump source of moderate power, to a level $N_{2,i}$ of inversion that is several times higher than

Fig. 7.18 Q-switching: evolution of resonator losses, inversion, and photon number

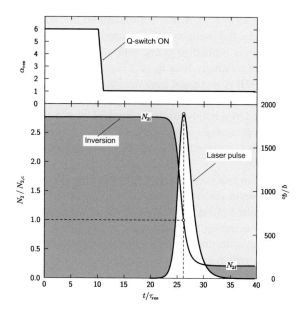

the critical inversion $N_{2,c}$; during the pump process, laser emission is inhibited by introducing an additional internal loss into the cavity. Once the target inversion is reached, the additional resonator loss is removed, leaving the resonator with a gain that exceeds the losses by far. Because of the excess gain, the photon number increases avalanche-like within a few resonator round trips and most of the energy stored in the gain medium is converted into laser light (Fig. 7.18). When the inversion is consumed to below $N_{2,c}$, the photon number drops to zero within few round trips. The left over, residual inversion is not exploited and decays by spontaneous emission.

The process of loss- or Q-switching can be induced by external control (active Q-switching), usually with an electro-optic switch (Pockels cell, Sect. 2.3.4), or by introducing a saturable absorber (Sect. 6.1.5) into the cavity; in this passive Q-switching scheme, the laser starts to oscillate when the gain is high enough to compensate the total losses, and bleaches the saturable absorber by absorption saturation (Degnan 1995). Passive Q-switching is less efficient than active Q-switching, but its implementation is extremely simple and reliable.

In a Q-switched laser, energy is accumulated in the gain medium over a relatively long time and released in form of a pulse that is many orders of magnitude shorter than the pumping time. Consequently, only gain materials with long upper state life time are suitable as Q-switched lasers. Assuming a step like onset of the pump process, the inversion, according to Eq. (7.1), is building up as

$$N_2(W_p, t) = NW_p\tau_2(1 - e^{-t/\tau_2}), \tag{7.41}$$

as long as $q = 0$. While the maximum inversion obtainable at a given pump rate is $NW_p\tau_2$, the pumping gets less and less efficient with time: in practice, the pump duration is limited to $\approx 2\tau_2$, when $\approx 86\%$ of the maximum inversion is reached. With its relatively long τ_2 of $230\,\mu$s, neodymium doped yttrium-aluminum-garnet (Nd:YAG) is a popular gain medium for such lasers. In a typical pump volume V_g of a few cm^3 and a Nd-concentration of some 10^{-4} (i.e., 1 out of 10^4 yttrium ions is replaced by a neodymium ion, equivalent to a Nd ion density of 1.38×10^{18} cm^{-3}), an energy storage of $V_g N_{2,i}\hbar\omega$ of 1 J can be realized; assuming a pulse duration of some ns (several round trips), this corresponds to an optical peak power of several 100 mW.

For an analysis of (active) Q-switching (Wagner and Lengyel 1963), we assume a four-level system that has been pumped to the initial value $N_{2,i}$; the duration of the emitted pulse is assumed to be so short that pumping as well as spontaneous emission during the pulse can be neglected; we also assume a perfect overlap between gain and mode volume. Using Eq. (7.10), rate equations (7.1) and (7.2) can be written as

$$\frac{dq}{dt} = q\sigma c N_{2,c} \left(N_2/N_{2,c} - 1 \right) \tag{7.42}$$

$$\frac{dN_2}{dt} = -\sigma c N_2 \frac{q}{V_g}. \tag{7.43}$$

We eliminate t by dividing the first equation by the second,

$$dq = -V_g \left(1 - N_{2,c}/N_2 \right) dN_2 \tag{7.44}$$

and integrate from the initial inversion $N_{2,i}$ to some arbitrary value N_2

$$q(N_2) = V_g \left[N_{2,i} - N_2 - N_{2,c} \ln(N_{2,i}/N_2) \right]. \tag{7.45}$$

The residual inversion $N_{2,f}$ is defined by $q(N_{2,f}) = 0$ and given by the (transcendent) equation

$$N_{2,i} - N_{2,f} = N_{2,c} \ln(N_{2,i}/N_{2,f}). \tag{7.46}$$

The optical extraction efficiency is the ratio of the exploited to initial inversion,

$$\eta_q = \frac{N_{2,i} - N_{2,f}}{N_{2,i}} = \frac{\ln(N_{2,i}/N_{2,f})}{(N_{2,i}/N_{2,c})}; \tag{7.47}$$

Figure 7.19 shows this parameter as a function of $N_{2,i}/N_{2,c}$; for $N_{2,i}/N_{2,c} > 3$, the efficiency exceeds 90%.

Fig. 7.19 Extraction efficiency η_q and pulse duration τ_p of a Q-switched laser as a function of the excess inversion $N_{2,i}/N_{2,c}$

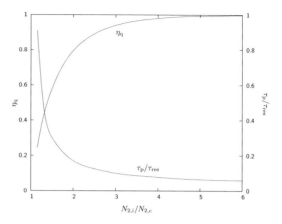

Since q is growing as long as $N_2 > N_{2,c}$, Eq. (7.45) also allows us to estimate the peak photon number q_{max}

$$q_{max} = q(N_{2,c}) = V_g N_{2,c} \left[\left(\frac{N_{2,i}}{N_{2,c}} \right) - 1 - \ln \left(\frac{N_{2,i}}{N_{2,c}} \right) \right]. \qquad (7.48)$$

For $N_{2,i}/N_{2,c} \gg 1$ we obtain

$$q_{max} = V_g N_{2,i}; \qquad (7.49)$$

in this regime, practically the entire inversion is converted to photons, and the process happens so quickly that most of the photons have not yet left the resonator when the inversion is consumed (Fig. 7.19).

In a typical Q-switched laser, the coupling losses dominate the total losses, so that

$$P_{out,max} \approx \hbar\omega q_{max}/\tau_{res}. \qquad (7.50)$$

The energy balance $\tau_p P_{out,max} \approx V_g(N_{2,i} - N_{2,f})\hbar\omega$ allows an estimate of the pulse duration τ_p

$$\tau_p = \frac{\eta_q \left(\frac{N_{2,i}}{N_{2,c}} \right)}{\left[\left(\frac{N_{2,i}}{N_{2,c}} \right) - 1 - \ln \left(\frac{N_{2,i}}{N_{2,c}} \right) \right]} \tau_{res}. \qquad (7.51)$$

The dependence of τ_p on the excess inversion $N_{2,i}/N_{2,c}$ is also shown in Fig. 7.19: for large values of $N_{2,i}N_{2,c}$, the residual inversion is very small and the pulse decays after reaching its peak with the decay time τ_{res}; the leading slope of the pulse grows with a rate that is $(N_{2,i}/N_{2,c} - 1)$ times faster [see Eq. (7.42)]. For small excess inversion, $N_{2,i}/N_{2,c} < 2$, the growth is slower, and the decay also takes longer because of the relatively large residual gain.

Q-switching allows producing "giant" pulses of nanosecond duration. The repetition rate of these pulses is limited by the long-lasting pumping process to typically less than several thousand pulse per second. Thermal limitations may further reduce the repetition rate drastically. The pulse duration is dominated by the resonator life time and usually much longer than the gain bandwidth would allow.

7.3.2 Mode Locking

According to Eq. (3.148), the relative gain bandwidth $\Delta\omega/\omega_0$ required to support a pulse duration of τ_p is essentially given by the ratio T/τ_p, where T is the oscillation period of the laser radiation. To obtain the shortest possible pulses from a given gain medium, as many resonator modes as possible must oscillate. According to Eq. (4.82), the frequency spacing between adjacent modes is $\Delta\omega_r = \pi c/d$; assuming that the gain medium supports a total number of $2N + 1$ modes (implying a gain bandwidth of $\Delta\omega_g = (2N + 1)\Delta\omega_r$), the electric field of the superimposed modes is

$$E(t) = \sum_{n=-N}^{N} E_{0,n} e^{j[(\omega_0 + n\Delta\omega_r)t + \phi_n]}, \qquad (7.52)$$

where $E_{0,n} e^{j\phi_n}$ is the complex amplitude of the nth mode and ω_0 denotes the frequency of the central mode (the phase ϕ_n is somewhat arbitrary because it depends on the choice of the time zero point—the transformation $t' := t - \tau$ changes the phase to $\phi'_n = \phi_n + (\omega_0 + n\Delta\omega_r)\tau$).

As the phase *difference* between the modes varies with time, the fields of the individual modes interfere with each other in a time varying fashion and $E(t)$ fluctuates in a quasi-random, yet periodic pattern (note that $E(t + t_{rep}) = E(t)$, where $t_{rep} = 2\pi/\Delta\omega_r = 2d/c$). Figure 7.20a shows the output power of such a superposition of modes. If (and only if) there exists a point in time where all modes are in phase (which is unlikely if the modes oscillate independently), the total field reaches the maximum possible value of $\sum_{n=-N}^{N} E_{0,n}$ (Fig. 7.20b); any process that establishes such a phase correlation is called "mode locking." In the following discussion, we choose the time axis such that the peak appears at $t = 0$ (and at any integer multiple of t_{rep}).

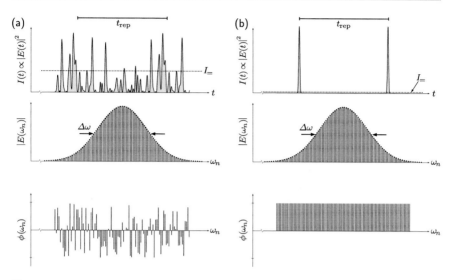

Fig. 7.20 Amplitude and phase distribution as well as output power of a typical multimode laser: (a) uncorrelated mode phases, (b) mode locked operation; note the different power scales

If we assume, for the sake of simplicity, equal amplitudes $E_{0,n} = E_0$, the peak field is $E(0) = (2N+1)E_0$. Under mode locked conditions, the phases of all modes at $t = 0$ are equal, $\phi_n = \phi_0$. The total field then is

$$E(t) = E_0 e^{j(\omega_0 t + \phi_0)} \sum_{n=-N}^{N} e^{jn\Delta\omega_r t}. \tag{7.53}$$

The first exponential factor is the carrier oscillation, the second factor constitutes the periodic envelope

$$\sum_{n=-N}^{N} e^{jn\Delta\omega_r t} = \frac{1 - e^{j(2N+1)\Delta\omega_r t}}{1 - e^{j\Delta\omega_r t}} e^{-jN\Delta\omega_r t} = \frac{\sin[(2N+1)\Delta\omega_r t/2]}{\sin(\Delta\omega_r t/2)}; \tag{7.54}$$

the resulting intensity is shown in Fig. 7.21.

Note that the field *energy* contained in one fluctuation period is—independent of the phase distribution—given by the sum of the individual mode energies $\propto (2N+1)E_0^2 t_{\text{rep}}$. Consequently, a mode locked pulse with the peak power $\propto [(2N+1)E_0]^2$ can only last for a duration of $t_{\text{rep}}/(2N+1) = 2\pi/\Delta\omega_g$, so that we can estimate the pulse duration as

$$\tau_p \approx \frac{t_{\text{rep}}}{2N+1}; \tag{7.55}$$

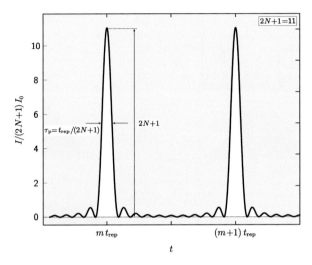

Fig. 7.21 Superposition of $2N + 1$ phase locked modes of equal amplitude; I_0 is the intensity of one of the modes, $(2N + 1)I_0$ is the averaged total intensity of all modes

for the remainder of the fluctuation cycle, the field must more or less vanish for reasons of energy conservation. This estimate is confirmed by Eq. (7.54), as the time between the two minima next to the peak is equal to $2t_{\text{rep}}/(2N + 1)$.

7.3.2.1 Active Mode Locking

Mode locking can be achieved extrinsically (active mode locking) or intrinsically by self-organization (passive mode locking). Active mode locking relies on the modulation of resonator gain or loss with a frequency Ω_m equal to the mode spacing. As a result of the net gain modulation, the amplitudes of the modes are also modulated; for the field of the nth mode we can write

$$(E_0 + \Delta E_0 \cos \Omega_m t)e^{j\omega_n t} = E_0 e^{j\omega_n t} + \Delta E_0 e^{j(\omega_n + \Omega_m)t} + \Delta E_0 e^{j(\omega_n - \Omega_m)t} \qquad (7.56)$$

where E_0 is the mode amplitude without modulation and ΔE_0 is a measure of the modulation depth; as one can see, amplitude modulation is equivalent to the formation of phase locked side bands at $\omega_n \pm \Omega_m$. If the modulation frequency is tuned to the mode spacing $\Omega_m = \Delta\omega_r$, each mode produces a "cross talk" into the two adjacent resonator modes. Starting from the central mode, the side bands "seed" the oscillation of the adjacent modes and finally all $2N + 1$ oscillate such that they are all in phase at $t = 0$ and any further multiple of t_{rep}.

Technically, loss modulation can be implemented using an acousto-optic modulator (Sect. 8.5) consisting of a transparent medium and an ultrasonic transducer that produces a standing acoustic wave between two parallel surfaces of the medium. The acoustic wave produces a spatial modulation of the refractive index, forming a diffraction grating. When positioned inside the laser resonator, the laser modes experience losses due to diffraction. As the standing wave actually vanishes twice per acoustic cycle [Eq. (8.237)], the diffraction losses are modulated at twice the acoustic frequency.

Semiconductor lasers can be directly gain modulated through the supply current. To increase the number of modes within the gain bandwidth (which, due to the short cavity length is very small), external resonators with appropriate length are often used.

7.3.2.2 Passive Mode Locking

Most mode locked lasers are modulated intrinsically, exploiting saturation or other intensity dependent effects (Haus et al. 2000; Ippen et al. 1972; Spence et al. 1991). As we have seen above [Eq. (7.53)], multimode operation inevitably produces periodic fluctuations of the laser power. In the time domain, mode locking is established if the fluctuation consists of essentially one dominating peak per round trip, with small or ideally no fluctuations in between. This state can be reached in a self-organized way if the round trip net gain is intensity dependent in such a way that it increases with intensity and is less than 1 for low intensities. Starting from random, periodic fluctuations, the fluctuation peak is amplified preferentially, and after a number of round trips develops into an isolated pulse that contains most of the available energy, while the small fluctuations are attenuated and finally extinguished. The "surviving" pulse tends to be further shortened by the same process, since the low intensity wings of the pulse are also suppressed.

Such an intensity dependent net gain can be realized by inserting a saturable absorber into the resonator. According to Eq. (6.82), low intensities experience higher losses, while intensity peaks bleach the absorber and reduce the loss. Consequently, peak fluctuations experience higher net gain at the expense of low intensity fluctuations.

Saturable absorbers need a certain time to recover from saturation, because the excited atoms need time (upper level life time τ_2) to return to the ground state. If this time is longer than the pulse duration (slow saturable absorber), only the leading edge of the pulse is shaped by the absorber (Fig. 6.13). The trailing edge of a pulse can be shaped by the saturation of the *gain*, however (Fig. 6.12): a peak fluctuation consumes a large fraction of the inversion, leaving only little gain for the trailing part of the pulse. The interplay of (slow) saturable absorber and (slow) saturable gain is capable of generating bandwidth limited laser pulses, provided that the recovery time is shorter than the pulse repetition time τ_{rep} (Fig. 7.22).

In Sect. 6.2.3, we have discussed saturation effects of short pulses. The critical parameter of an absorber or amplifier is the saturation energy fluence Eq. (6.102). For a mode locked laser to work properly, the pulse energy fluence must approximately match the saturation fluence, which can be achieved by designing the waist diameter of the laser mode by proper choice of the curvature of the resonator mirrors.

A very efficient, and virtually instantaneous intensity dependent loss mechanism relies on the nonlinear optical Kerr lens whose refractive power is proportional to the momentary incident power [Eq. (3.53)]. The combination of such a lens with

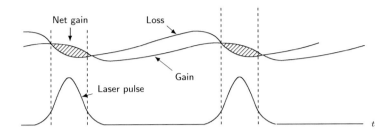

Fig. 7.22 Passive mode locking by saturation of gain and loss

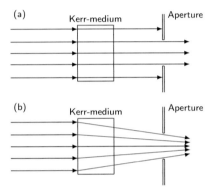

Fig. 7.23 Kerr lens mode locking: (**a**) at low power, self-focusing is negligible, resulting in high losses at the aperture; (**b**) Kerr lensing induced by high power reduces the losses at the aperture

a circular aperture (Fig. 7.23) introduces low losses at high power, because the focussed mode is transmitted through the aperture, while at low power, most of the (unfocussed) mode is clipped. This process shapes both, the leading and the trailing edge of the pulse. Pulses as short as few femtoseconds can be produced in this way.

7.3.3 Carrier Envelope Phase, CEP

To achieve the shortest and highest pulse possible, all modes must have the same phase at a certain instance of time; since the intensity is proportional to EE^*, the actual value of this common phase has no influence on the pulse shape, as an inspection of Eq. (7.53) shows. As can be seen from this equation, however, the common phase ϕ_0 is equivalent to a time shift between the pulse envelope and the carrier oscillation and is therefore called carrier envelope phase (CPE). If $\phi_0 = 0$, for example, the peak of the envelope coincides with a crest of the carrier oscillation (the physical field is the real part of its complex representation); $\phi_0 = \pi/2$ implies that the field actually vanishes at the peak of the envelope. Figure 7.24 shows two

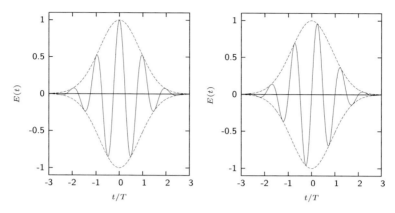

Fig. 7.24 Two ultrashort pulses with different carrier envelope phase: cosine-pulse (*left*), sine-pulse (*right*)

pulses with identical Gaussian envelope, but different CEP. For obvious reasons, pulses with $\phi_0 = 0$ or $\pi/2$ are called cosine- or sine-pulses, respectively. During propagation in a dispersive medium, the CEP changes continuously because the velocity of the envelope (the group velocity) usually differs from the phase velocity of the carrier.

The carrier envelope phase plays an important role in the nonlinear optics of few-cycle pulses, such as tunnel ionization or harmonic generation, because these effects depend on the electric *field*, in contrast to saturation effects or the Kerr effect that depend on the intensity (envelope).

Because of their insensitivity to the CEP, the mode locking mechanisms described above cannot control or stabilize the carrier envelope phase. Pulses from such mode locked lasers therefore have an unknown, and usually time varying CEP. CEP-stabilization is possible employing field dependent nonlinear optical processes, however.

7.4   Atomic and Molecular Lasers

Because of the many requirements on a practical gain medium, the variety of commercially important laser materials is rather limited (see Table 7.1). Table 7.2 summarizes performance data of some of the most popular lasers; the output parameters (average power, peak power, pulsed operation, etc.) are essentially determined by the gain medium. In particular, the resonant nature of the stimulated emission process implies that lasers are usually not tunable over a wide range of frequencies. Among the exceptions are the Ti:sapphire laser (tunable between 700 and 1100 nm), organic molecular lasers (so-called dye lasers), and, to a certain degree, semiconductor lasers.

Table 7.1 Properties of popular laser materials: $\Delta\omega_g$ and $\Delta\lambda_g$ denote the gain bandwidth; values of γ_0 refer to typical operating conditions; Rh6G is the organic dye rhodamine 6G in a methanol solution; data adapted from Bass (2010)

Medium	λ [nm]	σ [cm^2]	τ_2 [s]	$\Delta\omega_g$ [s^{-1}]	$\Delta\lambda_g$ [nm]	γ_0 [cm^{-1}]
Ruby (Cr:sapphire)	694	2×10^{-20}	3×10^{-3}	4×10^{11}	0.1	0.1
Nd:YAG	1064	7×10^{-19}	2×10^{-4}	7×10^{11}	0.4	0.1
Nd:glass	1050–1080	5×10^{-20}	3×10^{-4}	2×10^{13}	12	0.03
Er:glass	1550	6×10^{-21}	1×10^{-2}	3×10^{13}	30	0.03
Ti:sapphire	700–1100	4×10^{-19}	3×10^{-6}	9×10^{14}	400	0.2
HeNe	632.8	3×10^{-13}	3×10^{-7}	1×10^{10}	2×10^{-3}	0.002
Ar$^+$	488	5×10^{-12}	1×10^{-8}	1×10^{10}	1×10^{-3}	0.005
HeCd	441.6	8×10^{-14}	7×10^{-7}	1×10^{10}	1×10^{-3}	0.003
Cu-vapor	510.5	8×10^{-14}	5×10^{-7}	1×10^{10}	1×10^{-3}	0.05
CO$_2$	10600	2×10^{-16}		4×10^{8}		0.008
Excimer (ArF)	193	3×10^{-16}	9×10^{-9}	6×10^{13}	1.5	0.03
Rh6G	550–610	1×10^{-16}	5×10^{-9}	3×10^{14}	50	2.8
AlGaAs	720–850	1×10^{-15}	1×10^{-9}	6×10^{13}	20	10^{3}
InGaAsP	1000–1650	1×10^{-15}	1×10^{-9}		20	10^{3}
InGaN	380–515	1×10^{-15}	1×10^{-9}		10	10^{3}

Table 7.2 Important types of lasers: cw/continuous wave, p/pulsed, P_{avg} average power, P_p peak power, τ_p pulse duration, η_1 overall efficiency, FL/flashlamp, GD/gas discharge, LINAC/linear accelerator; the free electron laser (FEL), is included in the Table for comparison, although it is not a laser in the proper sense

Laser	λ [nm]		P_{avg} [W]	P_p [W]	τ_p [ns]	η_1 (%)	Pump
Ruby	694	p	1	10^{7}	10		FL
Nd:YAG	1064	cw	10–200			0.5	FL
Nd:YAG		p	10	10^{7}	10–1000	1–3	FL
Ti:sapphire	700–1100	cw	1–10				Laser
HeNe	632.8	cw	10^{-3}			0.05	GD
Ar$^+$	488	cw	10–100			0.05	GD
HeCd	441.6	cw	10^{-1}			0.1	GD
Cu-vapor	510.5	p	40	10^{5}	20	1–2	GD
CO$_2$	10600	cw	10^{4}			10–20	GD
Excimer	198	p	500	10^{7}	10	1	GD
Rh6G	550–610	cw	10			0.05	Laser
Rh6G		p	1	10^{6}	10–1000		Laser
AlGaAs	720–850	cw/p	0.001–1			40	DC
InGaAsP	1000–1650	cw/p	0.001–0.1			40	DC
InGaN	350, 405, 470, 515	cw/p	0.001–1			40 DC	
FEL	1–10^{6}	p		10^{9}	10^{3}–10^{-3}		LINAC

Fig. 7.25 Emission spectra
of Nd:YAG and Nd:Glass

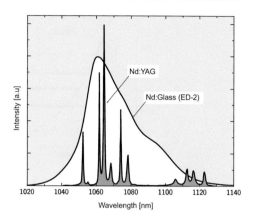

Nd:YAG

Nd:Glass (ED-2)

7.4.1   Atomic Solid State Lasers

Isolated atoms and ions constitute an important class of laser materials. Atomic
gases (e.g., noble gases or metal vapors) are pumped electrically by gas discharge;
alternatively, atoms are built into transparent solid host materials such as glass or
various crystals, and pumped optically. These laser are called solid state lasers, in
distinction from semiconductor lasers.

The historically important ruby laser, for example, is a solid state laser that relies
on Cr^{3+}-ions in a sapphire (Al_2O_3) host. One of the most popular solid state laser
materials is neodymium doped yttrium-aluminum-garnet ($Y_3Al_5O_{12}$), abbreviated
as Nd:YAG; YAG is also an excellent host for other rare earth ions (ytterbium,
erbium, and holmium) because of its outstanding thermal and mechanical properties.
As already mentioned, Nd:YAG is an important gain medium for Q-switched lasers
due to its long upper state life time. The gain bandwidth of Nd:YAG is relatively
small (Fig. 7.25) and does not support mode locked generation of pulses shorter
than 100 ps. Nd:glass, however, with its significant inhomogeneous line broadening,
allows the generation of pulse durations below 400 fs. Nd:glass can be produced in
large slabs and is therefore used for very high energy lasers, such as used in nuclear
fusion experiments. The glass matrix also allows the production of high quality
optical fibers, providing the gain medium for fiber lasers.

Another rare earth-based gain material of outstanding importance is erbium
doped glass which exhibits broadband gain at around 1.55 μm wavelength, where
silica fibers exhibit minimum losses. The relevant laser transition occurs between
the $^4I_{13/2}$ and the $^4I_{15/2}$ state of the Er^{3+} ion (Fig. 7.26). Erbium doped fiber
amplifiers (EDFAs, Sect. 5.3.5) constitute the backbone of long distance fiber optical
networks (Desurvire 2001). Pumping is achieved with semiconductor lasers at 980
and 1480 nm. Due to the interaction of the ions with the local field of the glass host,

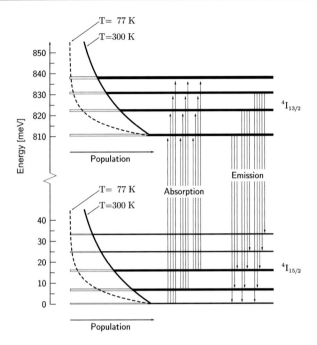

Fig. 7.26 Laser transitions of Er:glass at 1550 nm between sub-levels of the excited $^4I_{13/2}$ and the $^4I_{15/2}$ ground state; the occupation of the levels follows Boltzmann statistics (shown for two different temperatures)

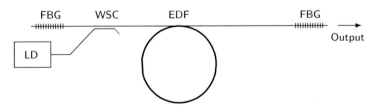

Fig. 7.27 Fully integrated fiber laser. The gain medium is a rare earth doped glass fiber (EDF), the mirrors are fiber Bragg gratings (FBGs) as described in Sect. 5.3.3; the pump light is supplied by a laser diode (LD) with a fiber-output ("pig tail"), coupled to the resonator by a dichroic waveguide coupler (WSC)

the gain bandwidth is substantial (>50 nm). At room temperature, all sub-levels of the $^4I_{15/2}$ ground state are populated so that the EDFA is effectively a three-level system requiring a significant pump power to exhibit gain (Fig. 5.35).

Rare earth doped fiber amplifiers are also used to build fiber lasers (Fig. 7.27). Fiber lasers are very compact and stable, have a very high efficiency and can deliver

Fig. 7.28 Energy levels of a vibronic solid state laser medium (titanium:sapphire); absorption and emission bands consist of a large number of vibrational sub-levels; compare Fig. 7.29

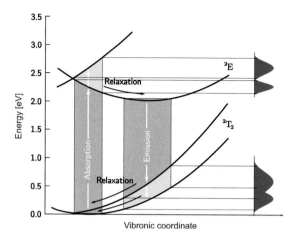

TEM$_{00}$ output beams. Because of their large gain bandwidth, they allow mode locked operation and can serve as tunable sources. Equipped with narrow band fiber gratings as mirrors, they can operate in single longitudinal mode, with a linewidth of 1 kHz or less. In comparison to bulk lasers, fiber lasers can be cooled very efficiently and do not suffer from various thermo-optical problems so that they are also well suited for high power continuous wave (cw) operation, delivering output powers of several 100 W. One of the limitations of fiber lasers results from the small mode cross section leading to very high intensities that can damage the output facet of the fiber. Q-switched operation of fiber lasers is possible, but self-focusing and optical damage limit the pulse energy to mJ.

Transition metal ions such as Cr^{3+} and Ti^{3+} in crystalline hosts (Al_2O_3) show broad bands of levels resulting from vibrations of the ions in the host material (Fig. 7.28). Ti-sapphire with an emission band between 680 and 1070 nm is a prominent gain material for tunable as well as ultrashort pulse lasers. Ti-sapphire can be pumped by frequency doubled Nd:YAG lasers.

Organic molecules, usually in a liquid host material, also show broad bands of vibrational and rotational levels; their transition wavelength can be customized over a broad range by chemical synthesis (Fig. 7.29).

Pumping of solid state and liquid lasers is provided by flashlamps (Fig. 7.30), especially if operated in the Q-switched mode (Sect. 7.3.1), or by other lasers. For cw-operation of solid state lasers, semiconductor laser pumping is advantageous over flashlamp pumping: not only is the efficiency of semiconductor lasers among

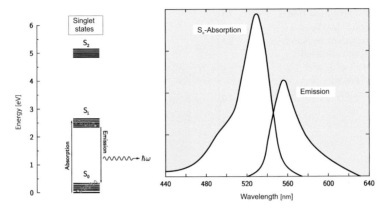

Fig. 7.29 Energy levels of an organic dye-molecule in a liquid solvent; for each electronic state, there is a range of vibrational and rotational sub-levels that form quasi-continuous bands. Absorption happens between the lowest levels of the S_0 state to upper levels of the S_1 state; emission takes place between the lowest levels of the S_1 state to upper levels of the S_0 state and is therefore red-shifted in comparison to absorption. Dye molecules thus act as four-level gain media

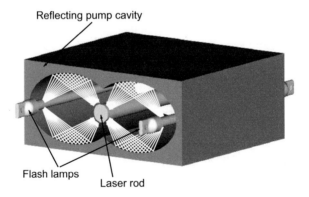

Fig. 7.30 Flash lamp pumped laser amplifier; the flashlamp is positioned parallel to the rod-like laser medium, and surrounded by a cylindrical reflecting cavity that optimizes the energy transfer and the homogeneity of the inversion density. Cooling is provided by water flowing through the cavity

the highest of all electric light sources, the wavelength of the pump source can also be precisely matched to the absorption band of the respective laser medium (Fig. 7.31).

7.4.2 Gas Lasers

Gaseous gain media can be pumped by electric discharge (Fig. 7.32). Important examples are Argon, Helium–Neon, CO_2, and mixtures of noble gases with

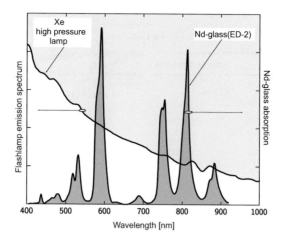

Fig. 7.31 Emission spectrum of a flash lamp, and absorption spectrum of Nd:glass; the poor spectral overlap is responsible for low pump efficiency

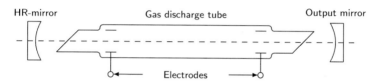

Fig. 7.32 Schematic of a gas laser: the windows of the discharge tube are mounted under Brewster's angle to avoid reflections; HR is a high reflection mirror, the second mirror serves as coupling mirror

halogens (excimers). Helium–Neon lasers emit at various wavelengths in the IR (1.15 and 3.39 μm) and in the visible at 632.8 and 543 nm (Fig. 7.33). The amplifying atom is Neon, Helium serves as intermediate medium that is excited into a long living 2S state by electron impact and transfers the energy resonantly to the Ne atoms. The return of the Ne atoms from the lower laser level to the ground state requires collisions with the walls of the discharge tube, limiting the diameter of the tube and consequently the output power. Argon-ion lasers can deliver several 10 W of cw radiation with excellent beam quality; argon atoms are ionized in a first step and then excited by a second electron impact. Because of this two step excitation, high current densities (kAcm^{-2}) are required. Argon-ion lasers emit in the visible, at 488 nm and 514 nm. The high temperature of the Argon-ion plasma (3000 K) is responsible for a considerable Doppler broadening that allows mode locked operation with a pulse duration of ≈150 ps. With the exception of the CO_2 laser (which emits at 10.6 μm and can deliver 100 kW of optical radiation), gas lasers suffer from very low overall efficiency and are replaced by solid state or semiconductor lasers wherever possible.

Excimer lasers (see, e.g., Basting and Marowsky 2005) rely on electronic transitions in two-atomic molecules (dimers) formed by an excited noble gas atom

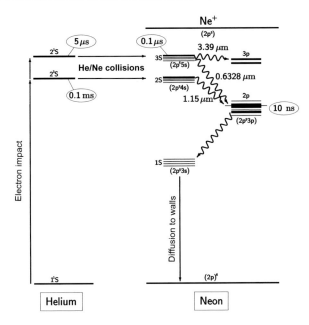

Fig. 7.33 Energy levels and transitions of a HeNe laser

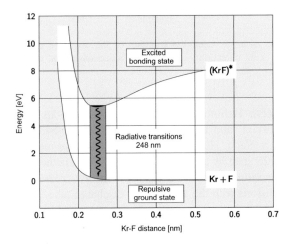

Fig. 7.34 Energy levels of an excimer; since the ground state is repulsive, there is practically no absorption at the emission wavelength

(argon, krypton, or xenon) and a halogen atom (fluorine or chlorine); the excitation of the noble gas is reached by electron impact in a discharge tube. After radiative transition to the ground state, the two constituents are driven apart by the repulsive potential of the noble gas atom (Fig. 7.34), so that the "ground state" is always unoccupied. Excimer lasers are important because of their emission wavelength in the UV (XeF 351 nm, XeCl 309 nm, KrF 248 nm, ArF 193 nm).

7.5   Semiconductor Lasers

Semiconductor laser amplifiers exhibit very large gain coefficients because of the high density of contributing electrons; they can be pumped directly by electron injection at the junction between differently doped semiconductors (pn-junction, Fig. 7.35). When such a laser diode is biased in the forward direction, electrons from the n-doped region are injected into the interfacial zone where they co-exist with holes injected from the p-region until they recombine via spontaneous or stimulated transitions. To reach amplification, the forward voltage must exceed the band gap, because the bias voltage determines the offset between the Fermi levels in the two sections of the diode. The resulting gain coefficient γ is large enough for an amplifier of about 100 μm length to support laser operation. The laser resonator is usually constituted by the two end facets of the semiconductor crystal (Fig. 7.36); due to the large refractive index (GaAs: ≈ 3.5) the reflectance is relatively large $R = [(n - 1)/(n + 1)]^2 \approx 0.31$, and low power semiconductor lasers are often operated without additional reflective coating.

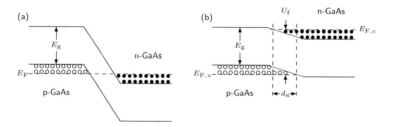

Fig. 7.35 Junction between degenerate p- and n-semiconductors: (**a**) without bias, (**b**) with forward bias voltage U_f

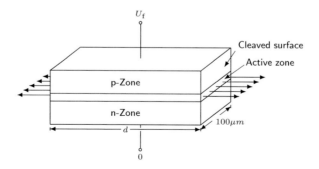

Fig. 7.36 Schematic of a semiconductor laser diode; the end facets serve as mirrors

The region of the diode that provides optical gain is called active zone; its volume is given by the product of the pn-interfacial area A and the thickness d_a of the active zone, which in the case of a simple pn-junction is equal to the diffusion length of the carriers (several μm). To establish a relation between the forward current of the diode and the carrier density n_e in the active zone, we assume an injection current density of j; the number of carriers transported per unit time into the zone then is equal to $\eta_i jA/e$ where the internal quantum efficiency η_i is defined as the fraction of injected electrons that actually reach the active zone (the others being lost by recombination). The recombination rate inside the active zone is $n_e A d_a/\tau_{rec}$, so that in equilibrium

$$j = e\frac{n_e d_a}{\eta_i \tau_{rec}}; \tag{7.57}$$

note that the current density required to obtain a certain carrier density is proportional to d_a. To reach a typical transparency carrier density of $1.5 \times 10^{18}\,\text{cm}^{-3}$, for example, a current density of $48\,\text{kAcm}^{-2}$ is needed if one assumes $\eta_i = 0.5$, a recombination time of $2\,\text{ns}$ and a thickness $d_a = 2\,\mu\text{m}$. Further assuming an interface area A of $200{\times}10\,\mu\text{m}^2$, this corresponds to a relatively high forward current of $1\,\text{A}$.

According to Eqs. (6.136) and (7.57), the gain of a semiconductor amplifier of length d is $e^{\alpha_0(j/j_{tr}-1)d}$, where $j_{tr} = e n_{tr} d_a/\eta_i \tau_{rec}$ is the current density corresponding to the transparency carrier density n_{tr}. If we assume mirror reflectivities R_1 and R_2 and express internal losses (due to free carrier absorption, absorption in regions adjacent to the active zone, and scattering at inhomogeneities) by the internal loss coefficient α_i, the threshold condition is

$$R_1 R_2 e^{[\alpha_0(j_c/j_{tr}-1)-\alpha_i]2d} = 1, \tag{7.58}$$

or

$$\alpha_0\left(\frac{j_c}{j_{tr}} - 1\right) = -\frac{\ln R_1}{2d} - \frac{\ln R_2}{2d} + \alpha_i, \tag{7.59}$$

where the threshold current density j_c (also called critical current density) is

$$j_c = j_{tr}\frac{2d\alpha_0 + 2d\alpha_i - \ln R_1 - \ln R_2}{2d\alpha_0}, \tag{7.60}$$

and the threshold current is

$$J_c = j_c A. \tag{7.61}$$

Fig. 7.37 Output power of a heterostructure-laser; the emission below threshold is due to luminescence (i.e., spontaneous emission)

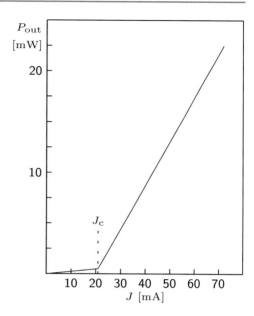

Since the critical carrier density is also the stationary operating carrier density, a current J exceeding the threshold current injects $\eta_i(J - J_c)/e$ carriers into the active zone that are deexcited by stimulated emission so as to conserve the stationary carrier density. The optical power internally generated is thus

$$P_i = \eta_i \frac{J - J_c}{e} \hbar\omega, \tag{7.62}$$

of which, according to Eq. (7.59), the fraction

$$\eta_m = \frac{-\ln R_1}{2d\alpha_i - \ln(R_1 R_2)} \tag{7.63}$$

is coupled out of the laser. The optical output power is therefore a linear function of the driving current

$$P_{out} = \eta_m \eta_i \frac{(J - J_c)}{e} \hbar\omega, \tag{7.64}$$

as shown in Fig. 7.37.

The electric input power of the diode is $P_p = JU_f$, where U_f is the forward voltage. The differential efficiency $\eta_1 = dP_{out}/dP_p$ is therefore

$$\eta_1 = \eta_m\eta_i\frac{\hbar\omega}{eU_f} \approx \eta_m\eta_i. \tag{7.65}$$

The approximation is reasonable since eU_f is equal to the difference of the quasi-Fermi levels and usually not much bigger than $\hbar\omega$. The product $\eta_m\eta_i$ is the external quantum efficiency; typical values for commercial semiconductor lasers are 50 %; far above the threshold, the total efficiency approaches the differential efficiency.

7.5.1 Heterostructure Lasers

Since the fraction U_fJ_c of the input power does not contribute to the output power, it is important to keep the threshold current low. One way to achieve this goal is to reduce the thickness d_a of the active zone by embedding the active zone between layers of higher band gap that provide barriers against carrier diffusion (Fig. 7.38). Because of the existence of two interfaces between different materials, such a structure is called double heterostructure; it is produced by epitaxial growth (molecular

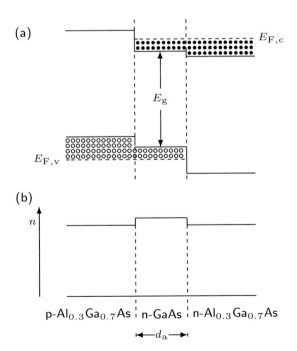

Fig. 7.38 Band diagram (**a**) and refractive index (**b**) of a double heterostructure laser diode

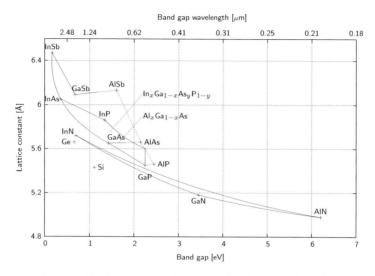

Fig. 7.39 Band gap and lattice constants of important III-V-compound semiconductors; compounds on *horizontal lines* are lattice matched; *dashed lines* indicate indirect band gaps, not suitable for laser operation

beam epitaxy) and requires matching lattice constants of the materials. Popular materials are various compositions of III-V-compounds such as $In_{1-x}Ga_xAs_{1-y}P_y$ and $Al_xGa_{1-x}As$, or $In_{1-x}Ga_xN$ and $Al_{1-x}Ga_xN$. The stoichiometric parameters x, y determine the electronic and optical properties of the compound, in particular the band gap (Fig. 7.39) and the refractive index. $Al_xGa_{1-x}As$ is a direct semiconductor in the range $0 \leq x < 0.38$, with the band gap and refractive index approximately given by $E_g[eV] \approx 1.42 + 1.30 x$ and $n \approx 3.5 - 0.71 x$.

Double heterostructure lasers exhibit several advantages over homostructure laser diodes:

- reduction of the thickness of the active zone from about $2 \mu m$ to 100–$200 nm$ results in a reduction of the threshold current by approximately the same factor;
- since the semiconductor material adjacent to the active zone has a lower refractive index, the sandwich structure acts as a waveguide for the laser light;
- because of its higher band gap, the semiconductor material adjacent to the active zone does not absorb the laser radiation so that internal losses are reduced.

The thickness of the active zone of double heterostructure lasers is typically $100 nm$, typical threshold currents are 10–$15 mA$. Further thinning of the active zone below $100 nm$ reduces the overlap between the laser mode and the gain medium and results in an increasing threshold current.

7.5.2 Quantum Well Lasers

If d_a is reduced below 20 nm, effects of quantum confinement come into play, as outlined in Sect. 6.3.4. In particular, quantum wells have a reduced density of states, which makes it easier to drive them into inversion and reach threshold for laser operation. Because of the low thickness of the active zone, epitaxial growth is possible even if the materials are not exactly lattice matched; a mismatch of up to several percent is tolerable. InGaAs-quantum wells, for example, can be grown between AlGaAs-layers. The strain that is induced by the mismatch also can modify the band structure significantly, an effect that is exploited in the technology of strained lattice quantum well lasers.

To improve the overlap of the gain region with the laser mode, several quantum wells can be stacked upon each other (multi-quantum wells, MQWs). MQW-lasers have threshold currents as low as 0.5 mA. They also show narrow bandwidth (10 MHz) and a reduced sensitivity to temperature.

7.5.3 Performance and Technology

Important semiconductor laser materials are AlGaAs (0.75–0.87 μm) and InGaAsP (1.1–1.6 μm). Gallium nitride based semiconductor lasers have become very popular since they operate in the visible up to the near UV (Fig. 7.39) and can be used for display applications and for high capacity optical storage.

Figure 7.40 shows the cross section of a typical heterostructure laser. The sandwich structure provides optical guiding in the direction of the current flow. In the lateral direction, the active zone is limited by the width of the injection electrode, providing gain-guiding of the laser mode; there is, however, no wave guiding in this direction because the refractive index of the active zone is lowered by the increased carrier density (see Sect. 6.3.5).

Gain guided semiconductor lasers usually show multiple transverse modes. Transverse single mode operation can be achieved by embedding the active zone laterally in a low index material (index guiding, Fig. 7.41).

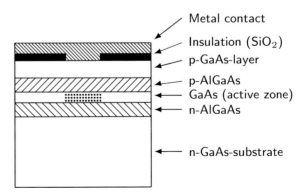

Fig. 7.40 Cross section of an oxide insulated stripe laser; the active region is defined by the electric current flow, i.e., by the insulation layer. Typical dimensions of the active zone are $0.2 \times 5 \times 100\,\mu m^3$

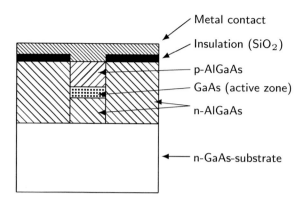

Fig. 7.41 Cross section of a buried heterostructure laser; the active zone is surrounded by semiconductor material of higher band gap and lower refractive index, forming a nonabsorbing waveguide

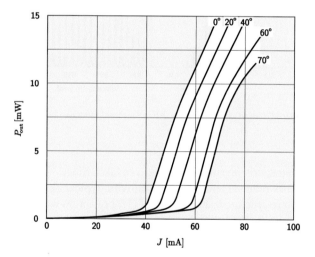

Fig. 7.42 Performance of a semiconductor laser at different operating temperatures

The performance parameters of semiconductor lasers, in particular the threshold current and emission wavelength, are highly temperature sensitive (Fig. 7.42). This is mostly due to the temperature dependence of the Fermi factor Eq. (6.134); in addition, the efficiency of the potential barriers of the heterostructure is reduced at elevated temperatures. Finally, the probability of Auger recombinations in which the energy of excited carriers is transferred to other carriers instead of photons, and eventually lost to lattice vibrations increases rapidly with temperature. The threshold current follows the empirical equation

$$J_c(T) \propto e^{T/T_0}, \tag{7.66}$$

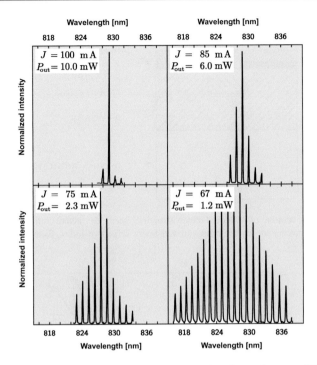

Fig. 7.43 Normalized output spectra of an AlGaAs-semiconductor laser at different driving currents

where T_0 is a characteristic parameter of a given laser. Since $\frac{\mathrm{d}J_c(T)}{\mathrm{d}T} \propto 1/T_0$, a higher value of T_0 implies reduced temperature sensitivity; typical T_0 values are $70\,^{\circ}\mathrm{C}$ for conventional heterostructure lasers, and $250\,^{\circ}\mathrm{C}$ for quantum well lasers.

The temperature drift of the emission wavelength of heterostructure lasers is typically $0.3\,\mathrm{nm}\,\mathrm{K}^{-1}$. This allows tuning laser diodes by heating or cooling; stable operation requires temperature stabilization.

Because of the short resonator length, the mode spacing $\Delta\omega_r = c\pi/d$ [Eq. (4.82)] is very substantial (expressed in terms of wavelength, 0.1–0.5 nm). The large gain bandwidth (Fig. 6.24) nevertheless provides gain for many axial modes. Near the threshold, these modes can be observed in the output spectrum (Fig. 7.43). Due to the fast intraband transitions, semiconductors behave like homogeneously broadened gain media, so that well above threshold the number of modes is reduced by mode competition; single mode operation is frustrated by spatial hole burning in standard heterostructure lasers, however.

Longitudinal single mode operation, which is required by many applications, can be obtained by different means. One is to employ an external resonator with a frequency selective element, such as a grating; the semiconductor amplifier itself is antireflection (AR) coated to avoid any additional resonances. Alternatively,

Fig. 7.44 Diagram of a
distributed-Bragg
reflector-laser (DBR) (**a**),
distributed feedback-laser
(DFB) (**b**), and DFB-laser
with a phase jump (**c**)

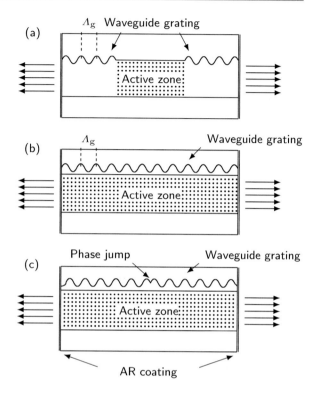

single mode selection is provided by a frequency selective laser structure (see, e.g.,
Kogelnik and Shank 1972); Fig. 7.44 shows three possible configurations.

In a so-called distributed-Bragg reflector-laser (DBR), the output facets are AR
coated and the laser resonator is formed by two integrated waveguide gratings that
serve as narrow band mirrors supporting only one mode (compare Sect. 5.3.3). In a
distributed feedback-laser (DFB), the active zone itself is corrugated longitudinally
to provide a feedback between the forward and backward travelling laser mode.
Such a laser can also be viewed as a waveguide grating with integrated gain.
Somewhat counterintuitively, such lasers do not oscillate at the Bragg wavelength of
the grating, but rather at *two* frequencies at the edges of the stop band. This can be
understood by an inspection of Fig. 5.28: the intracavity power distribution, which is
responsible for stimulated emission, is resonantly enhanced at the edges of the stop
band, while there is no such enhancement at the Bragg wavelength. Single frequency
operation can be achieved by introducing a $\lambda/4 = \Lambda_g/2$ spatial phase jump in the
waveguide grating as shown in Fig. 5.31, giving rise to resonant enhancement and
single longitudinal mode operation at the Bragg wavelength with narrow (MHz)
bandwidth.

In addition to these edge emitting lasers, there are also laser structures that emit
in the growth direction of the chip, so-called vertical cavity surface emitting lasers
(VCSEL, Fig. 7.45). The mirrors are multilayer reflectors integrated by growing

Fig. 7.45 Vertical cavity
surface emitting laser
(VCSEL); layers of high and
low refractive index serve as
resonator mirrors; laser
emission is orthogonal to the
substrate plane

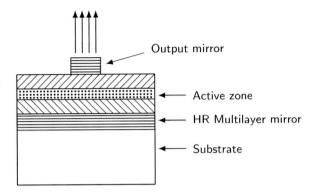

alternating layers of high and low refractive index. The active zone is very short
and requires high reflectance mirrors to keep the threshold low; the short resonator
results in very large mode spacing and thus to single mode operation.

Edge emitting lasers usually exhibit a strongly divergent and astigmatic output
beam, since the lateral dimensions of the active zone are very small and different
from each other [see Eq. (3.19)]. Typical values are 20–30° in the direction of
current transport and several degrees in the plane normal to it. By the use of
astigmatic collimating optics, the output beam can be rendered cylindrical. The
small dimensions of the laser mode also result in very high intensities at the output
facets; the onset of optical damage of the facets (damage threshold $10^9\,\mathrm{Wm^{-2}}$)
limits the output power of a single stripe laser diode to about 150 mW. Optical
damage is also responsible for the immediate destruction of laser diodes by supply
current spikes or by external reflections of the output light which are amplified in
the resonator; the latter problem can be avoided by a Faraday isolator (Sect. 2.4.2.1).

Heterostructure lasers can be produced by liquid phase epitaxy (LPE) allowing
for high growth rates (10 nm/s) and cost-effective large scale production. For the
production of quantum wells lasers, molecular beam epitaxy (MBE) is used, which
makes controlled layer by layer growth with very low defect densities possible.
Other commercial growth technologies are chemical vapor deposition (CVD) and
metal-organic chemical vapor deposition (MOCVD).

7.6   Free Electron Lasers*

Free electron lasers (FELs) are tunable sources of coherent radiation based upon
an oscillatory motion of high energy electrons in a spatially periodic, stationary
magnetic field. FELs do not rely on stimulated emission in the sense of Sect. 6.2, but
resemble travelling wave vacuum tubes or amplification schemes such as Brillouin
amplification (Sect. 8.3.6). Because of the high electron energy needed, they require
electron accelerators to operate and are large scale facilities. FELs can be described
in a purely electrodynamic framework including a relativistic equation of motion of
the electrons. Nonetheless, terms such as spontaneous or stimulated emission are

Fig. 7.46 Helical undulator: relativistic electrons (energy $m_e \gamma c_0^2$) are deflected by a helical series of magnets (north- and south-poles marked by N an S, respectively), resulting in a helical trajectory (*dotted line*)

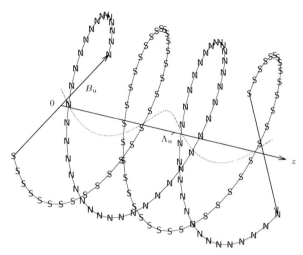

common in this context, because early theoretical treatments have been based on relativistic quantum electrodynamics.

The magnetic field for the electron deflection is provided by a periodic array of magnets (so-called undulator, Fig. 7.46). The velocity of the electrons and the spatial period Λ_u of the undulator determine the oscillation frequency. In the following discussion (that relies on Saldin 2000), we assume, for convenience, a helical undulator where the absolute value of the magnetic field is constant. In practice, most undulators are planar, however. The Lorentz force acting on the electrons induces a transverse acceleration and results in a helical electron trajectory.

We use a complex notation for periodically varying transverse quantities such as the magnetic field

$$\mathbf{B}_u = \mathrm{Re}\left[\tilde{\mathbf{B}}_{u,\perp}\right], \tag{7.67}$$

$$\tilde{\mathbf{B}}_{u,\perp} = B_u \begin{bmatrix} 1 \\ j \end{bmatrix} e^{-jK_u z} \tag{7.68}$$

$$K_u := 2\pi/\Lambda_u. \tag{7.69}$$

The Lorentz force acting on an electron propagating along the z-axis with velocity $\mathbf{v}$ (and axial component v_z) is given by

$$\tilde{\mathbf{F}}_L = -e\mathbf{v} \times \tilde{\mathbf{B}}_\perp = je v_z \tilde{\mathbf{B}}_\perp. \tag{7.70}$$

The relativistic equation of motion (see, e.g., Jackson 1999) in lab-coordinates then is

$$m_e \gamma \frac{d\tilde{\mathbf{v}}_\perp}{dt} = \tilde{\mathbf{F}}_L, \tag{7.71}$$

where

$$\gamma = 1/\sqrt{1 - \beta^2}, \quad \beta = v/c_0. \tag{7.72}$$

After substituting $dz = v_z \, dt$ and integration we obtain the transverse velocity

$$\tilde{\mathbf{v}}_\perp = -c_0 \frac{K}{\gamma} \begin{bmatrix} 1 \\ j \end{bmatrix} e^{-jK_u z} \tag{7.73}$$

with the dimensionless undulator-parameter

$$K := \frac{\Lambda_u e B_u}{2\pi m_e c_0} \approx 0.93 B_u[\text{T}] \Lambda_u[\text{cm}] \tag{7.74}$$

that represents the ratio of the undulator vector potential $e B_u c_0 \Lambda_u$ to the electron rest energy $m_e c_0^2$ and is typically on the order of 1. The transverse velocity amplitude can then be expressed as

$$v_\perp = c_0 \frac{K}{\gamma} =: c_0 \theta_s, \tag{7.75}$$

where $\theta_s = v_\perp/c_0 = K/\gamma \ll 1$ is approximately equal to the angle between the electron trajectory and the z-axis. With $\beta_z := v_z/c_0$, $\gamma_z := 1/\sqrt{1 - \beta_z^2}$ and $v_z^2 + v_\perp^2 = v^2$ we obtain the useful relation

$$\gamma^2 = \gamma_z^2 (1 + K^2). \tag{7.76}$$

7.6.1   "Spontaneous" Emission

In lab-coordinates, the electron oscillates at

$$\omega_u = 2\pi \frac{v_z}{\Lambda_u} \approx 2\pi \frac{c_0}{\Lambda_u}; \tag{7.77}$$

in the reference frame moving with the electron, the frequency, due to Lorentz time contraction (Sect. 2.4.3) is equal to $\omega_u \gamma_z$. An observer looking towards the electron along the z-axis detects an electromagnetic field oscillating at the Doppler shifted frequency [Eq. (2.197)]

$$\omega_0 = \omega_u \gamma_z \sqrt{\frac{1 + \beta_z}{1 - \beta_z}} \approx 2\gamma_z^2 \omega_u, \tag{7.78}$$

corresponding to a wavelength of

$$\lambda_0 = \frac{\Lambda_u}{2\gamma_z^2}. \tag{7.79}$$

The light emitted by an individual electron consists of $N_u = l_u/\Lambda_u$ cycles, where l_u is the length of the undulator; the duration of the emitted pulse is therefore $\tau_p = 2\pi N_u/\omega_0$. The shape of the power spectrum is given by the absolute square of the Fourier transform of the rectangular envelope $\text{rect}(t/\tau_p)e^{-j\omega_0 t}$, and is proportional to

$$S(\Delta\omega/\omega_0) \propto \frac{\sin^2(N_u \pi \Delta\omega/\omega_0)}{(N_u \pi \Delta\omega/\omega_0)^2} \tag{7.80}$$

with $\Delta\omega = \omega - \omega_0$ (Fig. 7.47); the FWHM bandwidth (normalized to ω_0) is $0.8895/N_u$.

Another electron (of same kinetic energy) passing the undulator produces the same pulse, but with a relative phase shift that depends on the difference of entrance

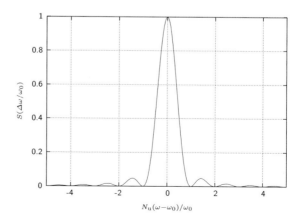

Fig. 7.47 Normalized emission spectrum of a single electron (or a bunch of uncorrelated electrons of identical energy) passing an undulator with N_u periods

time. An ensemble of uncorrelated electrons produces incoherent radiation with the power spectrum given above. This radiation is usually called spontaneous emission; the center frequency of the spectrum can be tuned by adjusting the electron velocity according to Eq. (7.78).

7.6.2 Light-Electron Coupling and Amplification

Let us now consider a (circularly polarized) light wave

$$\tilde{\mathbf{E}}_\perp(z,t) = E_0 \begin{bmatrix} 1 \\ -j \end{bmatrix} e^{-j(k_L z - \omega t)}, \tag{7.81}$$

co-propagating with the electron; note that the helicity is opposite to the electron trajectory and E_0 is the absolute value of the electric field at any time, $E_0 = \left| \text{Re}\left[\tilde{\mathbf{E}}_\perp(0,t) \right] \right|$; the corresponding intensity is E_0^2/Z_0. Due to the interaction of the moving electron (which constitutes a current) and the electric field, the electron energy $\mathcal{E}$ changes at the rate

$$\frac{d\mathcal{E}}{dt} = -e\mathbf{v}_\perp \cdot \mathbf{E}_\perp; \tag{7.82}$$

with $dz = v_z\, dt$ and $\mathbf{a} \cdot \mathbf{b} = \frac{1}{4}(\tilde{\mathbf{a}} + \tilde{\mathbf{a}}^*) \cdot (\tilde{\mathbf{b}} + \tilde{\mathbf{b}}^*) = (1/2)\text{Re}\left[\tilde{\mathbf{a}} \cdot \tilde{\mathbf{b}} + \tilde{\mathbf{a}} \cdot \tilde{\mathbf{b}}^* \right]$ we obtain

$$\frac{d\mathcal{E}}{dz} = -eE_0\theta_s \cos\psi, \tag{7.83}$$

where

$$\psi = K_u z + k_L z - \omega t + \psi_0 \tag{7.84}$$

$$= K_u z + k_L z - \frac{\omega}{v_z}z + \psi_0 \tag{7.85}$$

is the phase difference between $\mathbf{v}_\perp(z,t)$ and $\mathbf{E}_\perp(z,t)$; ψ_0 is the phase at entry and assumed to be statistically distributed. The differential energy exchange is thus proportional to the normalized transverse velocity θ_s, the electric field amplitude E_0 and the cosine of ψ that determines sign and amount of the transfer. If ψ varies strongly over the interaction length, the sign of the energy transfer changes several times and the integrated energy exchange is small or zero. A necessary condition for a significant integrated exchange (be it positive or negative) is that ψ varies only very little

$$\frac{d\psi}{dz} = \left(K_u + \frac{\omega}{c_0} - \frac{\omega}{v_z} \right) \approx 0; \tag{7.86}$$

with $K_u = 2\pi/\Lambda_u$, and $\omega = 2\pi c_0/\lambda_L$ we obtain the synchronism condition

$$\frac{\Lambda_u}{v_z} \approx \frac{\lambda_L}{c_0 - v_z}; \tag{7.87}$$

during the time Λ_u/v_z that it takes the electron to pass one undulator period, the light wave, travelling faster by the amount $c_0 - v_z$, acquires a lead of one wavelength in respect to the electron. Expressed in terms of the wavelength and assuming $\beta_z \approx 1$, we obtain

$$\lambda_L = \Lambda_u \frac{1 - \beta_z}{\beta_z} = \Lambda_u \frac{1 - \beta_z^2}{\beta_z(1 + \beta_z)} \approx \frac{\Lambda_u}{2\gamma_z^2} \tag{7.88}$$

which is equal to the wavelength of the "spontaneous" radiation Eq. (7.79). With Eq. (7.76) and $v_z \approx c_0$ we can formulate the synchronism condition as

$$\lambda_L \approx \Lambda_u \frac{1 + K^2}{2\gamma^2}. \tag{7.89}$$

When the synchronism condition is met, $\psi(z)$ in Eq. (7.83) remains equal to the initial value ψ_0, and $\cos\psi_0$ determines whether the electron gains or loses energy (by acceleration or deceleration) during the interaction. For an ensemble of uncorrelated electrons, the time of entry (measured in reference to the light wave), and therefore ψ_0, is statistically distributed and one has to evaluate the angular average over all input phases, defined as

$$\langle . \rangle = \frac{1}{2\pi} \int_0^{2\pi} (.) \, d\psi_0; \tag{7.90}$$

since $\langle \cos\psi_0 \rangle = 0$, an ensemble of uncorrelated electrons experiences, for statistical reasons, no net energy exchange; for the average energy transfer to be non-vanishing, the synchronism condition must be slightly violated. We write Eq. (7.86) as

$$\frac{d\psi}{dz} = K_u + \frac{\omega}{c_0} - \frac{\omega}{v_z(\mathcal{E})} \approx \underbrace{K_u + \frac{\omega}{c_0} - \frac{\omega}{v_z(\mathcal{E}_0)}}_{D} + \frac{\omega}{v_z^2(\mathcal{E}_0)} \frac{dv_z}{d\mathcal{E}} \Delta\mathcal{E}, \tag{7.91}$$

where $\mathcal{E}_0$ is the electron energy at entry, $\Delta\mathcal{E} = \mathcal{E} - \mathcal{E}_0$ is the energy exchange and D is the deviation from synchronism at entry. With Eq. (7.76), we can express the electron energy $\mathcal{E} = m_e c_0^2 \gamma$ as $m_e c_0^2 \gamma_z \sqrt{K^2 + 1}$ to obtain $d\mathcal{E}/dv_z \approx \gamma_z^2 \mathcal{E}_0/c_0$ since $v_z \approx c_0$. In combination with Eq. (7.83), we find the set of coupled differential equations

$$\frac{d\psi}{dz} = D + \frac{\omega}{c_0 \gamma_z^2 \mathcal{E}_0} \Delta\mathcal{E} \tag{7.92}$$

$$\frac{d\Delta\mathcal{E}}{dz} = -eE_0 \theta_s \cos\psi \tag{7.93}$$

that describes the interplay between light field and electrons.

We eliminate $\Delta\mathcal{E}$ and introduce the normalized coordinate $\zeta := z/l_u$ and field $u := (e\omega\theta_s E_0 l_u^2)/(c_0 \gamma_z^2 \mathcal{E}_0)$ and obtain

$$\frac{d^2\psi}{d\zeta^2} + u\cos\psi = 0, \tag{7.94}$$

which is formally equivalent to the equation of a pendulum, indicating that the phase swings around the values $\pm\pi/2$: the reason for this oscillation is that the energy transfer changes the electron velocity and thus the phase $\psi(z)$; in phase space, the electrons are attracted to $\psi = \pm\pi/2$ where the energy exchange is zero, in the same way a pendulum is attracted to the vertical by gravitation.

With the initial conditions $\psi(0) = \psi_0$ and $\frac{d\psi}{dz}|_{z=0} = D$, equivalent to $\frac{d\psi}{d\zeta}|_{\zeta=0} = Dl_u =: D'$, integration of Eq. (7.94) yields the phase as a function of ζ

$$\psi(\zeta) = \psi_0 + D'\zeta + \Delta\psi(\zeta, \psi_0), \tag{7.95}$$

where

$$\Delta\psi(\zeta, \psi_0) = -\int_0^\zeta d\zeta' \int_0^{\zeta'} u_0 \cos(\psi_0 + D'\zeta'')\, d\zeta'' \tag{7.96}$$

$$= \frac{u_0}{D'^2} \left[\cos(\psi_0 + D'\zeta) - \cos\psi_0\right] + \frac{u_0\zeta}{D'} \sin\psi_0 \tag{7.97}$$

is the pendulum component, ψ_0 the initial phase and $D'\zeta$ the accumulated phase slip. We now can integrate Eq. (7.83) over the undulator length to evaluate the total energy exchange of one electron

$$\Delta\mathcal{E} = -eE_0 \theta_s l_u \int_0^1 \cos[\psi_0 + D'\zeta + \Delta\psi(\zeta, \psi_0)]\, d\zeta. \tag{7.98}$$

We restrict ourselves to small phase excursions $\Delta\psi(\zeta, \psi_0) \ll 1$ (implying $\sin \Delta\psi(\zeta, \psi_0) \approx \Delta\psi(\zeta, \psi_0)$, $\cos \Delta\psi(\zeta, \psi_0) \approx 1$) and use the identity $\cos(a+b) = \cos a \cos b - \sin a \sin b$ to obtain

$$\Delta\mathcal{E} \approx -eE_0\theta_s l_u \int_0^1 \cos(\psi_0 + D'\zeta) - \Delta\psi(\zeta, \psi_0)\sin(\psi_0 + D'\zeta)\,d\zeta; \qquad (7.99)$$

after integration over ζ and averaging over the initial phase (assumed to be uniformly distributed between $\psi_0 = 0 \ldots 2\pi$), we arrive at the average exchange per electron of

$$\langle \Delta\mathcal{E} \rangle = eE_0\theta_s l_u u_0 f(D') = \frac{\theta_s^2 \omega l_u^3 e^2 E_0^2}{c_0^3 \gamma_z^2 \gamma m_e} f(D') \qquad (7.100)$$

with

$$f(D') = \frac{2}{D'^3}\left[1 - \cos D' - (D'/2)\sin D'\right] \qquad (7.101)$$

$$= -\frac{d}{2\,dD'}\left(\frac{\sin^2(D'/2)}{(D'/2)^2}\right). \qquad (7.102)$$

If we denote the electron flux density entering the undulator with j_0, the increase of the light intensity is given by $(j_0/e)\langle \Delta\mathcal{E} \rangle$. The small signal gain, defined here as the ratio of intensity increment to input intensity E_0^2/Z_0, can be expressed as

$$\frac{(j_0/e)\langle \Delta\mathcal{E} \rangle}{E_0^2/Z_0} = f(D')64\sqrt{2}\pi^2 \frac{j_0}{I_A}\frac{K^2\sqrt{\Lambda_u}\lambda^{3/2}}{(1+K^2)^{3/2}}N_u^3, \qquad (7.103)$$

where $I_A = 4\pi c_0^3\varepsilon_0 m_e/e \approx 17\,\text{kA}$ is the so-called Alfven-current. Using Eq. (7.86) in the form $(1/c_0 - 1/v_z) = -K_u/\omega_0$, D' can be expressed in terms of $\omega - \omega_0$,

$$D' = l_u\left(K_u + \frac{\omega}{c_0} - \frac{\omega}{v_z}\right) = -\frac{\omega - \omega_0}{\omega_0}l_u K_u = -2\pi N_u\frac{\omega - \omega_0}{\omega_0}, \qquad (7.104)$$

so that the gain profile $f(D')$ Eq. (7.102) turns out to be proportional to the derivative of Eq. (7.80),

$$f(D') \propto \frac{d}{d\Delta\omega}\left(\frac{\sin^2(N_u\pi\Delta\omega/\omega_0)}{(N_u\pi\Delta\omega/\omega_0)^2}\right). \qquad (7.105)$$

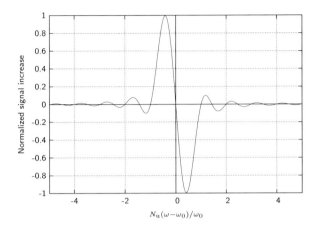

Fig. 7.48   FEL-gain as a function of frequency; N_u is the number of undulator periods

Figure 7.48 shows the dependence of the gain on the detuning $\omega - \omega_0$. As outlined above, the gain vanishes at $D' = 0$, and, for equivalent reasons, at $D' = 2m\pi$. The most efficient energy transfer happens at approximately $D' = \pm\pi$.

A requirement for the efficient operation of an FEL amplifier is that the electron energy distribution is narrow, since the velocity determines the degree of detuning. Moreover, the electron velocity decreases due to the gain process, so that the gain also decreases along the undulator length. This can be compensated by reducing the undulator period along the undulator axis (tapered undulator). An FEL-amplifier can be converted into an oscillator by embedding it into an optical resonator. Alternatively, an FEL can be seeded by an external coherent light source.

7.7   Summary

Lasers are not the only, but certainly the most important sources of coherent optical radiation. The fundamentals of laser operation have been laid in Sect. 4.3 and Chap. 6; here we describe various types of lasers and their mode of operation. We analyze the stationary operation of laser oscillators, exemplified by four and three-level atomic lasers as well as semiconductor lasers. We discuss mode competition and selection, and the impact of inhomogeneous and homogeneous line broadening, respectively, on laser performance. We also discuss the fact that the theoretical spectral width of the laser modes is not zero but given by the Schawlow–Townes limit that takes spontaneous emission of the laser medium into the laser mode into account.

Non-stationary laser operation is theoretically interesting and practically important. We describe the trajectory in phase space that a laser has to pass before reaching stationary equilibrium; the relaxation oscillations that precede the stationary state are analyzed and the frequency response of pump modulated lasers is derived. Gain modulation, mode locking, and Q-switching are modeled.

The operation and technology of semiconductor lasers is described in detail; important atomic lasers are presented with their main operating features. Finally, the free electron laser is presented, which, however, resembles a vacuum electronic tube rather than a conventional that relies on stimulated emission; the synchronicity condition, which is basically a phase matching condition between the undulating electrons and the electromagnetic field, is derived and the gain as a function of signal frequency is estimated.

7.8   Problems

1. A four-level system (Nd:YAG) has a stimulated emission cross section of 7×10^{-19} cm^2 and a spontaneous life time of $\tau_2 = 230$ μs at 1064 nm wavelength. The length of the laser rod (and the resonator) is 4 cm, its diameter 4 mm. Assume two laser mirrors with transmissions $T_1 = 0$, $T_2 = 0.4$; internal losses amount to 2 % per round trip. Calculate the threshold inversion. Assuming ideal conditions (complete absorption of the pump power etc.), calculate the threshold pump power (pump wavelength 800 nm). Finally, calculate the output power as a function of pump power assuming a pump efficiency (compared to the ideal case) of 10 % .

2. Using the results of Problem 6, Sect. 6.5, calculate numerically the gain coefficient of a GaAs laser diode at room temperature as a function of the injection current, assuming that the active zone has the dimensions (thickness × width × length) $1 \times 5 \times 100$ μm^3. Assuming a quantum efficiency of 1, what is the threshold current for this laser (the cleaved surfaces of the GaAs crystal serve as mirrors, neglect internal losses)? Calculate the output power as a function of the current. For the properties of GaAs, refer to Table 6.1.

3. Derive the formula for the optimal output coupler and present the results graphically. Discuss the case $\alpha_i = 0$.

4. Integrate numerically the rate equations (7.42) and (7.43) of a Q-switched laser using appropriate discrete Δt-intervals and a finite number of initial photons. Assume a Nd-YAG rod of 6 mm diameter and 70 mm length and an output mirror with a transmission of 80 %; set the mode volume equal to the rod volume and neglect internal losses. Assume different ratios $N_i/N_{2,c}$ of the initial inversion to the critical inversion and determine the laser pulse duration and extraction efficiency. Compare these results with the approximative solutions shown in Fig. 7.19. See Table 7.1 for the properties of Nd:YAG.

References and Suggested Reading

Bachmann, F., Loosen, P., & Poprawe, R. (2007). *High power diode lasers: Technology and applications* (Vol. 128). New York: Springer.

Bass, M. (Ed.). (2010). *Handbook of optics*. New York: McGraw-Hill.

Basting, D., & Marowsky, G. (Eds.). (2005). *Excimer laser technology*. New York: Springer.

Botez, D., & Scifres, D. R. (Eds.). (2005). *Diode laser arrays*. New York: Cambridge University Press.

Capasso, F. (1990). *Physics of quantum electron devices*. New York: Springer.

Cerullo, G., Longhi, S., Nisoli, M., Stagira, S., Svelto, O. Saldin, E., Schneidmiller, E.V., Yurkov, M.V. (2001). *Problems in laser physics*. New York: Springer.

Chow, W. W., & Koch, S. W. (1999). *Semiconductor-laser fundamentals*. New York: Springer.

Connelly, M. J. (2002). *Semiconductor optical amplifiers*. New York: Springer.

Corzine, S. W., Coldren, L. A., & Mashanovitch, M. L. (2012). *Diode lasers and photonic integrated circuits*. New York: Wiley.

Degnan, J. J. (1995). Optimization of passively Q-switched lasers. *IEEE Journal of Quantum Electronics, 31*(11), 1890–1901.

Desurvire, E. (2001). *Erbium doped fiber amplifiers*. New York: Wiley.

Digonnet, M. J. F. (2001). *Rare earth doped fiber lasers and amplifiers*. Boca Raton: CRC Press.

Duling, I. N. (Ed.). (2006). *Compact sources of ultrashort pulses*. New York: Cambridge University Press.

Faist, J. (2011). *Quantum cascade lasers*. London: Oxford University Press.

Fermann, M. E., & Hartl, I. (2009). Ultrafast fiber laser technology. *IEEE Journal of Selected Topics in Quantum Electronics, 15*(1), 191–206.

Ghafouri-Shiraz, H. (1995). *Fundamentals of laser diode amplifiers*. New York: Wiley.

Haus, H. A. (2000). Mode-locking of lasers. *IEEE Journal of Selected Topics in Quantum Electronics, 6*(6), 1173–1185.

Ippen, E. P., Shank, C. V., & Dienes, A. (1972). Passive mode locking of the cw dye laser. *Applied Physics Letters, 21*(8), 348–350.

Jackson, J. D. (1999). *Classical electrodynamics*. New York: Wiley.

Koechner, W. (2006). *Solid-state laser engineering*. New York: Springer.

Kogelnik, H., & Shank, C. V. (1972). Coupled-wave theory of distributed feedback lasers. *Journal of Applied Physics, 43*(5), 2327–2335.

Meschede, D. (2007). *Optics, light and lasers*. New York: Wiley.

Mollenauer, L. F., & White, J. C. (Eds.). (1992). *Tunable lasers*. Berlin: Springer.

Nakamura, S., Pearton, S., & Fasol, G. (2013). *The blue laser diode: The complete story*. New York: Springer.

Numai, T. (2004). *Fundamentals of semiconductor lasers*. Berlin: Springer.

Paschotta, R. (2009). *Encyclopedia of laser physics and technology*. New York: John Wiley.

Quimby, R. S. (2006). *Photonics and lasers*. New York: Wiley.

Saldin, E. L., et. al. (2000). *The physics of free electron lasers*. Berlin: Springer.

Saleh, B. E., & Teich, M. C. (2007). *Fundamentals of photonics*. New York: Wiley.

Siegman, A. E. (1986). *Lasers*. Mill Valley, CA: University Science Books.

Spence, D. E., Kean, P. N., & Sibbett, W. (1991). 60-fsec pulse generation from a self-mode-locked Ti: Sapphire laser. *Optics Letters, 16*(1), 42–44.

Svelto, O. (2010). *Principles of lasers*. New York: Plenum Press.

Träger, F. (2007). *Springer handbook of lasers and optics*. New York: Springer.

Wagner, W. G., & Lengyel, B. A. (1963). Evolution of the giant pulse in a laser. *Journal of Applied Physics, 34*(7), 2040–2046.

Yariv, A., & Yeh, P. (2006). *Photonics*. Oxford: Oxford University Press.

Young, M. (2000). *Optics and lasers*. Berlin: Springer.

Nonlinear Optics and Acousto-Optics

8

Nonlinear optics deals with optical phenomena that result from the dependence of the optical susceptibility on the electromagnetic field. Exemplary effects are the intensity-dependent propagation index, the electro-optic effects, and parametric effects such as frequency mixing, harmonic generation, or parametric amplification. In a more general sense, gain and absorption saturation are also nonlinear optical effects, but they are usually not treated in terms of susceptibilities. Acousto-optic effects are related to nonlinear optical effects in the sense that the susceptibility is influenced by acoustic fields; since the mathematical treatment is very similar, they are included in this chapter.

8.1 Nonlinear Susceptibility

The polarization response of a material on the electric field $\mathbf{E}$ is not a strictly linear function; to account for that, we can write it as Taylor expansion

$$\mathbf{P} = \sum_i \mathbf{P}^{(i)} = \varepsilon_0 \chi^{(1)} \mathbf{E} + \varepsilon_0 \chi^{(2)} \mathbf{E}\mathbf{E} + \varepsilon_0 \chi^{(3)} \mathbf{E}\mathbf{E}\mathbf{E} + \dots, \tag{8.1}$$

where $\chi^{(1)}$ is the "linear" susceptibility, while the nonlinear contributions are represented by the "nonlinear susceptibilities" $\chi^{(i)}$ ($i \geq 2$) of order i. The total polarization is the sum of the "linear" polarization $\mathbf{P}^{(1)} = \varepsilon_0 \chi^{(1)} \mathbf{E}$ and "nonlinear polarizations" $\mathbf{P}^{(i)} = \varepsilon_0 \chi^{(i)} \mathbf{E}^i$ of order i. Second (third) order effects are also called quadratic (cubic).

For a rough estimate of the magnitude of these nonlinearities, we can assume that at fields comparable to the inner atomic electric field E_{at}, polarizations of different order are of about the same magnitude, so that $\varepsilon_0 \chi^{(i)} E_{at}^i \approx \varepsilon_0 \chi^{(1)} E_{at}$. Since $\chi^{(1)}$ is of the order of unity, this estimate yields $\chi^{(i)} \approx 1/E_{at}^{i-1}$. Taking as a reference the field

© Springer International Publishing Switzerland 2016
G.A. Reider, *Photonics*, DOI 10.1007/978-3-319-26076-1_8

of a hydrogen nucleus at a distance equal to Bohr's radius ($a_0 = 5.3 \times 10^{-11}$ m), $E_{\text{at}} \approx e/4\pi\varepsilon_0 a_0^2 \approx 5 \times 10^{11}$ Vm^{-1}, we obtain reasonable agreement with experimental values, with $\chi^{(2)}$ ranging between 10^{-13} and 10^{-10} V^{-1}m, and $\chi^{(3)}$ ranging between 10^{-23} and 10^{-18} V^{-2}m^2.

To get a feeling for the optical intensities required to generate significant nonlinear polarizations, we assume a field of 10^8 Vm^{-1}, which is less than $10^{-3}E_{\text{at}}$, and use Eq. (1.71) to obtain a value of 10^{13} Wm^{-2}; with a very tightly focused cw 10 W laser, such intensities can be reached. Usually, pulsed lasers with much higher peak powers are used to produce nonlinear optical effects.

The (non)linear susceptibility is a tensor, usually given in cartesian representation, as known from the treatment of wave propagation in anisotropic media (Sect. 2.3). Using Einstein's convention, (8.1) can be expressed as

$$P_i = \varepsilon_0 \chi_{ij}^{(1)} E_j + \varepsilon_0 \chi_{ijk}^{(2)} E_j E_k + \varepsilon_0 \chi_{ijkl}^{(3)} E_j E_k E_l + \dots \qquad (8.2)$$

Symmetry has a strong impact on tensors; certain symmetries actually rule out particular effects because all elements of the relevant tensor vanish. For example, the nonlinear optical susceptibility of second (even) order is zero in centrosymmetric materials. This becomes obvious if we look at the quadratic polarization induced by an electric field in a centrosymmetric medium which is invariant under the operation of inversion; inversion changes the sign of polar vectors: $\mathbf{E} \rightarrow -\mathbf{E}$ and $\mathbf{P}^{(2)} \rightarrow -\mathbf{P}^{(2)}$. On the other hand, $\mathbf{P}^{(2)}(-\mathbf{E}) = \varepsilon_0 \chi^{(2)} \mathbf{E}\mathbf{E} = \mathbf{P}^{(2)}(\mathbf{E})$. Centrosymmetry forces the quadratic polarization and thus the second order nonlinear susceptibility to vanish, $\chi^{(2)} = 0$.

In a more general way, invariance under a certain symmetry operation implies that the transformed tensor is equal to the original; in the formulation of Eq. (2.116), $m'_{ijk} = m_{ijk}$. These relations give rise to a number of equations between the tensor elements (note that m'_{ijk} is a linear combination of all m_{ijk}). Inversion, for example, is represented by the transformation matrix Eq. (2.3.1.1) $A_{ij} = -\delta_{ij}$, so that $\chi_{ijk}^{(2)}{}' = (-1)^3 \chi_{ijk}^{(2)}$; for centrosymmetric media we obtain

$$-\chi_{ijk}^{(2)} = \chi_{ijk}^{(2)} = 0. \qquad (8.3)$$

Other symmetries (such as mirror planes or two-, three-, four-, or sixfold rotations), also reduce the number of non-vanishing elements or establish linear relations between them. Table 8.1 shows the structure of the $\chi^{(2)}$-tensor for some important point groups.

If the field driving the nonlinear polarization is monochromatic, the sequence of fields in Eq. (8.2) is irrelevant and we have the additional symmetry $\chi_{ijk}^{(2)} = \chi_{ikj}^{(2)}$. The pair of interchangeable indices is sometimes contracted into a single index

Table 8.1 Non-zero elements of $\chi^{(2)}$ for selected point groups

Point group	$\chi^{(2)}_{ijk}$	$\chi^{(2)}_{i\xi}$
$\bar{4}2m$	$\chi^{(2)}_{123} = \chi^{(2)}_{213}(=)\chi^{(2)}_{132} = \chi^{(2)}_{231}$	$= \chi^{(2)}_{14}$
	$\chi^{(2)}_{312} = \chi^{(2)}_{321}$	$= \chi^{(2)}_{36}$
$\bar{4}3m$	$\chi^{(2)}_{123} = \chi^{(2)}_{213}(=)\chi^{(2)}_{132} = \chi^{(2)}_{231}$	$= \chi^{(2)}_{14}$
mm2	$\chi^{(2)}_{311}$	$= \chi^{(2)}_{31}$
	$\chi^{(2)}_{322}$	$= \chi^{(2)}_{32}$
	$\chi^{(2)}_{333}$	$= \chi^{(2)}_{33}$
	$\chi^{(2)}_{131}(=)\chi^{(2)}_{113}$	$= \chi^{(2)}_{15}$
	$\chi^{(2)}_{223}(=)\chi^{(2)}_{232}$	$= \chi^{(2)}_{24}$
4mm	$\chi^{(2)}_{131} = \chi^{(2)}_{232} = \chi^{(2)}_{113} = \chi^{(2)}_{223}$	$= \chi^{(2)}_{15}$
	$\chi^{(2)}_{311} = \chi^{(2)}_{322}$	$= \chi^{(2)}_{31}$
	$\chi^{(2)}_{333}$	$= \chi^{(2)}_{33}$
3m	$\chi^{(2)}_{131} = \chi^{(2)}_{232}(=)\chi^{(2)}_{113} = \chi^{(2)}_{223}$	$= \chi^{(2)}_{15}$
	$\chi^{(2)}_{222} = -\chi^{(2)}_{211} = -\chi^{(2)}_{112} = -\chi^{(2)}_{121}$	$= \chi^{(2)}_{22}$
	$\chi^{(2)}_{311} = \chi^{(2)}_{322}$	$= \chi^{(2)}_{31}$
	$\chi^{(2)}_{333}$	$= \chi^{(2)}_{33}$
32	$\chi^{(2)}_{111} = -\chi^{(2)}_{122} = -\chi^{(2)}_{221} = -\chi^{(2)}_{212}$	$= \chi^{(2)}_{11}$
	$\chi^{(2)}_{123} = -\chi^{(2)}_{213}(=)\chi^{(2)}_{132} = -\chi^{(2)}_{231}$	$= \chi^{(2)}_{14}$

Table 8.2 Piezoelectric index contraction

jk	$\rightarrow$	ξ
11	$\rightarrow$	1
22	$\rightarrow$	2
33	$\rightarrow$	3
23, 32	$\rightarrow$	4
13, 31	$\rightarrow$	5
12, 21	$\rightarrow$	6

according to the piezoelectric contraction, Table 8.2, $\chi^{(2)}_{ijk} = \chi^{(2)}_{i\xi}$; examples are $\chi^{(2)}_{21}$ instead of $\chi^{(2)}_{211}$ and $\chi^{(2)}_{14}$ instead of $\chi^{(2)}_{123} = \chi^{(2)}_{132}$.

8.1.1 Frequency Mixing

An important consequence of the nonlinear response is the generation of sum and difference frequencies, also termed frequency mixing (Fig. 8.1). Let us assume an input field containing two distinct frequencies

$$\mathbf{E}(\mathbf{x}, t) = \frac{1}{2}\left[\tilde{\mathbf{E}}(\mathbf{x}, \omega_1)e^{j\omega_1 t} + \tilde{\mathbf{E}}(\mathbf{x}, \omega_2)e^{j\omega_2 t} + c.c.\right]; \tag{8.4}$$

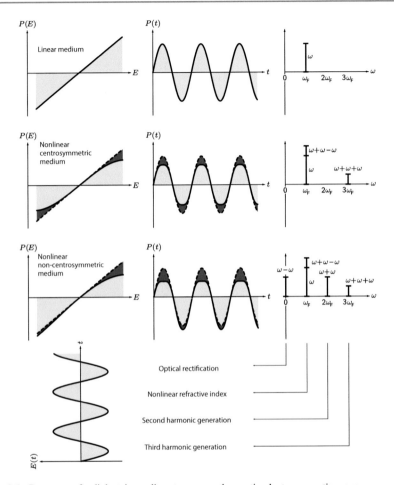

Fig. 8.1 Response of a dielectric medium to a monochromatic electromagnetic wave

the quadratic nonlinear polarization then assumes the form

$$\mathbf{P}^{(2)}(\mathbf{x}, t) = \varepsilon_0 \chi^{(2)} \mathbf{E}(\mathbf{x}, t) \mathbf{E}(\mathbf{x}, t)$$

$$= \varepsilon_0 \frac{1}{4} \chi^{(2)} \Big[\tilde{\mathbf{E}}(\mathbf{x}, \omega_1) \tilde{\mathbf{E}}(\mathbf{x}, \omega_1) e^{j 2\omega_1 t}$$

$$+ \tilde{\mathbf{E}}(\mathbf{x}, \omega_2) \tilde{\mathbf{E}}(\mathbf{x}, \omega_2) e^{j 2\omega_2 t}$$

$$+ 2\tilde{\mathbf{E}}(\mathbf{x}, \omega_1) \tilde{\mathbf{E}}(\mathbf{x}, \omega_2) e^{j(\omega_1 + \omega_2) t}$$

$$+ 2\tilde{\mathbf{E}}(\mathbf{x}, \omega_1) \tilde{\mathbf{E}}^*(\mathbf{x}, \omega_2) e^{j(\omega_1 - \omega_2) t}$$

$$+ \tilde{\mathbf{E}}(\mathbf{x}, \omega_1) \tilde{\mathbf{E}}^*(\mathbf{x}, \omega_1)$$

$$+ \tilde{\mathbf{E}}(\mathbf{x}, \omega_2) \tilde{\mathbf{E}}^*(\mathbf{x}, \omega_2) + c.c. \Big]. \qquad (8.5)$$

This can be written as

$$
\begin{aligned}
\mathbf{P}^{(2)}(\mathbf{x}, t) = \frac{1}{2} & \left[\tilde{\mathbf{P}}(\mathbf{x}, 2\omega_1) e^{j2\omega_1 t} + c.c. \right] \\
+ \frac{1}{2} & \left[\tilde{\mathbf{P}}(\mathbf{x}, 2\omega_2) e^{j2\omega_2 t} + c.c. \right] \\
+ \frac{1}{2} & \left[\tilde{\mathbf{P}}(\mathbf{x}, \omega_1 + \omega_2) e^{j(\omega_1 + \omega_2)t} + c.c. \right] \\
+ \frac{1}{2} & \left[\tilde{\mathbf{P}}(\mathbf{x}, \omega_1 - \omega_2) e^{j(\omega_1 - \omega_2)t} + c.c. \right] \\
+ \frac{1}{2} & \left[\tilde{\mathbf{P}}(\mathbf{x}, 0) + c.c. \right]
\end{aligned}
\tag{8.6}
$$

and contains frequency components at $2\omega_i$ (second harmonic)

$$
\tilde{\mathbf{P}}(\mathbf{x}, 2\omega_i) = \frac{1}{2} \varepsilon_0 \chi^{(2)} \tilde{\mathbf{E}}(\mathbf{x}, \omega_i) \tilde{\mathbf{E}}(\mathbf{x}, \omega_i),
\tag{8.7}
$$

at the sum frequency $\omega_1 + \omega_2$

$$
\tilde{\mathbf{P}}(\mathbf{x}, \omega_1 + \omega_2) = \varepsilon_0 \chi^{(2)} \tilde{\mathbf{E}}(\mathbf{x}, \omega_1) \tilde{\mathbf{E}}(\mathbf{x}, \omega_2),
\tag{8.8}
$$

at the difference frequency $\omega_1 - \omega_2$

$$
\tilde{\mathbf{P}}(\mathbf{x}, \omega_1 - \omega_2) = \varepsilon_0 \chi^{(2)} \tilde{\mathbf{E}}(\mathbf{x}, \omega_1) \tilde{\mathbf{E}}^*(\mathbf{x}, \omega_2)
\tag{8.9}
$$

and finally a dc-component at $\omega = 0$

$$
\tilde{\mathbf{P}}(\mathbf{x}, 0) = \frac{1}{2} \varepsilon_0 \chi^{(2)} \left[\tilde{\mathbf{E}}(\mathbf{x}, \omega_1) \tilde{\mathbf{E}}^*(\mathbf{x}, \omega_1) + \tilde{\mathbf{E}}(\mathbf{x}, \omega_2) \tilde{\mathbf{E}}^*(\mathbf{x}, \omega_2) \right].
\tag{8.10}
$$

This last term is proportional to the intensity of the respective field and is equivalent to optical rectification (Fig. 8.2). Note that the factor $\frac{1}{2}$ in Eq. (8.7) accompanies the second harmonic components, but is missing in the sum and difference components; the reason is that mixed terms show up twice in the calculation.

Let us further assume plane waves $\tilde{\mathbf{E}}(\mathbf{x}, \omega_i) = \tilde{\mathbf{E}}(\omega_i) e^{-j\mathbf{k}_i \cdot \mathbf{x}}$ so that the driving field is

$$
\mathbf{E}(\mathbf{x}, t) = \frac{1}{2} \left[\tilde{\mathbf{E}}(\omega_1) e^{-j(\mathbf{k}_1 \cdot \mathbf{x} - \omega_1 t)} + \tilde{\mathbf{E}}(\omega_2) e^{-j(\mathbf{k}_2 \cdot \mathbf{x} - \omega_2 t)} + c.c. \right].
\tag{8.11}
$$

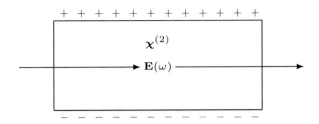

Fig. 8.2 Optical rectification in a quadratic nonlinear medium

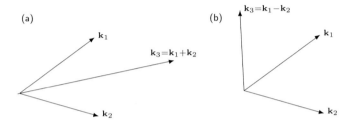

Fig. 8.3 Phase matching diagram for **(a)** sum frequency and **(b)** difference frequency generation

The sum frequency component of the nonlinear polarization then is

$$\mathbf{P}^{(2)}_{\omega_1+\omega_2}(\mathbf{x}, t) = \frac{1}{2}\left[\varepsilon_0 \chi^{(2)}\tilde{\mathbf{E}}(\omega_1)\tilde{\mathbf{E}}(\omega_2)e^{-j[(\mathbf{k}_1+\mathbf{k}_2)\cdot\mathbf{x}-(\omega_1+\omega_2)t]} + c.c.\right]. \quad (8.12)$$

This is a planar polarization wave with the wave vector $\mathbf{k}_1 + \mathbf{k}_2$, that can serve as a source term for an electromagnetic wave

$$\mathbf{E}_{\omega_3}(\mathbf{x}, t) = \frac{1}{2}\left[\tilde{\mathbf{E}}(\omega_3)e^{-j(\mathbf{k}_3\cdot\mathbf{x}-\omega_3 t)} + c.c.\right] \quad (8.13)$$

with the frequency ω_3

$$\omega_3 = \omega_1 + \omega_2. \quad (8.14)$$

For the coupling between the source Eq. (8.12) and the field Eq. (8.13) to be efficient, the two waves must have a constant phase relation in space, implying the equality of the wave vectors (Fig. 8.3a)

$$\mathbf{k}_3 = \mathbf{k}_1 + \mathbf{k}_2; \quad (8.15)$$

this equation is known as phase matching condition.

An analogue condition applies to difference frequency generation

$$\omega_3 = \omega_1 - \omega_2 \tag{8.16}$$

with the polarization density

$$\mathbf{P}^{(2)}_{\omega_1 - \omega_2}(\mathbf{x}, t) = \frac{1}{2}\left[\varepsilon_0 \chi^{(2)}\tilde{\mathbf{E}}(\omega_1)\tilde{\mathbf{E}}^*(\omega_2)e^{-j[(\mathbf{k}_1 - \mathbf{k}_2)\cdot\mathbf{x} - (\omega_1 - \omega_2)t]} + c.c.\right] \tag{8.17}$$

(Fig. 8.3b) in the form

$$\mathbf{k}_3 = \mathbf{k}_1 - \mathbf{k}_2. \tag{8.18}$$

Nonlinear optical processes can also be understood in a photon picture: in this framework, sum frequency generation (SFG) is merging of two photons of energy $\hbar\omega_1$ and $\hbar\omega_2$, respectively, to a new one of energy $\hbar\omega_3$ (Fig. 8.4a). The phase matching condition can be interpreted as momentum conservation, since the momentum of a photon is equal to $\hbar\mathbf{k}$.

While the total energy and momentum of the participating photons is conserved, the total number of photons is not. Assuming a common direction of propagation of all fields, the photon flux in SFG must obey the equations $dF_{\omega_3}/dz = -dF_{\omega_1}/dz = -dF_{\omega_2}/dz$ since one photon at ω_1 and ω_2, respectively, is annihilated to produce one ω_3-photon. In terms of intensity $I_{\omega_i} = \hbar\omega_i F_{\omega_i}$, this implies

$$\frac{dI_{\omega_3}}{\omega_3\,dz} = -\frac{dI_{\omega_1}}{\omega_1\,dz} = -\frac{dI_{\omega_2}}{\omega_2\,dz}. \tag{8.19}$$

These relations are called Manley–Rowe relations and will be derived from purely electrodynamic arguments later [Eq. (8.80)].

Difference frequency generation can be understood as "splitting" of a photon of energy $\hbar\omega_3$ into two with energies $\hbar\omega_1$ and $\hbar\omega_2$ (Fig. 8.4b). The Manley–Rowe relations are, of course, also valid in this case.

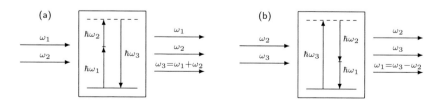

Fig. 8.4 Photon diagram of (**a**) sum and (**b**) difference frequency generation

In the mixing processes described, two input fields produce a new field, so that actually three fields are present, giving rise to a three wave mixing process where each field interacts with one of the others to produce a polarization at the frequency of the third one. Depending on the relative intensity of the fields, it is common to distinguish different processes, albeit physically they are all the same; in the following listing of quadratic effects, it is understood that $\omega_3 - \omega_1 - \omega_2 = 0$:

- In SFG, I_{ω_1} and I_{ω_2} are large and the goal is to produce efficiently a field at ω_3.
- Second harmonic generation (SHG) is a degenerate case of SFG with $\omega_1 = \omega_2 =:$ ω called the fundamental and $\omega_3 = 2\omega$ the second harmonic.
- In parametric amplification, a weak input signal at ω_1 interacts with a strong pump wave at ω_3, splitting pump photons into pairs of photons; one of this pair is a replica of the signal photon and enhances the signal; the second photon, at the difference frequency ω_2 contributes to a new wave (called "idler") not present at the entrance. The idler wave, however, also interacts with the pump to enhance itself and the signal wave. Because of this mutual enhancement, both waves are amplified in a quasi-exponential fashion. In contrast to the amplification by stimulated emission, this gain process in principle works at any signal frequency below the pump frequency.
- In the process of parametric frequency conversion, a weak signal field at ω_1 interacts with a strong pump wave at ω_2 to shift the signal frequency (usually in the mid-IR) to ω_3, where it can be conveniently detected by quantum detectors. According to the Manley–Rowe relations, the number of output photons cannot exceed the number of signal photons; ideally, the conversion is one-to-one.

8.1.2 Anharmonic Oscillator

Modeling the linear susceptibility as a response of harmonically oscillating electrons [Eq. (2.51)] provided important insights in the nature of light–matter interaction. We now extend this simple model to the (quadratic) nonlinear susceptibility by adding a quadratic term to the restoring force term[1]

$$m_e\ddot{x} + b\dot{x} + ax + Dx^2 = -eE(t). \qquad (8.20)$$

To solve this nonlinear differential equation, we treat the quadratic term as a small perturbation, $|Dx^2| \ll |ax|$ which is neglected in a first step of iteration. The driving field is assumed to have two frequencies, $E(t) = E_{\omega_1}(t) + E_{\omega_2}(t)$, with

$$E_{\omega_{1,2}}(t) = \frac{1}{2}\left[\tilde{E}(\omega_{1,2})e^{j\omega_{1,2}t} + c.c.\right]. \qquad (8.21)$$

---

[1] Introducing higher order nonlinear restoring force terms allows, in a similar fashion, estimating nonlinear susceptibilities of corresponding order.

According to Eqs. (2.52) and (2.56), the electron displacement can be expressed as

$$\tilde{x}(\omega_{1,2}) = -\chi^{(1)}(\omega_{1,2}) \frac{\varepsilon_0}{n_e e} \tilde{E}(\omega_{1,2}). \tag{8.22}$$

In a second step, we use this displacement to calculate the nonlinear restoring force component

$$Dx^2(t) = \frac{1}{4} D \left[\tilde{x}(\omega_1) e^{j\omega_1 t} + \tilde{x}(\omega_2) e^{j\omega_2 t} + c.c. \right]^2 \tag{8.23}$$

which contains, among other frequency components, a sum frequency term

$$\frac{1}{2} \left[D\tilde{x}(\omega_1)\tilde{x}(\omega_2) e^{j\omega_3 t} + c.c. \right] \tag{8.24}$$

that has no counterpart in the remaining Eq. (8.20). Next, we adjust the motion of the electrons by adding a small sum frequency component $x_{\omega_3}(t) = \frac{1}{2} \left[\tilde{x}(\omega_3) e^{j\omega_3 t} + c.c. \right]$ such that the linear force component $m_e \ddot{x}_{\omega_3} + b\dot{x}_{\omega_3} + ax_{\omega_3}$ compensates the nonlinear force term at ω_3

$$\begin{aligned}
\tilde{x}(\omega_3) &= \frac{-D\tilde{x}(\omega_1)\tilde{x}(\omega_2)}{m_e[(\omega_0^2 - \omega_3^2) + j\omega_3 \Gamma]} \\
&= \frac{-D\varepsilon_0^3}{n_e^3 e^4} \chi^{(1)}(\omega_1)\chi^{(1)}(\omega_2)\chi^{(1)}(\omega_3)\tilde{E}(\omega_1)\tilde{E}(\omega_2),
\end{aligned} \tag{8.25}$$

where Eqs. (8.22) and (2.56) have been used.

The complex amplitude of the nonlinear polarization density is then

$$\tilde{P}(\omega_3) = -n_e e\tilde{x}(\omega_3) \tag{8.26}$$

and comparison with Eq. (8.8) in the form

$$\tilde{P}(\omega_3) = \varepsilon_0 \chi^{(2)} \tilde{E}(\omega_1)\tilde{E}(\omega_2), \tag{8.27}$$

allows us to express the nonlinear susceptibility as

$$\chi^{(2)}(\omega_3; \omega_1, \omega_2) = \frac{D\varepsilon_0^2}{n_e^2 e^3} \chi^{(1)}(\omega_1)\chi^{(1)}(\omega_2)\chi^{(1)}(\omega_3). \tag{8.28}$$

This is Miller's rule, which states that the nonlinear susceptibility is proportional to the product of the three linear susceptibilities at the frequencies involved. In

Table 8.3 Nonlinear susceptibility $\chi_{i\xi}^{(2)}$ [10^{-12} mV^{-1}] of selected materials; symmetry allowed but very small components are not included

Material		Symmetry	$\chi_{i\xi}^{(2)}$	Transparency range [µm]	n_o, n_e, n_z
KDP	KH$_2$PO$_4$	$\bar{4}$2m	$\chi_{14}^{(2)} = 0.8$	0.22–1.50	1.494, 1.495
			$\chi_{36}^{(2)} = 0.9$		
KTP	KTiOPO$_4$	mm2	$\lvert\chi_{31}^{(2)}\rvert = 13$	0.35–4.50	1.737, 1.745, 1.829
			$\lvert\chi_{32}^{(2)}\rvert = 10$		
			$\lvert\chi_{33}^{(2)}\rvert = 27$		
			$\lvert\chi_{15}^{(2)}\rvert = 12$		
			$\lvert\chi_{24}^{(2)}\rvert = 15$		
BBO	BaB$_2$O$_4$	3m	$\lvert\chi_{22}^{(2)}\rvert = 3.2$	0.19–3.00	1.655, 1.542
			$\lvert\chi_{31}^{(2)}\rvert = 0.2$		
Lithium niobate	LiNbO$_3$	3m	$\chi_{22}^{(2)} = 5.2$	0.40–5.00	2.232, 2.150
			$\chi_{31}^{(2)} = -9.7$		
			$\chi_{33}^{(2)} = -88$		
Gallium arsenide	GaAs	$\bar{4}$3m	$\chi_{14}^{(2)} = 270$	>0.9	3.491
α-Quartz	SiO$_2$	32	$\lvert\chi_{11}^{(2)}\rvert = 0.7$	>0.18	1.544, 1.553
			$\lvert\chi_{14}^{(2)}\rvert = 0.006$		

particular, it states that $\chi^{(2)}$ shows resonant enhancement if $\chi^{(1)}$ is resonant at any of the three frequencies.

It turns out, moreover, that the prefactor $D\varepsilon_0^2/n_e^2e^3$ in Eq. (8.28) has about the same value ($\approx -0.3 \times 10^{-12}$ mV^{-1}) for a wide variety of dielectrics and semiconductors (see Problem 1). Since $\chi^{(1)} = n^2 - 1$, this implies that media with large refractive index also exhibit a large nonlinear susceptibility (compare Table 8.3). The large $\chi^{(2)}$-values of lithium niobate ($n = 2.2$, $\chi^{(1)} = 3.8$) and gallium arsenide ($n = 3.3$, $\chi^{(1)} = 9.8$), compared to KDP ($n = 1.5$, $\chi^{(1)} = 1.25$) are consistent with this rule, for example.

8.2   Second Order Processes

8.2.1   Second Harmonic Generation

In Sect. 8.1.1, we have calculated the nonlinear polarization induced by plane fundamental waves. The various frequency components of this nonlinear polarization wave are also plane waves with a wave vector that is either the sum or the difference of the fundamental wave vectors. We now want to evaluate the electromagnetic field that is radiated by this polarization wave, for the important example of SHG. We start with the wave equation (1.17)

$$-\nabla^2\mathbf{E} + \mu_0\frac{\partial^2(\varepsilon_0\mathbf{E} + \mathbf{P})}{\partial t^2} = 0, \tag{8.29}$$

where according to Eq. (1.8), $\mathbf{D} = \varepsilon_0 \mathbf{E} + \mathbf{P}$. With $\mathbf{P} = \mathbf{P}^{(1)} + \mathbf{P}^{(2)}$ and $\varepsilon_0 \mathbf{E} + \mathbf{P}^{(1)} = \varepsilon_0 \varepsilon \mathbf{E}$ we obtain

$$\nabla^2 \mathbf{E} - \frac{\varepsilon}{c_0^2} \frac{\partial^2 \mathbf{E}}{\partial t^2} = \mu_0 \frac{\partial^2 \mathbf{P}^{(2)}}{\partial t^2}. \tag{8.30}$$

The term on the right-hand side is the time derivative of the nonlinear polarization current density, which is the source term for the second harmonic electromagnetic field.

Assuming the fields to propagate along the z-direction with the transverse field components

$$E_{\omega,i}(\mathbf{x}, t) = \frac{1}{2} \left[\tilde{E}_i(\omega) e^{-j(k_\omega z - \omega t)} + c.c. \right], \quad i = 1, 2 \tag{8.31}$$

and

$$E_{2\omega,i}(\mathbf{x}, t) = \frac{1}{2} \left[\tilde{E}_i(2\omega) e^{-j(k_{2\omega} z - 2\omega t)} + c.c. \right], \quad i = 1, 2, \tag{8.32}$$

the operator ∇^2 in Eq. (8.30) reduces to $\partial^2/\partial z^2$ and

$$\frac{\partial^2}{\partial z^2} \tilde{E}_i(2\omega) e^{-j(k_{2\omega} z - 2\omega t)}$$

$$= \left[\frac{\partial^2}{\partial z^2} \tilde{E}_i(2\omega) - 2jk_{2\omega} \frac{\partial}{\partial z} \tilde{E}_i(2\omega) - k_{2\omega}^2 \tilde{E}_i(2\omega) \right] e^{-j(k_{2\omega} z - 2\omega t)}$$

$$\approx \left[-2jk_{2\omega} \frac{\partial}{\partial z} \tilde{E}_i(2\omega) - k_{2\omega}^2 \tilde{E}_i(2\omega) \right] e^{-j(k_{2\omega} z - 2\omega t)}, \tag{8.33}$$

where we have neglected the second order spatial derivative assuming that $\tilde{E}_i(2\omega)$ changes slowly on the length scale of the wavelength, $|\partial^2 \tilde{E}_i(2\omega)/\partial z^2| \ll |k_{2\omega} \partial \tilde{E}_i(2\omega)/\partial z|$ [slowly varying envelope approximation, compare Eq. (3.3)].

For the second order time derivatives we obtain, using Eq. (1.27)

$$-\frac{\varepsilon_{2\omega} \partial^2}{c_0^2 \partial t^2} \tilde{E}_i(2\omega) e^{-j(k_{2\omega} z - 2\omega t)} = \varepsilon_{2\omega} \frac{4\omega^2}{c_0^2} \tilde{E}_i(2\omega) e^{-j(k_{2\omega} z - 2\omega t)} =$$

$$= k_{2\omega}^2 \tilde{E}_i(2\omega) e^{-j(k_{2\omega} z - 2\omega t)} \tag{8.34}$$

and

$$\mu_0 \frac{\partial^2}{\partial t^2} \tilde{P}_i(2\omega) e^{-j(2k_\omega z - 2\omega t)} = -4\mu_0 \omega^2 \tilde{P}_i(2\omega) e^{-j(2k_\omega z - 2\omega t)}; \tag{8.35}$$

substitution of Eqs. (8.33)–(8.35) in Eq. (8.30) yields

$$-2jk_{2\omega}\frac{\partial}{\partial z}\tilde{E}_i(2\omega)e^{-jk_{2\omega}z} = -4\mu_0\omega^2\tilde{P}_i(2\omega)e^{-j2k_{\omega}z}. \tag{8.36}$$

With the source term Eq. (8.7)

$$\tilde{P}_i(2\omega) = \frac{1}{2}\varepsilon_0\chi^{(2)}_{ijk}\tilde{E}_j(\omega)\tilde{E}_k(\omega), \tag{8.37}$$

Equation (8.36) assumes the form

$$\frac{\partial}{\partial z}\tilde{E}_i(2\omega) = -\frac{j\omega}{2c_0n_{2\omega}}\chi^{(2)}_{ijk}\tilde{E}_j(\omega)\tilde{E}_k(\omega)e^{j\Delta kz}, \tag{8.38}$$

where $k_{2\omega} = 2\omega n_{2\omega}/c_0$, $\mu_0\varepsilon_0 = 1/c_0^2$, and

$$\Delta k := k_{2\omega} - 2k_{\omega} \tag{8.39}$$

is the deviation from the phase matching condition. Integration of Eq. (8.38) with the boundary condition $\tilde{\mathbf{E}}(2\omega)|_{z=0} = 0$ yields

$$\tilde{E}_i(2\omega)\Big|_{z=l} = -\frac{j\omega}{2c_0n_{2\omega}}\chi^{(2)}_{ijk}\tilde{E}_j(\omega)\tilde{E}_k(\omega)\frac{e^{j\Delta kl}-1}{j\Delta k}, \quad i = 1, 2, \tag{8.40}$$

where it was tacitly assumed that the conversion efficiency from the fundamental to the SH-field is so small that the fundamental amplitude remains practically constant.

We now write the fundamental field $\tilde{E}_i(\omega) = \tilde{E}(\omega)e_i$ as a product of the scalar amplitude $\tilde{E}(\omega)$ and a unit vector e_i; then, $\chi^{(2)}_{ijk}\tilde{E}_j(\omega)\tilde{E}_k(\omega) = \chi^{(2)}_{ijk}e_je_k\tilde{E}\tilde{E}$. Equation (1.71) relates the complex amplitude $\tilde{E}$ to the intensity I

$$I = \frac{n\tilde{E}\tilde{E}^*}{2Z_0} \qquad \tilde{E}\tilde{E}^* = \frac{2Z_0I}{n}, \tag{8.41}$$

so that we can express the fundamental field in terms of its intensity; with the identity $|e^{jx} - 1|^2 = 2(1 - \cos x) = 4\sin^2(x/2)$ we obtain for the SH-intensity

$$I_{2\omega} = I_{\omega}^2\frac{\omega^2Z_0l^2}{2c_0^2n_{2\omega}n_{\omega}^2}\sum_{i=1,2}\left|\chi^{(2)}_{ijk}e_je_k\right|^2\left[\frac{\sin(\Delta kl/2)}{\Delta kl/2}\right]^2. \tag{8.42}$$

Fig. 8.5 Second harmonic (SH) power as a function of interaction length for different degrees of phase mismatch $|\Delta k|$; a is an arbitrary scaling factor

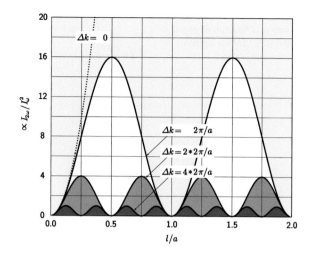

For a given phase mismatch Δk, the SH-intensity is a periodic function of z

$$I_{2\omega}(l) \propto \frac{\sin^2(\Delta kl/2)}{(\Delta k)^2},$$ (8.43)

with maxima at $l = 2\pi(m + \frac{1}{2})/|\Delta k|$ (Fig. 8.5); the spatial period

$$l_c = \frac{2\pi}{|\Delta k|} = \frac{2\pi}{|k_{2\omega} - 2k_\omega|} = \frac{\lambda_0}{2|n_{2\omega} - n_\omega|}$$ (8.44)

is called, somewhat misleadingly, coherence length, not to be confused with the same expression from Sect. 4.4.1. The first intensity maximum is reached at $l = l_c/2$, which is, for obvious reasons, also the maximum useful crystal length. Note that the maximum intensity is proportional to l_c^2, or $1/|n_{2\omega} - n_\omega|^2$.

Alternatively, if the phase mismatch is varied for a given crystal length, we obtain

$$I_{2\omega}(l) \propto \left[\frac{\sin(\Delta kl/2)}{\Delta kl/2}\right]^2.$$ (8.45)

This function has a central maximum at $\Delta k = 0$ and maxima of higher order whose height decays with the square of the order (Fig. 8.6). The phase matched intensity follows from Eq. (8.42)

$$I_{2\omega} = I_\omega^2 \frac{\omega^2 Z_0 l^2}{2c_0^2 n_{2\omega} n_\omega^2} \sum_{i=1,2} \left|\chi_{ijk}^{(2)} e_j e_k\right|^2.$$ (8.46)

Fig. 8.6 Second harmonic power as a function of phase mismatch $|\Delta k|$ for a given interaction length l

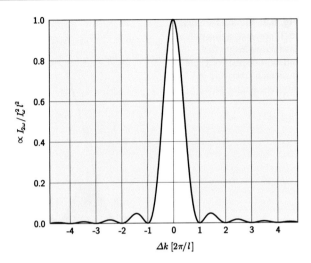

8.2.1.1 SHG of Gaussian Beams

According to Eq. (8.42), the SH conversion efficiency $I_{2\omega}/I_\omega$ for plane waves is proportional to the fundamental intensity and to the square of the interaction length under phase matched conditions. In practice, nonlinear processes are driven by laser beams of a certain power and it seems natural to increase the conversion efficiency by focusing the fundamental beam as tightly as possible. Apart from limits due to optical damage, however, there is a tradeoff between the intensity in the beam waist and the useful interaction length within the nonlinear medium, because the beam diverges with an angle that is inversely proportional to the beam waist diameter.

To get a more quantitative picture, we consider a Gaussian beam, with an intensity profile Eq. (3.23)

$$I_\omega(r) = \frac{2P_\omega}{\pi w_{0,\omega}^2} e^{-2r^2/w_{0,\omega}^2}, \tag{8.47}$$

where P_ω is the beam power and $w_{0,\omega}$ the beam waist. If the interaction length l is so short that the beam profile does not widen significantly, the phase matched SH-intensity according to Eq. (8.42) is

$$I_{2\omega} = \frac{\omega^2 Z_0 l^2}{2c_0^2 n_{2\omega} n_\omega^2} \frac{4P_\omega^2}{\pi^2 w_{0,\omega}^4} \sum_{i=1,2} \left| \chi_{ijk}^{(2)} e_j e_k \right|^2 e^{-2r^2/w_{0,2\omega}^2}, \tag{8.48}$$

which is also Gaussian but with a reduced waist $w_{0,2\omega} = w_{0,\omega}/\sqrt{2}$. A measure for the useful interaction length is the confocal distance $2z_0 = w_{0,\omega}^2 k_\omega = w_{0,\omega}^2 n_\omega \omega/c_0$ [Eq. (3.12)]. If we shape our beam such that the confocal range matches the length of the nonlinear crystal, $l = 2z_0$ or $w_{0,\omega}^2 = l c_0/n_\omega \omega$, the resulting SH power

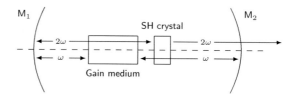

Fig. 8.7 Intracavity SHG: mirror M_1 is high reflecting at ω and 2ω, the coupling mirror M_2 is high reflecting at ω and transparent at 2ω

$P_{2\omega} = I_{2\omega}(0)\pi w_{0,2\omega}^2/2$ is approximately

$$P_{2\omega} = P_\omega^2 \frac{\omega^3 Z_0 l}{2\pi c_0^3 n_{2\omega} n_\omega} \sum_{i=1,2} \left| \chi_{ijk}^{(2)} e_j e_k \right|^2$$

$$= P_\omega^2 \frac{4\pi^2 Z_0 l}{\lambda_0^3 n_{2\omega} n_\omega} \sum_{i=1,2} \left| \chi_{ijk}^{(2)} e_j e_k \right|^2 ; \qquad (8.49)$$

SH generation optimized in this way thus increases only *linearly* with the crystal length.

The SH conversion efficiency $P_{2\omega}/P_\omega$ is proportional to the fundamental beam power; with a typical value of $\chi^{(2)} = 10^{-12}\,\mathrm{mV}^{-1}$ and a crystal length of $l = 1\,\mathrm{cm}$, Eq. (8.49) yields a conversion efficiency of $10^{-4}P_\omega$ at a fundamental wavelength of $\lambda_0 = 1.064\,\mu\mathrm{m}$ (Nd:YAG laser). The overall efficiency can be greatly improved by placing the doubling crystal inside the cavity of the laser that provides the fundamental beam (Fig. 8.7).

8.2.2   Phase Matching

8.2.2.1 Birefringent Phase Matching

The frequency dependence (dispersion) of the propagation index of the nonlinear crystal is responsible for the phase mismatch Eq. (8.44) and the finite coherence length $l_c = \lambda_0/2|n_{2\omega} - n_\omega|$. Because of the large frequency difference between fundamental and SH, $|n_{2\omega} - n_\omega|$ is usually substantial (several %), and the coherence length amounts to not more than some ten wavelengths.

One possible way to achieve phase matching is to exploit the natural birefringence present in many nonlinear materials (Fig. 8.8). According to Eq. (2.145), the propagation index of the extraordinary wave depends on the angle θ between the wave vector and the optical axis

$$\frac{1}{n^2(\theta)} = \frac{\cos^2\theta}{n_o^2} + \frac{\sin^2\theta}{n_e^2}. \qquad (8.50)$$

To obtain phase matching, the propagation direction and the polarization of the fundamental and the SH wave, respectively, is chosen such that the propagation index of the ordinary wave at ω matches that of the extraordinary wave at 2ω (or *vice versa*) (Fig. 8.8). In this configuration, phase matching requires $n_{2\omega}(\theta) = n_{\omega,o}$,

Fig. 8.8 Phase matching in a uniaxial crystal

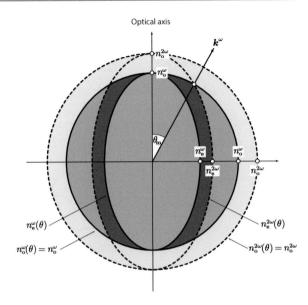

or

$$\frac{1}{n_{\omega,o}^2} = \frac{\cos^2\theta}{n_{2\omega,o}^2} + \frac{\sin^2\theta}{n_{2\omega,e}^2};\tag{8.51}$$

this defines the phase matching angle θ_m

$$\cos^2\theta_m = \frac{n_{\omega,o}^{-2} - n_{2\omega,e}^{-2}}{n_{2\omega,o}^{-2} - n_{2\omega,e}^{-2}}.\tag{8.52}$$

The compensation of dispersion by birefringence is possible only if the birefringence is larger than the dispersion, since there is no intersection between the k-surfaces at ω and 2ω otherwise. One problem with this scheme is that in birefringent materials, the Poynting vector (the direction of energy transport) also depends on the polarization state, so that the fundamental and the SH beam tend to separate spatially [Fig. 2.33]. Only if $\theta_m = 90°$, this effect can completely be avoided (90°-phase matching). Since the propagation index is temperature dependent, 90°-phase matching can be realized in some cases by heating the crystal.

8.2.2.2 Quasi-Phase Matching

A very powerful alternative is to modify the nonlinear susceptibility periodically (Fig. 8.9) with a spatial period equal to l_c; $\chi^{(2)}$ can then be written as a Fourier series

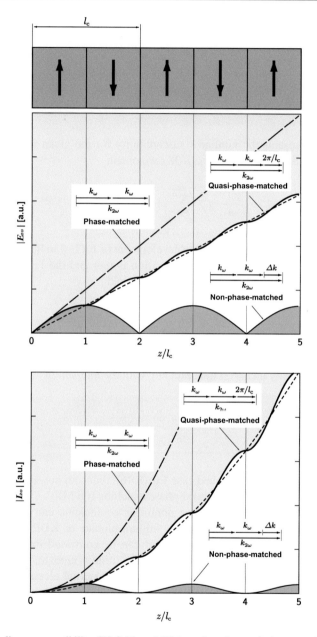

Fig. 8.9 Nonlinear susceptibility, SH-field, and SH-intensity of a quasi-phase matched (periodically poled) frequency doubler

$$\chi^{(2)}(z) = \chi_0^{(2)} \sum_{m=-\infty}^{\infty} F_m e^{-j2m\pi/l_c z}$$

$$= \chi_0^{(2)} \sum_{m=-\infty}^{\infty} F_m e^{-jm\Delta kz}, \tag{8.53}$$

where $\chi_0^{(2)}$ represents the nonlinear susceptibility for the given fundamental field configuration. Substituted into Eq. (8.38), we obtain

$$\frac{\partial}{\partial z} \tilde{E}_i(2\omega) = -\frac{j\omega}{2c_0 n_{2\omega}} \chi_0^{(2)} \sum_{m=-\infty}^{\infty} F_m \tilde{E}_j(\omega) \tilde{E}_k(\omega) e^{j(1-m)\Delta kz}; \tag{8.54}$$

Figure 8.10 shows the corresponding development of $\tilde{E}_i(2\omega)$ in the complex plane.

For $m = 1$, the phase matching condition is met and the Fourier component $\chi_0^{(2)} F_1$ is the source of a linearly increasing SH-field

$$\frac{\partial}{\partial z} \tilde{E}_i(2\omega) = -\frac{j\omega}{2c_0 n_{2\omega}} \chi_0^{(2)} F_1 \tilde{E}_j(\omega) \tilde{E}_k(\omega); \tag{8.55}$$

the other Fourier components are responsible for a superimposed spatial oscillation. The output intensity increases quadratically with the interaction length

$$I_{2\omega} = F_1^2 I_\omega^2 l^2 \frac{\omega^2 Z_0}{2c_0^2 n_{2\omega} n_\omega^2} \sum_{i=1,2} \left| \chi_{ijk}^{(2)} e_j e_k \right|^2, \tag{8.56}$$

as in the perfectly phase matched case Eq. (8.46), reduced, however, by the factor F_1^2. This technique is known as quasi-phase matching (QPM).

Periodic structures of alternating nonlinear coefficients can be produced by "poling" of ferroelectric media such as lithium niobate or KDP. These materials exist in two metastable configurations that can be converted into each other by a strong (20 kV/mm) dc-electric field pulse. In lithium niobate, for example, the metal ions change sites under the influence of an external electric field in the z-direction (Fig. 8.11), converting the initial configuration into its mirror image that has nonlinear coefficients $\chi_{311}^{(2)}$ and $\chi_{333}^{(2)}$ of same magnitude but opposite sign. An initially homogeneous crystal is transformed into a periodic structure of ferroelectric domains by applying lithographically a periodic electrode structure to the surface of the crystal which is removed after the poling process. Since the first Fourier component of the resulting rectangular $\chi^{(2)}$-modulation is $F_1 = 2/\pi$, the resulting the SH-intensity is smaller by a factor of $(2/\pi)^2 = 0.4$ than a perfectly phase matched output. This disadvantage is compensated by the possibility to pick a fundamental field configuration (propagation direction and polarization state) that

Fig. 8.10 Locus of the SH amplitude in the complex plane: phase mismatch results in a closed circular loop (*dashed line*) with a curvature proportional to the phase mismatch; quasi-phase matching periodically alters the sign of the source term, providing monotonic amplitude growth (*solid line*). The two amplitudes shown refer to the same propagation distance in the crystal, the *dashed arrow* representing the phase mismatched signal, the *solid* one the quasi-phase matched result

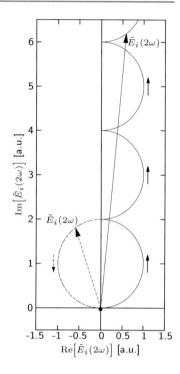

optimizes the effective $\chi_0^{(2)}$ and avoids the walk-off of the fundamental and SH Poynting vectors.

8.2.3 Optical Parametric Amplification

Another important quadratic effect is parametric amplification of a signal at frequency ω_s in the presence of a strong pump wave at ω_p ($\omega_p > \omega_s$). It is a special case of difference frequency generation and produces an additional wave at $\omega_i = \omega_p - \omega_s$ (the so-called idler wave). The process is called parametric because it can be understood as a modulation of the system parameter χ at the frequency ω_p; the generation of the idler results from a beating of the signal with the pump frequency and *vice versa*. Under phase matched conditions, the beat wave adds coherently to the signal or idler wave, respectively, resulting in amplification (optical parametric amplification, OPA).

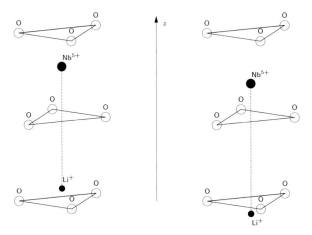

Fig. 8.11 Two stable lattice configurations of LiNbO$_3$ with opposite values of $\chi_{33}^{(2)}$ and $\chi_{31}^{(2)}$; they can be converted into each other by applying a dc-electric field pulse along the z-axis

For a simplified treatment, we assume in the following that all wave vectors are collinear and treat the fields as scalars; we use the complex amplitudes

$$\tilde{P}(\omega_s) = \varepsilon_0 \chi^{(2)} \tilde{E}(\omega_p) \tilde{E}^*(\omega_i)$$

$$\tilde{P}(\omega_i) = \varepsilon_0 \chi^{(2)} \tilde{E}(\omega_p) \tilde{E}^*(\omega_s)$$

$$\tilde{P}(\omega_p) = \varepsilon_0 \chi^{(2)} \tilde{E}(\omega_s) \tilde{E}(\omega_i). \tag{8.57}$$

The interaction of the three waves is described by a set of three coupled differential equations which can be easily derived along the lines of Eq. (8.38) (slowly varying envelope approximation):

$$\frac{d\tilde{E}(\omega_s)}{dz} = -\frac{j\omega_s}{2c_0 n_{\omega_s}} \chi^{(2)} \tilde{E}(\omega_p) \tilde{E}^*(\omega_i) e^{-j(k_p - k_i - k_s)z}$$

$$\frac{d\tilde{E}(\omega_i)}{dz} = -\frac{j\omega_i}{2c_0 n_{\omega_i}} \chi^{(2)} \tilde{E}(\omega_p) \tilde{E}^*(\omega_s) e^{-j(k_p - k_s - k_i)z}$$

$$\frac{d\tilde{E}(\omega_p)}{dz} = -\frac{j\omega_p}{2c_0 n_{\omega_p}} \chi^{(2)} \tilde{E}(\omega_s) \tilde{E}(\omega_i) e^{-j(k_s + k_i - k_p)z}. \tag{8.58}$$

For convenience, we introduce a normalized field amplitude $\tilde{A}_i$ such that $\tilde{A}_i\tilde{A}_i^*$ equals the photon flux F_{ω_i}

$$\tilde{A}_i\tilde{A}_i^* = F_{\omega_i} = \frac{n_{\omega_i}\tilde{E}(\omega_i)\tilde{E}^*(\omega_i)}{2Z_0\hbar\omega_i}, \tag{8.59}$$

or

$$\tilde{A}_i := \frac{\tilde{E}(\omega_i)}{\sqrt{2Z_0\hbar\omega_i/n_{\omega_i}}}. \tag{8.60}$$

Equations (8.58) then assume the form

$$\frac{d\tilde{A}_s}{dz} = -j\kappa\tilde{A}_i^*\tilde{A}_p e^{-j\Delta k z} \tag{8.61}$$

$$\frac{d\tilde{A}_i}{dz} = -j\kappa\tilde{A}_s^*\tilde{A}_p e^{-j\Delta k z} \tag{8.62}$$

$$\frac{d\tilde{A}_p}{dz} = -j\kappa\tilde{A}_s\tilde{A}_i e^{j\Delta k z}, \tag{8.63}$$

where $\Delta k = k_p - k_i - k_s$ and the coupling factor κ is given by

$$\kappa = \chi^{(2)}\sqrt{\frac{\omega_s\omega_i\omega_p}{n_{\omega_s}n_{\omega_i}n_{\omega_p}}}\frac{\sqrt{2\hbar Z_0}}{2c_0}. \tag{8.64}$$

For the further treatment, we neglect pump depletion ($d\tilde{A}_p/dz \approx 0$) and assume $\tilde{A}_p$ to be positive and real, $\tilde{A}_p = |\tilde{A}_p|$ (which can always be arranged by proper choice of time zero). Under phase matched conditions ($\Delta k = 0$) Eqs. (8.61) and (8.62) have the form

$$\frac{d\tilde{A}_s}{dz} = -j\kappa'\tilde{A}_i^* \tag{8.65}$$

$$\frac{d\tilde{A}_i}{dz} = -j\kappa'\tilde{A}_s^* \tag{8.66}$$

where κ' is the normalized coupling coefficient

$$\kappa' = \kappa\tilde{A}_p = \kappa\sqrt{I_{\omega_p}/\hbar\omega_p} = \chi^{(2)}\sqrt{\frac{\omega_s\omega_i}{n_{\omega_s}n_{\omega_i}n_{\omega_p}}\frac{Z_0 I_{\omega_p}}{2c_0^2}}. \tag{8.67}$$

Taking the derivative of Eq. (8.65) and substituting $d\tilde{A}_i^* / dz$ from Eq. (8.66) we obtain

$$\frac{d^2\tilde{A}_s}{dz^2} - \kappa'^2 \tilde{A}_s = 0 \tag{8.68}$$

and in analogous way

$$\frac{d^2\tilde{A}_i}{dz^2} - \kappa'^2 \tilde{A}_i = 0. \tag{8.69}$$

The solutions of these equations are linear combinations of $e^{\kappa' z}$ and $e^{-\kappa' z}$,

$$\tilde{A}_{s,i}(z) = \left(a_{s,i}^+ e^{\kappa' z} + a_{s,i}^- e^{-\kappa' z} \right). \tag{8.70}$$

To determine the coefficients $a_s^{\pm}$, we use the boundary conditions $\tilde{A}_s(0) = a_s^+ + a_s^-$ and $\tilde{A}_i(0) = 0$ from which follows, using Eq. (8.65), $a_s^+ - a_s^- = 0$ so that

$$\tilde{A}_s(l) = \tilde{A}_s(0) \cosh \kappa' l; \tag{8.71}$$

$\tilde{A}_i$ follows from Eq. (8.65)

$$\tilde{A}_i(l) = -j\tilde{A}_s^*(0) \sinh \kappa' l. \tag{8.72}$$

The corresponding photon flux densities are

$$F_{\omega_s}(l) = F_{\omega_s}(0) \cosh^2 \kappa' l \tag{8.73}$$

$$F_{\omega_i}(l) = F_{\omega_s}(0) \sinh^2 \kappa' l. \tag{8.74}$$

The quasi-exponential growth pertains as long as pump depletion is negligible (Fig. 8.12). With a pump intensity of 10^7 Wcm^{-2}, $\omega_{s,i} \approx 10^{15}$ s^{-1}, $n = 2$, and $\chi^{(2)} = 10^{-11}$ mV^{-1}, the gain coefficient is $\kappa' \approx 0.5$ cm^{-1} and the (power) gain after 2 cm is equal to $\cosh^2(2) = 2.25$.

Inclusion of the phase mismatch term modifies Eqs. (8.68) and (8.69) to

$$\frac{d^2\tilde{A}_s}{dz^2} + j\Delta k \frac{d\tilde{A}_s}{dz} - \kappa'^2 \tilde{A}_s = 0 \tag{8.75}$$

$$\frac{d^2\tilde{A}_i}{dz^2} + j\Delta k \frac{d\tilde{A}_i}{dz} - \kappa'^2 \tilde{A}_i = 0, \tag{8.76}$$

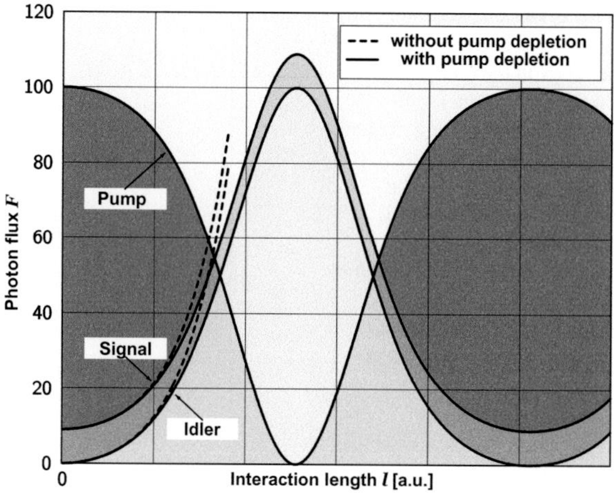

Fig. 8.12 Signal, idler, and pump photon flux density in a phase matched optical parametric amplifier as a function of the interaction length; after complete consumption of the pump, signal and idler interact to regenerate the pump wave by sum frequency generation. *Dotted lines* show the result Eq. (8.71), valid for negligible pump depletion

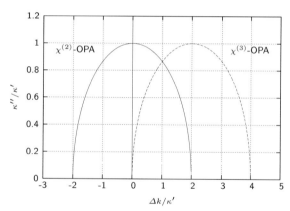

Fig. 8.13 Normalized gain coefficient κ''/κ' as a function of the phase mismatch $\Delta k/\kappa'$ for an OPA based on second or third order nonlinearity, respectively

with the solutions

$$\tilde{A}_{s,i}(z) = \left(a_{s,i}^{+}e^{\kappa''z} + a_{s,i}^{-}e^{-\kappa''z}\right)e^{-j(\Delta k/2)z}, \qquad (8.77)$$

where the modified gain coefficient κ'' is given by

$$\kappa'' := \sqrt{\kappa'^2 - (\Delta k/2)^2}; \qquad (8.78)$$

note that only real values of κ'' provide exponential growth. In contrast to SHG, where any phase mismatch results in an oscillatory dependence of the output power on the interaction length, parametric amplification allows for a certain mismatch $|\Delta k| < 2|\kappa'|$ (Fig. 8.13). Outside this interval, the output signals oscillate along the propagation distance. An inspection of Fig. 8.14 explains the transition from

Fig. 8.14 Locus of the signal and idler amplitudes of an OPA in the complex plane. In the *upper panel*, the straight arrows along the real and imaginary axis, respectively, refer to the phase matched situation, while the spirals describe quasi-exponential growth for small phase mismatch $|\Delta k| < 2|\kappa'|$. *Lower panel*: excessive phase mismatch $|\Delta k| > 2|\kappa'|$ results in a closed loop locus and oscillatory output power; the numbers indicate equally spaced points along the propagation direction

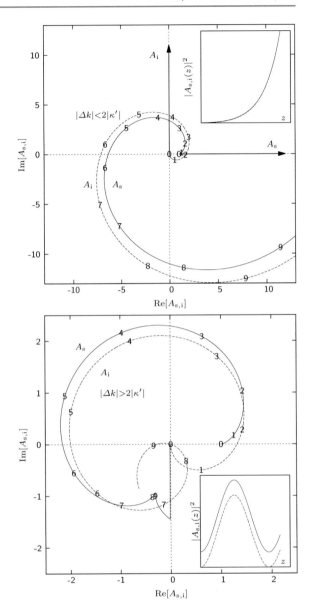

the amplifying to the oscillating regime. Phase mismatch results in a curved locus of $\tilde{A}_{s,i}(z)$ in the complex plane; different from SHG, where the modulus of the differential field increment $|dA|$ is constant and the locus, if mismatched, is a circle (Fig. 8.10), $|d\tilde{A}_{s,i}|$ is initially growing and the trajectory is a spiral within the gain interval. Outside the gain regime, the curvature is so strong that the idler returns periodically to zero and the signal to its initial value.

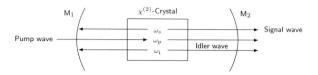

Fig. 8.15 Optical parametric oscillator (OPO); mirror M_1 is transparent at ω_p and high reflecting at $\omega_{s,i}$, while M_2 is high reflecting at ω_p and partially transmitting at ω_s

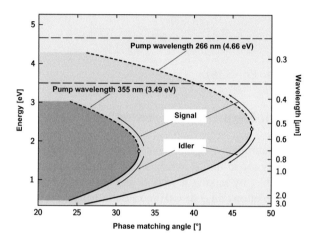

Fig. 8.16 Tuning of an OPO (pump wavelength 355 or 266 nm, respectively) by tilting the nonlinear crystal: at a given phase matching angle, a distinct pair of signal and idler frequencies starts to oscillate

In the absence of an input signal, the parametric process can also start from so-called parametric fluorescence, the equivalent of spontaneous emission; the emitted photon pairs are "entangled" and can be used for quantum cryptography. In combination with a resonator, a parametric amplifier can operate as oscillator (optical parametric oscillator, OPO). To realize an OPO, the nonlinear medium is placed in a resonator with mirrors that have high reflectance at ω_i (Fig. 8.15); the pump radiation is usually pulsed. Like a laser, an OPO starts to emit coherent radiation only above a certain threshold, where the parametric gain compensates the resonator losses. The oscillator can be frequency tuned by changing the phase matching angle (Fig. 8.16).

8.2.3.1 Manley–Rowe Relations

Equations (8.61)–(8.63) also allow deriving the Manley–Rowe relations Eq. (8.19) from purely electromagnetic arguments: multiplication of Eq. (8.61) with $\tilde{A}_s^*$, Eq. (8.62) with $\tilde{A}_i^*$, and the conjugate of Eq. (8.63) with $\tilde{A}_p$ yields

$$\frac{d\tilde{A}_s}{dz}\tilde{A}_s^* = \frac{d\tilde{A}_i}{dz}\tilde{A}_i^* = -\frac{d\tilde{A}_p^*}{dz}\tilde{A}_p; \qquad (8.79)$$

since $\tilde{A}_i \tilde{A}_i^*$ is the photon flux density at ω_i,

$$\frac{\mathrm{d}F_{\omega_i}}{\mathrm{d}z} = \frac{\mathrm{d}I_{\omega_i}}{\hbar\omega_i\,\mathrm{d}z} = \frac{\mathrm{d}(\tilde{A}_i \tilde{A}_i^*)}{\mathrm{d}z} = \frac{\mathrm{d}\tilde{A}_i}{\mathrm{d}z}\tilde{A}_i^* + \frac{\mathrm{d}\tilde{A}_i^*}{\mathrm{d}z}\tilde{A}_i. \tag{8.80}$$

Substitution of Eq. (8.79) and its conjugate, respectively, yields

$$\frac{\mathrm{d}F_{\omega_1}}{\mathrm{d}z} = \frac{\mathrm{d}F_{\omega_2}}{\mathrm{d}z} = -\frac{\mathrm{d}F_{\omega_3}}{\mathrm{d}z}, \tag{8.81}$$

which is equivalent to Eq. (8.19).

8.2.4 Parametric Frequency Conversion*

In the process of parametric frequency conversion (or frequency up-conversion), a signal at frequency ω_{ir} in the IR is converted to a frequency $\omega_{vis} = \omega_{ir} + \omega_p$, typically in the visible, by mixing it with an intense pump field at frequency ω_p. Neglecting pump depletion ($\mathrm{d}A_p/\mathrm{d}z \approx 0$) and assuming phase matching, the process is described by

$$\frac{\mathrm{d}\tilde{A}_{ir}}{\mathrm{d}z} = -\mathrm{j}\kappa\tilde{A}_{vis}\tilde{A}_p^* = -\mathrm{j}\kappa'\tilde{A}_{vis}$$

$$\frac{\mathrm{d}\tilde{A}_{vis}}{\mathrm{d}z} = -\mathrm{j}\kappa\tilde{A}_{ir}\tilde{A}_p = -\mathrm{j}\kappa'\tilde{A}_{ir}, \tag{8.82}$$

where $\tilde{A}_p$ is assumed to be real and positive and

$$\kappa' = \kappa\tilde{A}_p = \kappa\tilde{A}_p^* = \kappa\sqrt{I_{\omega_p}/\hbar\omega_p}. \tag{8.83}$$

The two equations Eq. (8.82) can be decoupled

$$\frac{\mathrm{d}^2\tilde{A}_{ir}}{\mathrm{d}z^2} + \kappa'^2\tilde{A}_{ir} = 0$$

$$\frac{\mathrm{d}^2\tilde{A}_{vis}}{\mathrm{d}z^2} + \kappa'^2\tilde{A}_{vis} = 0 \tag{8.84}$$

and, with the boundary condition $\tilde{A}_{vis}(0) = 0$, have the solution

$$\tilde{A}_{ir}(l) = \tilde{A}_{ir}(0)\cos\kappa'l$$

$$\tilde{A}_{vis}(l) = -\mathrm{j}\tilde{A}_{ir}(0)\sin\kappa'l, \tag{8.85}$$

Fig. 8.17 Photon flux density of the IR-signal (**a**) and upconverted wave (**b**) in a parametric upconverter

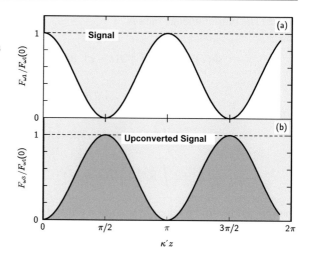

corresponding to the photon flux densities

$$F_{\omega_{ir}}(l) = F_{\omega_{ir}}(0) \cos^2 \kappa' l \qquad (8.86)$$

$$F_{\omega_{vis}}(l) = F_{\omega_{vis}}(0) \sin^2 \kappa' l. \qquad (8.87)$$

After the interaction distance $l = \pi/2\kappa'$, all signal photons are converted (Fig. 8.17). This technique is frequently used in IR-spectroscopy, where fast and highly sensitive quantum detectors are unavailable, for the detection of very small IR-signals.

8.2.5 Second Order Autocorrelation

The multiplicative capabilities of second order polarization can be used to obtain the intensity autocorrelation of optical light pulses. For this purpose, a quadratic nonlinear crystal is inserted in the output of a Michelson interferometer (output 1 in Fig. 4.1). A light pulse launched into the interferometer is split into two replicas that travel along the two legs; recombination at the beam splitter produces the superposition $\propto E(t) + E(t - \tau)$ of the two replicas, delayed in respect to each other by the time $\tau = 2\Delta s/c_0$, where Δs is the (adjustable) length difference of the legs. The nonlinear crystal produces a second harmonic field proportional to $[E(t) + E(t - \tau)]^2$; after removal of the fundamental radiation by a filter, a (slow) detector operating as integrator measures the SH pulse energy $\int \left| [E(t) + E(t - \tau)]^2 \right|^2 \, dt$.

For an analysis of the output, we adopt a complex notation $E \propto [A(t)e^{j\omega t} + A^*(t)e^{-j\omega t}]$ where the pulse envelope $A(t)$ is normalized such that $I(t) = A(t)A^*(t)$ is the momentary intensity; note that $A(t)$ is complex and can include a time dependent phase. The output fluence then is

$$\Phi(\tau) = \int I(t)^2 + I(t - \tau)^2 \, dt$$

$$+4 \int [I(t)I(t - \tau)] \, dt$$

$$+4 \int [I(t) + I(t - \tau)] \mathrm{Re} \left[A(t)A^*(t - \tau)e^{j\omega\tau} \right] dt$$

$$+2 \int \mathrm{Re} \left[A^2(t)(A^2(t - \tau))^* e^{j2\omega\tau} \right] dt; \qquad (8.88)$$

the integral has to be taken over several pulse durations. After multiplication with the beam cross section, this is the SH energy resulting from a single input pulse for a given delay time τ. To measure the entire correlation function, a train of identical pulses is launched into the interferometer and the output is recorded as a function of τ; the result for a Gaussian pulse is shown in Fig. 8.18.

The first term in Eq. (8.88) is a constant offset Φ_0 and can be used to normalize the output; $\Phi(\tau)/\Phi_0$ oscillates between ≈ 0 and 8. The second term represents the intensity autocorrelation. The third term in Eq. (8.88) is the amplitude auto-correlation multiplied by a τ-dependent factor, and the fourth term is the SH amplitude autocorrelation. Equation 8.88 is called interferometric autocorrelation; it is sensitive to a time varying phase of the amplitude, i.e., to a frequency chirp within the pulse (see Sect. 3.2.1.6).

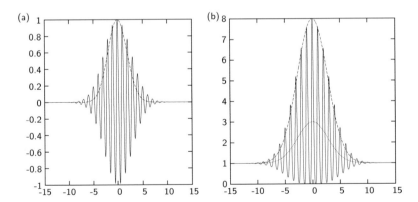

Fig. 8.18 A Fourier limited Gaussian pulse and its interferometric autocorrelation: (**a**) electric field (*solid line*) and intensity envelope (*dashed line*) of the pulse, (**b**) output of second order autocorrelator (*solid line*); note that the envelope of the interferometric autocorrelation deviates from the intensity autocorrelation (*dashed line*); interpretation of this deviation allows for an analysis of the possible chirp of the pulse. Averaging over the interference fringes provides the exact intensity autocorrelation, offset by 1 (*dotted line*)

If the delay τ is scanned so rapidly that the detector cannot resolve the oscillating terms three and four, taking the time average instead, the normalized output signal reduces to

$$\Phi(\tau)/\Phi_0 = 1 + 2\int [I(t)I(t-\tau)]\,d\tau, \qquad (8.89)$$

which is the exact intensity autocorrelation, offset by 1.

It is not possible to completely reconstruct the pulse envelope $A(t)$ from the intensity or interferometric autocorrelation, because there is an infinite manifold of wave functions yielding the same autocorrelation; with numerical means and physical intuition, however, pulses can be constructed that reproduce the measured interferometric autocorrelation function.

An extension of this technique is frequency resolved optical gating, where in addition to the SH pulse energy its spectrum is recorded as a function of the delay time (Trebino 2000). This allows for a complete temporal reconstruction of the electric field of a light pulse, with exception of the carrier envelope phase.

8.3 Third Order Processes

8.3.1 Third Harmonic Generation

The third order nonlinear susceptibility is a fourth rank tensor; the symmetry properties of the medium determine which of the coefficients are equal to zero or linear combinations of other components. In contrast to the second order susceptibility, all symmetry classes allow for non-zero elements; isotropic and centrosymmetric media such as glasses, gases, and liquids exhibit the non-vanishing components $\chi_{iiii}^{(3)}$, $\chi_{iijj}^{(3)}$, $\chi_{jiij}^{(3)}$, and $\chi_{ijij}^{(3)}$, $i = 1, 2, 3, i \neq j$. The following discussion is limited to such materials.

We first discuss the case of a monochromatic input field, linearly polarized along the x-axis. The nonlinear polarization is then parallel to the input field

$$P_1^{(3)}(t) = \varepsilon_0 \chi_{1111}^{(3)} E_1^3(t), \quad P_{2,3}^{(3)} = 0, \qquad (8.90)$$

and we can adopt a scalar formulation $\chi_{1111}^{(3)} =: \chi^{(3)}$

$$P^{(3)}(t) = \varepsilon_0 \chi^{(3)} E^3(t). \qquad (8.91)$$

With

$$E(t) = \frac{1}{2}\left[\tilde{E}(\omega)e^{j\omega t} + c.c.\right] \qquad (8.92)$$

we obtain

$$P^{(3)}(t) = \frac{1}{2}\left[\tilde{P}^{(3)}(\omega)e^{j\omega t} + c.c.\right] + \frac{1}{2}\left[\tilde{P}^{(3)}(3\omega)e^{j3\omega t} + c.c.\right],$$ (8.93)

where

$$\tilde{P}^{(3)}(\omega) = \frac{3}{4}\varepsilon_0\chi^{(3)}\tilde{E}(\omega)\tilde{E}(\omega)\tilde{E}^*(\omega)$$ (8.94)

$$\tilde{P}^{(3)}(3\omega) = \frac{1}{4}\varepsilon_0\chi^{(3)}\tilde{E}(\omega)\tilde{E}(\omega)\tilde{E}(\omega).$$ (8.95)

The polarization at 3ω (Fig. 8.19a) serves as a source term for third harmonic generation (THG) in a similar way as we have seen in SHG. THG is a possible process to produce coherent radiation in the UV, where lasers are difficult to operate. If the phase matching condition $k_{3\omega} = 3k_\omega$ is met, the TH field grows linearly (and the TH-power quadratically) with the propagation distance; the conversion efficiency scales quadratically with the input power.

One term that has no counterpart in quadratic nonlinearities is the nonlinear polarization at the *fundamental* frequency (Fig. 8.19b). This term gives rise to the intensity dependence of the propagation index, the so-called Kerr effect.

8.3.2 Optical Kerr Effect

8.3.2.1 Self Phase Modulation
Driven by a monochromatic field, the combined linear and cubic polarization at the fundamental frequency ω is

$$\tilde{P}(\omega) = \varepsilon_0\chi^{(1)}\tilde{E}(\omega) + \frac{3}{4}\varepsilon_0\chi^{(3)}\tilde{E}(\omega)\left[\tilde{E}(\omega)\tilde{E}^*(\omega)\right]$$

$$= \varepsilon_0\left[\chi^{(1)} + \frac{3\chi^{(3)}Z_0}{2n}I_\omega\right]\tilde{E},$$ (8.96)

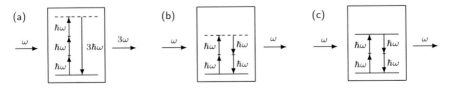

Fig. 8.19 Photon diagram of third harmonic generation (**a**), optical Kerr effect (**b**), and two-photon absorption (**c**)

where the electric field amplitude is expressed by $I = n\tilde{E}\tilde{E}^*/2Z_0$. This is equivalent to an effective susceptibility

$$\chi(I) = \chi_0 + \Delta\chi = \chi_0 + \chi^{(3)}\frac{3Z_0}{2n}I \tag{8.97}$$

where $\chi_0 = \chi^{(1)}$ is the susceptibility at very low intensities. The small nonlinear contribution to the susceptibility gives rise to a modified propagation index [Eq. (2.74)]

$$n(I) \approx n_0 + \frac{1}{2n_0}\Delta\chi(I) = n_0 + \frac{1}{2n_0}\frac{3\chi^{(3)}Z_0}{2n_0}I, \tag{8.98}$$

with $n_0 = \sqrt{\chi_0 + 1}$. This can be written as

$$n = n_0 + n_2 I \tag{8.99}$$

where

$$n_2 := \frac{3Z_0}{4n_0^2}\chi^{(3)} \tag{8.100}$$

is called nonlinear propagation index. Typical values for n_2 in glasses are between 10^{-20} and $10^{-18}\,\mathrm{m^2W^{-1}}$; silica ($SiO_2$), for example, has an n_2-value of $3.2 \times 10^{-20}\,\mathrm{m^2W^{-1}}$. According to Miller's rule Eq. (8.28), $\chi^{(3)}$ scales with $(n^2 - 1)^4$, and optically dense media consequently have relatively large n_2-values: lead glasses with a propagation index in the range of 2.4 exhibit n_2-values above $10^{-18}\,\mathrm{m^2W^{-1}}$.

The change of the propagation index results in a phase change

$$\Delta\phi = n_2 k_0 I d \tag{8.101}$$

of the wave, where d is the propagation distance. This effect is called self-phase modulation (SPM) and is proportional to the intensity of the field. Important manifestations of SPM are the Kerr lens (Sect. 3.1.3.4), spectral broadening (Sect. 3.2.2.1), and soliton propagation (Sect. 3.2.2.2).

8.3.2.2 Cross Phase Modulation
We now extend the discussion to polychromatic (yet linearly polarized) fields with discrete frequencies ω_i

$$E(t) = \frac{1}{2}\left[\sum_{i=1}^{m}\tilde{E}(\omega_i)e^{j\omega_i t} + c.c.\right]; \tag{8.102}$$

the resulting nonlinear polarization comprises a total of $(2m)^3$ terms of the form $\varepsilon_0\chi^{(3)}\tilde{E}^{(*)}(\omega_l)\tilde{E}^{(*)}(\omega_j)\tilde{E}^{(*)}(\omega_k)e^{j(\pm\omega_l\pm\omega_j\pm\omega_k)t}$, of which we only consider the terms oscillating at ω_1

$$\tilde{P}^{(3)}(\omega_1) = \frac{3}{4}\varepsilon_0\chi^{(3)}\left[\tilde{E}(\omega_1)\tilde{E}^*(\omega_1) + 2\sum_{i=2}^{m}\tilde{E}(\omega_i)\tilde{E}^*(\omega_i)\right]\tilde{E}(\omega_1)$$

$$=: \varepsilon_0\Delta\chi\tilde{E}(\omega_1). \tag{8.103}$$

The resulting propagation index is

$$n = n_0 + n_2 I_{\omega_1} + 2n_2\sum_{i=2}^{m} I_{\omega_i}. \tag{8.104}$$

In addition to the self-term Eq. (8.96), the susceptibility (and the propagation index) is also modified by a term proportional to twice the sum of all other field intensities; this effect is called cross phase modulation (XPM).

XPM can be used to influence light waves by light (all-optical devices). It also is responsible for cross talk in wavelength division multiplexing in optical communication fibers.

8.3.2.3 Nonlinear Polarization Rotation

Elliptically polarized light exhibits, in any cartesian coordinate system, two orthogonal field vector components, so that not only $\chi^{(3)}_{1111}$ but (in isotropic media) also $\chi^{(3)}_{iijj}$, $\chi^{(3)}_{ijji}$, and $\chi^{(3)}_{ijij}$, $i \neq j$, come into play; for symmetry reasons, $\chi^{(3)}_{iijj} + \chi^{(3)}_{ijij} + \chi^{(3)}_{ijji} = \chi^{(3)}_{iiii}$ and $\chi^{(3)}_{ijij}(\omega = \omega + \omega - \omega) = \chi^{(3)}_{iijj}$. Assuming propagation along z, $E_3 = 0$, and the polarization at the fundamental frequency is

$$\Delta\tilde{P}_i(\omega) = \frac{6}{4}\varepsilon_0\chi^{(3)}_{1122}\tilde{E}_i(\omega)\sum_{j=1}^{3}\tilde{E}_j(\omega)\tilde{E}_j^*(\omega) + \frac{3}{4}\varepsilon_0\chi^{(3)}_{1221}\tilde{E}_i^*(\omega)\sum_{j=1}^{3}\tilde{E}_j(\omega)\tilde{E}_j(\omega),$$

$$\tag{8.105}$$

which is equivalent to

$$\Delta\tilde{\mathbf{P}} = \frac{6}{4}\varepsilon_0\chi^{(3)}_{1122}(\tilde{\mathbf{E}}\cdot\tilde{\mathbf{E}}^*)\tilde{\mathbf{E}} + \frac{3}{4}\varepsilon_0\chi^{(3)}_{1221}(\tilde{\mathbf{E}}\cdot\tilde{\mathbf{E}})\tilde{\mathbf{E}}^*. \tag{8.106}$$

In general, this contribution to the polarization is not parallel to the driving field; this means that the nonlinear interaction modifies the polarization state during propagation. An exception is circularly polarized light that turns out to be an eigenstate of the isotropic Kerr medium. We thus use a circularly polarized base [Eq. (1.78)]

$$\tilde{\mathbf{E}} = E^+ \sigma^+ + E^- \sigma^-, \tag{8.107}$$

$$\Delta \tilde{\mathbf{P}} = \Delta P^+ \sigma^+ + \Delta P^- \sigma^- \tag{8.108}$$

where

$$\sigma^\pm = \frac{1}{\sqrt{2}} \begin{bmatrix} 1 \\ \pm j \end{bmatrix} \tag{8.109}$$

with the properties $(\sigma^\pm)^* = \sigma^\mp$, $\sigma^+ \cdot \sigma^+ = \sigma^- \cdot \sigma^- = 0$, and $\sigma^+ \cdot \sigma^- = \sigma^- \cdot \sigma^+ = 1$. In this base, the nonlinear polarization components are

$$\Delta P^+ = \frac{6}{4} \varepsilon_0 \left[\chi^{(3)}_{1122} (|E^+|^2 + |E^-|^2) E^+ + \chi^{(3)}_{1221} (E^+ E^-) E^{-*} \right]$$

$$= \frac{6}{4} \varepsilon_0 \left[\chi^{(3)}_{1122} (|E^+|^2 + |E^-|^2) + \chi^{(3)}_{1221} |E^-|^2 \right] E^+ =: \varepsilon_0 \Delta \chi^+ E^+,$$

$$\Delta P^- = \frac{6}{4} \varepsilon_0 \left[\chi^{(3)}_{1122} (|E^+|^2 + |E^-|^2) + \chi^{(3)}_{1221} |E^+|^2 \right] E^- =: \varepsilon_0 \Delta \chi^- E^-. \tag{8.110}$$

These relations are scalar, implying that the circularly polarized states $\sigma^\pm$ are propagation eigenstates (compare optically active media Sect. 2.4.1) with the propagation indices

$$n^\pm \approx n_0 + \frac{3Z_0}{2n_0^2} \left[\chi^{(3)}_{1122} (I^+ + I^-) + \chi^{(3)}_{1221} I^\mp \right], \tag{8.111}$$

where $I^\pm = n E^\pm E^{\pm *} / 2Z_0$. While the values of $E^\pm$ and thus the ellipticity Eq. (1.131) is not altered by the Kerr effect, the difference

$$n^+ - n^- = \frac{3Z_0}{2n_0^2} \chi^{(3)}_{1221} (I^- - I^+) \tag{8.112}$$

results, according to Eqs. (1.90) and (1.124), respectively, in a rotation of the polarization ellipse by the angle $\Delta\varphi = (n^+ - n^-) k_0 d/2$, where k_0 is the wave number; the effect is known as nonlinear polarization rotation. In combination with a polarizer, the (intensity-dependent) polarization rotation can be used as intensity discriminator, similar to the setup shown in Fig. 7.23, and is employed for mode locked operation of fiber lasers [see, e.g., Fermann and Hartl (2009)].

For linearly polarized light ($I^- = I^+ =: I_0/2$), the eigenstates are degenerate, and Eq. (8.111) reduces to Eq. (8.98) because of $\chi^{(3)}_{iijj} + \chi^{(3)}_{ijij} + \chi^{(3)}_{ijji} = \chi^{(3)}_{iiii}$. Linearly and circularly polarized light, respectively, experiences SPM without change of the polarization state.

8.3.3 Third Order Parametric Amplification

OPA is also possible in the absence of a quadratic nonlinearity $\chi^{(2)}$, i.e., in (centrosymmetric) media such as glass. This is of particular interest in glass fibers that can provide high pump intensities over a long interaction length.

Similarly to OPA in quadratic media, a strong pump wave (wave vector $\mathbf{k}_p$, frequency ω_p) interacts with a signal wave ($\mathbf{k}_s$, ω_s) and an idler wave ($\mathbf{k}_i$, ω_i). In the following we assume codirectional propagation of all waves (as it is the case in fibers) and linearly polarized waves (allowing a scalar treatment). For the three waves we set

$$E_{p,s,i}(z,t) = \frac{1}{2}\left[\tilde{A}_{p,s,i}(z)e^{-j(k_{p,s,i}z - \omega_{p,s,i}t)} + c.c.\right]. \tag{8.113}$$

The nonlinear polarization has components at $\omega_{p,s,i}$ with the complex amplitudes

$$\tilde{P}_s^{(3)}(z,\omega_s) = \frac{3}{4}\varepsilon_0\chi^{(3)}\left[2|\tilde{A}_p|^2\tilde{A}_s e^{-jk_s z} + \tilde{A}_p\tilde{A}_p\tilde{A}_i^* e^{-j(2k_p - k_i)z}\right]$$

$$\tilde{P}_i^{(3)}(z,\omega_i) = \frac{3}{4}\varepsilon_0\chi^{(3)}\left[2|\tilde{A}_p|^2\tilde{A}_i e^{-jk_i z} + \tilde{A}_p\tilde{A}_p\tilde{A}_s^* e^{-j(2k_p - k_s)z}\right]$$

$$\tilde{P}_p^{(3)}(z,\omega_p) = \frac{3}{4}\varepsilon_0\chi^{(3)}|\tilde{A}_p|^2\tilde{A}_p e^{-jk_p z}, \tag{8.114}$$

where we have tacitly assumed that

$$\omega_s + \omega_i = 2\omega_p \tag{8.115}$$

and terms proportional to $\tilde{A}_s\tilde{A}_i^*$ have been neglected as of second order (no pump depletion). Thus, signal and idler appear as symmetric side bands of the pump frequency,

$$\omega_{s,i} = \omega_p \pm (\omega_s - \omega_i)/2; \tag{8.116}$$

at the entrance of the medium, the idler wave amplitude is usually zero and is built up only during propagation.

Within the slowly varying envelope approximation [compare Eq. (8.38)], we obtain coupled differential equations for the development of the complex amplitudes

$$\frac{\partial}{\partial z}\tilde{A}_s = -j\frac{3\omega_s}{8c_0 n_{\omega_s}}\chi^{(3)}\left[2|\tilde{A}_p|^2\tilde{A}_s + \tilde{A}_p\tilde{A}_p\tilde{A}_i^* e^{-j\Delta k z}\right]$$

$$\frac{\partial}{\partial z}\tilde{A}_i = -j\frac{3\omega_i}{8c_0 n_{\omega_i}}\chi^{(3)}\left[2|\tilde{A}_p|^2\tilde{A}_i + \tilde{A}_p\tilde{A}_p\tilde{A}_s^* e^{-j\Delta k z}\right]$$

$$\frac{\partial}{\partial z}\tilde{A}_p = -j\frac{3\omega_p}{8c_0 n_{\omega_p}}\chi^{(3)}|\tilde{A}_p|^2\tilde{A}_p \tag{8.117}$$

with

$$\Delta k := 2k_p - k_s - k_i. \tag{8.118}$$

To simplify matters, we choose the time zero point such that $\tilde{A}_p$ is real and positive so that $|\tilde{A}_p|^2 = \tilde{A}_p^2$, and assume the frequencies $\omega_{s,i,p}$ to differ only slightly, which allows us to introduce a common coupling coefficient κ'

$$\kappa' := \frac{3\omega_p}{8c_0 n_{\omega_p}} \chi^{(3)} |\tilde{A}_p|^2; \tag{8.119}$$

this can be expressed, with the help of Eq. (8.100) and $I_p = n|\tilde{A}_p|^2/2Z_0$, as

$$\kappa' = n_2 I_p k_0. \tag{8.120}$$

Our system of equations then is

$$\frac{\partial}{\partial z} \tilde{A}_s = -j2\kappa' \tilde{A}_s - j\kappa' \tilde{A}_i^* e^{-j\Delta k z}$$

$$\frac{\partial}{\partial z} \tilde{A}_i = -j2\kappa' \tilde{A}_i - j\kappa' \tilde{A}_s^* e^{-j\Delta k z} \tag{8.121}$$

$$\frac{\partial}{\partial z} \tilde{A}_p = -j\kappa' \tilde{A}_p. \tag{8.122}$$

With the substitution

$$\tilde{A}_{s,i} =: \tilde{A}'_{s,i} e^{-j2\kappa' z}, \quad \tilde{A}_p =: \tilde{A}'_p e^{-j\kappa' z}, \tag{8.123}$$

the system can be cast in the form

$$\frac{\partial}{\partial z} \tilde{A}'_s = -j\kappa' \tilde{A}'^*_i e^{-j\Delta' k z} \tag{8.124}$$

$$\frac{\partial}{\partial z} \tilde{A}'_i = -j\kappa' \tilde{A}'^*_s e^{-j\Delta' k z} \tag{8.125}$$

$$\frac{\partial}{\partial z} \tilde{A}'_p = 0, \tag{8.126}$$

where

$$\Delta' k = \Delta k - 2\kappa'. \tag{8.127}$$

The transformation Eq. (8.123) incorporates pump-induced SPM and XPM [Eq. (8.104)] in the modified wave vectors

$$k'_{s,i} = k_{s,i} + 2\kappa', \quad k'_p = k_p + \kappa'.$$ (8.128)

Accordingly, the phase matching condition is changed to

$$\Delta'k = 2k'_p - k'_s - k'_i = 0.$$ (8.129)

Elimination of one of the two fields Eqs. (8.124) and (8.125) yields

$$\frac{\partial^2}{\partial z^2}\tilde{A}'_{s,i} + j\Delta'k\frac{\partial}{\partial z}\tilde{A}'_{s,i} - \kappa'^2\tilde{A}'_{s,i} = 0$$ (8.130)

with the solutions

$$\tilde{A}'_{s,i}(z) = \left(a^+_{s,i}e^{\kappa''z} + a^-_{s,i}e^{-\kappa''z}\right)e^{-j(\Delta'k/2)z},$$ (8.131)

where

$$\kappa'' := \sqrt{\kappa'^2 - (\Delta'k/2)^2}.$$ (8.132)

The coefficients $a^\pm_{s,i}$ are determined in the same way that led to Eq. (8.71) and result in

$$\tilde{A}'_s(z) = \tilde{A}'_s(0)\left(\cosh\kappa''z + j\frac{\Delta'k}{2\kappa''}\sinh\kappa''z\right)e^{-j(\Delta'k/2)z}.$$ (8.133)

The signal power is proportional to

$$|\tilde{A}'_s|^2(z) = |\tilde{A}'_s(0)|^2\left[1 + \left(1 - \frac{\Delta'k^2}{4\kappa''^2}\right)\sinh^2\kappa''z\right].$$ (8.134)

According to the Manley–Rowe relations Eq. (8.19) (and using $\omega_s \approx \omega_i$), the measure for the idler power is offset by $-|\tilde{A}'_s(0)|^2$:

$$|\tilde{A}'_i|^2(z) = |\tilde{A}'_s(0)|^2\left(1 - \frac{\Delta'k^2}{4\kappa''^2}\right)\sinh^2\kappa''z.$$ (8.135)

The above solutions are valid if κ'' is real, which implies [Eq. (8.132)]

$$\kappa'^2 - (\Delta k/2 - \kappa')^2 > 0 \qquad (8.136)$$

or

$$0 < \Delta k < 4\kappa' = 4n_2 k_0 I_p. \qquad (8.137)$$

Outside this range, one obtains oscillatory solutions. Figure 8.13 shows the normalized gain coefficient κ''/κ' as a function of $\Delta k/\kappa'$; the maximum gain coefficient appears at $\Delta k = 2\kappa'$

$$\kappa''_{\max} = \kappa' = n_2 k_0 I_p. \qquad (8.138)$$

With $n_2 = 3.2\times 10^{-20}\,\mathrm{m^2 W^{-1}}$, we obtain $\kappa''_{\max} \approx I_p \times 2\times 10^{-13}\,\mathrm{W^{-1}m}$ at a wavelength of $1\,\mu m$ for silica.

8.3.4 Two-Photon Absorption

The cubic susceptibility $\chi^{(3)}$, and thus the effective linear susceptibility Eq. (8.97), can assume complex values. According to Eq. (1.65), the imaginary part of the effective susceptibility is responsible for transfer of energy from the light field to the material (absorption); this happens whenever one of the involved frequencies coincides (within the relevant line width) with a resonance frequency of the material (Fig. 8.19c). If the SH (or SF) of the incoming waves is resonant with a transition, the process is called two-photon absorption (TPA); note that in the absence of linear absorption at ω, the medium is transparent at low intensities.

According to Eq. (2.75), the imaginary part of the susceptibility results in an imaginary component of the propagation index

$$\kappa \approx -\frac{3Z_0}{4n_0^2} \mathrm{Im}\left[\chi^{(3)}\right] I, \qquad (8.139)$$

which allows us to define a TPA coefficient [compare Eq. (2.71)]

$$\alpha_{\mathrm{TPA}} = 2\kappa k_0 = -k_0 \frac{3Z_0}{2n_0^3} \mathrm{Im}\left[\chi^{(3)}\right] I =: \beta_{\mathrm{TPA}} I. \qquad (8.140)$$

The total, effective absorption coefficient is

$$\alpha_{\mathrm{eff}} = \alpha + \beta_{\mathrm{TPA}} I, \qquad (8.141)$$

where β_{TPA} is usually given in [cm/GW]. Silicon, for example, exhibits a $\beta_{TPA}=1.5$ cm/GW at a wavelength of $1.064\,\mu$m. It is common to introduce a TPA cross section σ_{TPA} which is related to β_{TPA} by the volume density N of absorbing molecules:

$$\beta_{TPA} = N\sigma_{TPA}/\hbar\omega; \tag{8.142}$$

σ_{TPA} is frequently given in units of $1\,GM = 10^{-50}$ cm^{4}s (1 Goeppert-Mayer).
For TPA, the absorption "law" Eq. (2.70) does not apply; from

$$\frac{dI}{dz} = -\beta_{TPA}I^2 \tag{8.143}$$

we find

$$I(z) = \frac{I(0)}{1 + I(0)\beta_{TPA}z}; \tag{8.144}$$

if linear *and* TPA has to be taken into account, the attenuation is given by

$$I(z) = \frac{I(0)}{e^{\alpha z} + I(0)(\beta_{TPA}/\alpha)(e^{\alpha z} - 1)}, \tag{8.145}$$

as can be verified by differentiation.

TPA has a number of important applications that usually rely on the fact that in the focus of a laser beam the probability for TPA is strongly enhanced. Following the simultaneous absorption of two photons, the excited molecule can decay under spontaneous emission of a photon (two-photon fluorescence), a process that is used for scanning microscopy, where the focus of a pulsed laser is scanned in a three-dimensional fashion through the sample (for example, a biological tissue). Since the emitted fluorescence light originates only from the small focal volume of the tightly focused beam, three-dimensional pictures with a resolution below the fundamental wavelength can be obtained. In a somewhat similar technique, two-photon excited molecules can polymerize; within an originally liquid phase, solid features can be generated by scanning a laser focus through the medium (rapid prototyping). In semiconductors, TPA can be used to realize intensity correlators and power limiters, respectively, provided that $\hbar\omega < E_g < 2\hbar\omega$.

8.3.5 Raman Amplification

The polarizability of molecules generally depends on the molecular bond lengths. Provided that this dependence is linear, the susceptibility of an ensemble of molecules vibrating at Ω_v consequently exhibits a component oscillating at this frequency, and the polarization induced by a light field of frequency ω_p shows side bands at $\omega_p - \Omega_v$ ("Stokes line") and $\omega_p + \Omega_v$ ("anti-Stokes line"). The emission of frequency shifted light by the polarization side bands is called Raman scattering and provides a valuable spectroscopic tool because it allows determining the characteristic vibrational resonances of the molecule.

The vibrational resonance frequencies Ω_v of a molecule are much lower than typical electronic resonance frequencies since the vibrating atoms are much heavier than the electrons. Raman scattering is usually rather weak, because the thermal excitation of molecular vibrations is low at room temperature. However, the molecular vibrations can also be optically driven and become substantial. For this purpose, the molecular sample is exposed to a superposition of a pump wave at ω_p and a "signal" wave at ω_s such that the beat frequency $\omega_p - \omega_s$ is close to the vibrational resonance frequency Ω_v. The resulting molecular vibrations generate a Stokes side band $\omega_p - (\omega_p - \omega_s)$ of the pump wave that coincides with the signal frequency and is capable of amplifying the signal wave. This so-called stimulated Raman scattering is the basis of Raman amplification.

To understand the driving process, we start from the force acting on the molecules in the presence of an electromagnetic field. The dielectric energy density of a medium with susceptibility χ is $W = \varepsilon_0 \varepsilon E^2/2 = \varepsilon_0[\chi + 1](E^2/2)$, where $\chi(q)$ is assumed to depend on a representative intramolecular coordinate q. A displacement Δq results in a change of the energy density of $\Delta W = (dW/dq)\Delta q = \varepsilon_0(E^2/2)(d\chi/dq)\Delta q$. Thus, the force $F = \Delta W/\Delta q$ of the electromagnetic field on a molecule is

$$F = \varepsilon_0 \langle E^2/2 \rangle (d\xi/dq), \tag{8.146}$$

where $\xi = \chi/N$ is the polarizability and N is the volume density of molecules; the temporal average is taken over a cycle of the light field, which is short in comparison to the molecular vibrational period. Note that no static dipole moment is required for this interaction, so that symmetric molecules such as H_2 or O_2 can be Raman active.

The superposition

$$E(t) = \frac{1}{2}\left[\tilde{E}(\omega_p)e^{j\omega_p t} + \tilde{E}(\omega_s)e^{j\omega_s t} + c.c.\right] \tag{8.147}$$

produces an oscillating force

$$F(t) = \frac{1}{4}\varepsilon_0(d\xi/dq)\left[\tilde{E}(\omega_p)\tilde{E}^*(\omega_s)e^{j(\omega_p-\omega_s)t} + c.c.\right] \tag{8.148}$$

on the molecules. Describing the molecule as a linear oscillator driven by this force, we obtain the equation of motion

$$\ddot{q} + \Gamma_v \dot{q} + \Omega_v^2 q = F(t)/m \tag{8.149}$$

where m is the effective mass. With the ansatz $q(t) := \frac{1}{2}\left[\tilde{q}e^{j(\omega_p - \omega_s)t} + c.c.\right]$, we get the complex displacement amplitudes

$$\tilde{q} = \frac{\varepsilon_0 \tilde{E}(\omega_p)\tilde{E}^*(\omega_s)(\,d\xi/\,dq)}{2m\left[\Omega_v^2 - (\omega_p - \omega_s)^2 + j\Gamma_v(\omega_p - \omega_s)\right]}. \tag{8.150}$$

The displacement $q(t)$, multiplied with $N(\,d\xi/\,dq)$ yields the alternating component $\chi(t)$ of the susceptibility,

$$\chi(t) = \frac{1}{2}\left[\tilde{q}e^{j(\omega_p - \omega_s)t} + c.c.\right]N\frac{d\xi}{dq}. \tag{8.151}$$

The incoming pump field generates the cubic nonlinear polarization $\varepsilon_0\chi(t)E(t)$ that has a frequency component at ω_s

$$\tilde{P}^{(3)}(\omega_s) = \frac{\varepsilon_0^2 N(\,d\xi/\,dq)^2}{4m}\frac{\tilde{E}(\omega_p)\tilde{E}^*(\omega_p)\tilde{E}(\omega_s)}{\left[\Omega_v^2 - (\omega_p - \omega_s)^2 - j\Gamma_v(\omega_p - \omega_s)\right]}$$

$$=: \varepsilon_0 \Delta\chi\tilde{E}(\omega_s). \tag{8.152}$$

Depending on the relative phase, this polarization wave can amplify or attenuate the signal wave at ω_s. Note that phase matching is automatically provided in this process since $\mathbf{k}_p - \mathbf{k}_p + \mathbf{k}_s = \mathbf{k}_s$.

Comparison with Eq. (8.103) allows us to describe the stimulated Raman effect by a third order susceptibility

$$\chi_{Rn}^{(3)} := \frac{2\Delta\chi}{3\tilde{E}(\omega_p)\tilde{E}^*(\omega_p)} \tag{8.153}$$

with the complex value

$$\chi_{Rn}^{(3)} = \frac{\varepsilon_0 N(\,d\xi/\,dq)^2}{6m\left[\Omega_v^2 - (\omega_p - \omega_s)^2 - j\Gamma_v(\omega_p - \omega_s)\right]} = \text{Re}\left[\chi_{Rn}^{(3)}\right] + j\,\text{Im}\left[\chi_{Rn}^{(3)}\right], \tag{8.154}$$

so that the effective susceptibility at ω_s is

$$\chi = \chi_0 + \chi_{Rn}^{(3)}\frac{3Z_0}{n}I_p. \tag{8.155}$$

The real part of this susceptibility represents a XPM of the signal by the pump (Kerr effect Sect. 8.3.2), i.e., a change of the real part of the propagation index. The imaginary part plays a role similar to the imaginary part of the propagation index [see Eq. (2.75)]

$$\kappa \approx -\frac{3Z_0}{2n_0^2} \text{Im}\left[\chi_{\text{Rn}}^{(3)}\right] I_{\text{p}}. \tag{8.156}$$

Since, however,

$$\text{Im}\left[\chi_{\text{Rn}}^{(3)}\right] = \frac{\varepsilon_0 N (\,\mathrm{d}\xi/\,\mathrm{d}q)^2 \Gamma_{\text{v}}(\omega_{\text{p}} - \omega_{\text{s}})}{6m\left[(\Omega_{\text{v}}^2 - (\omega_{\text{p}} - \omega_{\text{s}})^2)^2 + \Gamma_{\text{v}}^2(\omega_{\text{p}} - \omega_{\text{s}})^2\right]} \tag{8.157}$$

is positive in the Stokes regime $\omega_{\text{s}} < \omega_{\text{p}}$, κ and accordingly the absorption coefficient Eq. (2.71) turn negative, which is equivalent to amplification of the signal wave with the Raman gain coefficient

$$\gamma_{\text{Rn}} := -2\kappa k_0 = k_{\text{s}}\frac{3Z_0}{n_0^3} \text{Im}\left[\chi_{\text{Rn}}^{(3)}\right] I_{\text{p}}. \tag{8.158}$$

The maximum gain is provided at $\omega_{\text{s}} \approx \omega_{\text{p}} - \Omega_{\text{v}}$ and the gain bandwidth is given by Γ_{v} (Fig. 8.20). The Raman gain coefficient for SiO_2 is $10^{-13} \text{W}^{-1}\text{m} I_{\text{p}}$, somewhat smaller than the OPA-value (Sect. 8.3.3); CS_2 shows a particularly large value (about 30 times larger than SiO_2). The Stokes shift $\hbar\Omega_{\text{v}}$ is 57 meV in SiO_2, the bandwidth is about 2 meV. Raman fiber amplifiers are attractive alternatives to EDFAs (Sect. 5.3.5), since they can amplify arbitrary wavelengths, and in particular the 1.55 μm telecommunication band (see, e.g., Islam 2004). They do not require

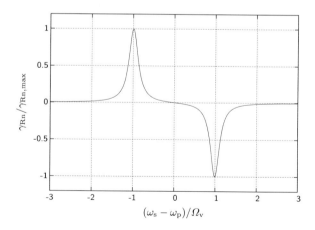

Fig. 8.20 Frequency dependence of the Raman gain; peak amplification is provided at $\omega_{\text{s}} \approx \omega_{\text{p}} - \Omega_{\text{v}}$

any modification (doping) of the transmission fiber; due to the small mode diameter, sufficient pump intensity can be obtained at a moderate pump power.

8.3.6 Brillouin Amplification

In contrast to the Raman effect, which is the interaction of light with the vibrations of isolated molecules, Brillouin scattering is the interaction with acoustic (compression) waves propagating in a medium. A compression wave is equivalent to a density wave, and eventually to a propagation index wave, because the propagation index depends on the density of the medium. Such a wave acts like a dielectric multilayer system that travels through the medium at the speed of sound. Light that is scattered by such a wave is frequency shifted because of this motion.

The frequency Ω and wave number K of an acoustic density wave

$$\rho = \rho_0 + \frac{1}{2}\left[\tilde{\rho}e^{-\mathrm{j}(Kz-\Omega t)} + c.c.\right] \tag{8.159}$$

obey the dispersion relation

$$\frac{\Omega}{K} = v_{\mathrm{ak}}, \tag{8.160}$$

where v_{ak} is the acoustic phase velocity with typical values of $10^3 \ldots 10^4$ m/s in solids and liquids. Since the acoustic phase velocity is many orders of magnitude smaller than the velocity of light, an acoustic wave with a wave number comparable to that of a light wave has a frequency that is many orders of magnitude smaller than the optical frequency.

Acoustic waves are always present in a medium in the form of thermally excited phonons. Scattering of light from these waves is known as spontaneous Brillouin scattering. Similar to molecular vibrations in Raman scattering, acoustic waves can also be driven by electromagnetic waves through the process of electrostriction.

To calculate the electrostrictive pressure p_e, we start from the electric contribution $\varepsilon_0 \chi E^2/2$ of the energy density [Eq. (1.52)]. Changing the density ρ (and thus the susceptibility χ) of the medium results in a change of the energy density by the amount

$$\frac{\varepsilon_0 E^2}{2}\frac{\partial \chi}{\partial \rho}\Delta\rho, \tag{8.161}$$

which is equal to the corresponding mechanical work per unit volume $p\frac{\Delta V}{V} = -p\frac{\Delta\rho}{\rho}$. The electrostrictive pressure is therefore proportional to the square of the electric field

$$p_e = -\frac{\varepsilon_0 E^2}{2} \frac{\rho \partial \chi}{\partial \rho} =: -\gamma_e \frac{\varepsilon_0 E^2}{2}, \tag{8.162}$$

where

$$\gamma_e = \frac{\rho \partial \chi}{\partial \rho} \approx \chi \tag{8.163}$$

is the electrostriction coefficient; the approximation is valid if we assume the susceptibility to be proportional to the density, so that $\partial\chi/\chi = \partial\rho/\rho$. A superposition of electromagnetic waves

$$E(z,t) = \frac{1}{2}\left[\tilde{E}(\omega_p)e^{-j(\mathbf{k}_p\cdot\mathbf{x}-\omega_p t)} + \tilde{E}(\omega_s)e^{-j(\mathbf{k}_s\cdot\mathbf{x}-\omega_s t)} + c.c.\right] \tag{8.164}$$

gives rise to a pressure field Eq. (8.162) that contains a component at the difference frequency $\omega_p-\omega_s$

$$\frac{1}{2}\left[\tilde{E}(\omega_p)\tilde{E}^*(\omega_s)e^{-j\left[(\mathbf{k}_p-\mathbf{k}_s)\cdot\mathbf{x}-(\omega_p-\omega_s)t\right]} + c.c.\right] \tag{8.165}$$

which can couple to acoustic phonons of frequency Ω, provided that the corresponding wave vector $\mathbf{k}_p - \mathbf{k}_s$ matches the acoustic wave vector $\mathbf{K}$. For co-propagating electromagnetic fields, this condition cannot be met, since $k_p - k_s \approx (\omega_p-\omega_s)/c \ll K = \Omega/v_{ak}$. We therefore assume counterpropagating signal and pump waves with wave vectors $\mathbf{k}_p = [0,0,k_p]$ and $\mathbf{k}_s = [0,0,-k_s]$ (Fig. 8.21), so that the phase matching condition is $K = k_p+k_s$; since the acoustic and optical wave numbers K and $k_{p,s}$ are now of the same order of magnitude, the acoustic frequency Ω must be smaller than the optical frequencies by a factor on the order of v_{ak}/c_0, implying $\omega_p \approx \omega_s$ and $k_p \approx k_s$. At a given pump frequency ω_p, the matching acoustic wave number can therefore be approximated by

$$K_{Bn} = k_p+k_s \approx 2\frac{\omega_p}{c_0}n, \tag{8.166}$$

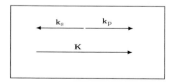

Fig. 8.21 Wave vectors of a pump wave with frequency ω_p, a counterpropagating signal wave (also called Stokes wave) with frequency $\omega_s < \omega_p$, and an acoustic wave of frequency Ω and wave vector $\mathbf{K}$ in a Brillouin amplifier

where n is the propagation index at ω_p. The corresponding acoustic frequency

$$\Omega_{Bn} = 2\omega_p \frac{v_{ak}}{c_0} n \qquad (8.167)$$

is called Brillouin frequency; in contrast to the frequency Ω_v characterizing the Raman effect, Ω_{Bn} is not a resonance frequency, but the acoustic frequency at which phase matching between the electrostrictive pressure wave and phonons is possible.

The wave equation of an externally driven acoustic compression wave can be cast in the form (see Starunov and Fabelinskii 1970)

$$\frac{\partial^2 \rho}{\partial t^2} - \Gamma \nabla^2 \frac{\partial \rho}{\partial t} - v_{ak}^2 \nabla^2 \rho = \nabla^2 p_e, \qquad (8.168)$$

where $\rho(z,t)$ is the local mass density and Γ denotes the acoustic damping coefficient; we simplify the equation by assuming that the complex amplitude is stationary, $\partial\tilde{\rho}/\partial t = 0$ and homogeneous, $\partial\tilde{\rho}/\partial z = 0$, which is reasonable since the strong attenuation of acoustic waves in the relevant frequency regime prevents the acoustic amplitude from building up during propagation. According to Eq. (8.162), the driving term in Eq. (8.168) is given by

$$\tilde{p}_e(\Omega) = -\gamma_e \varepsilon_0 \tilde{E}(\omega_p) \tilde{E}^*(\omega_s)/2. \qquad (8.169)$$

Substituting Eq. (8.159) with $K \approx K_{Bn}$ and using $v_{ak}^2 K^2 \approx \Omega_{Bn}^2$, we obtain

$$\left(-\Omega^2 + jK_{Bn}^2 \Gamma\Omega + \Omega_{Bn}^2\right)\tilde{\rho} = \frac{\gamma_e \varepsilon_0}{2} K_{Bn}^2 \tilde{E}(\omega_p)\tilde{E}^*(\omega_s) \qquad (8.170)$$

or

$$\tilde{\rho} = \frac{\gamma_e \varepsilon_0}{2} K_{Bn}^2 \frac{\tilde{E}(\omega_p)\tilde{E}^*(\omega_s)}{\Omega_{Bn}^2 - \Omega^2 + j\Omega\Gamma_{Bn}}, \qquad (8.171)$$

where $\Gamma_{Bn} = K_{Bn}^2 \Gamma$. With Eq. (8.163), this corresponds to a "susceptibility-wave"

$$\Delta\chi_e = \frac{1}{2}\left[\tilde{\rho}e^{-j(Kz-\Omega t)} + c.c.\right]\frac{\gamma_e}{\rho_0}. \qquad (8.172)$$

The interaction of the pump wave $E(\omega_p)$ with this susceptibility generates a cubic nonlinear polarization at ω_s

$$\tilde{P}^{(3)}(\omega_s) = \frac{\gamma_e^2 \varepsilon_0^2 K_{Bn}^2}{2\rho_0} \frac{\tilde{E}(\omega_p)\tilde{E}^*(\omega_p)\tilde{E}(\omega_s)}{\Omega_{Bn}^2 - \Omega^2 - j\Omega\Gamma_{Bn}}, \qquad (8.173)$$

which is formally equivalent to the stimulated Raman polarization Eq. (8.152). We therefore can introduce the complex nonlinear susceptibility

$$\chi_{\mathrm{Bn}}^{(3)} := \frac{\varepsilon_0 \gamma_{\mathrm{e}}^2 K_{\mathrm{Bn}}^2}{3\rho_0 \left[\Omega_{\mathrm{Bn}}^2 - (\omega_{\mathrm{p}} - \omega_{\mathrm{s}})^2 - j\Gamma_{\mathrm{Bn}}(\omega_{\mathrm{p}} - \omega_{\mathrm{s}}) \right]} \tag{8.174}$$

with the imaginary part

$$\mathrm{Im}\left[\chi_{\mathrm{Bn}}^{(3)}\right] = \frac{\varepsilon_0 \gamma_{\mathrm{e}}^2 K_{\mathrm{Bn}}^2 \Gamma_{\mathrm{Bn}}(\omega_{\mathrm{p}} - \omega_{\mathrm{s}})}{3\rho_0 \left[(\Omega_{\mathrm{Bn}}^2 - (\omega_{\mathrm{p}} - \omega_{\mathrm{s}})^2)^2 + \Gamma_{\mathrm{Bn}}^2(\omega_{\mathrm{p}} - \omega_{\mathrm{s}})^2 \right]} \tag{8.175}$$

and define, analog to Eq. (8.158), the Brillouin gain coefficient

$$\gamma_{\mathrm{Bn}} = k_{\mathrm{s}} \frac{3Z_0}{n_0^3} \mathrm{Im}\left[\chi_{\mathrm{Bn}}^{(3)}\right] I_{\mathrm{p}}; \tag{8.176}$$

the frequency dependence of the Brillouin gain coefficient follows Fig. 8.20, with Ω_{v} replaced by Ω_{Bn} and Γ_{v} by Γ_{Bn}.

The treatment of Raman and Brillouin amplification given above is valid only in the small signal approximation; in general, the back-action of the signal wave on the pump has to be included.

8.3.7 Phase Conjugation*

Consider the monochromatic signal wave

$$\mathbf{E}_{\mathrm{s}}(\mathbf{x}, t) := \mathrm{Re}\left[\tilde{\mathbf{E}}_{\mathrm{s}}(\mathbf{x}) \mathrm{e}^{\mathrm{j}\omega t} \right] = \frac{1}{2}\left[\tilde{\mathbf{E}}_{\mathrm{s}}(\mathbf{x}) \mathrm{e}^{\mathrm{j}\omega t} + c.c. \right], \tag{8.177}$$

where $\tilde{\mathbf{E}}_{\mathrm{s}}(\mathbf{x})$ is a solution of the Helmholtz equation (1.22). Then the so-called phase conjugate wave

$$\mathbf{E}_{\mathrm{c}}(\mathbf{x}, t) = \mathrm{Re}\left[\tilde{\mathbf{E}}_{\mathrm{s}}^*(\mathbf{x}) \mathrm{e}^{\mathrm{j}\omega t} \right] = \frac{1}{2}\left[\tilde{\mathbf{E}}_{\mathrm{s}}^*(\mathbf{x}, \omega) \mathrm{e}^{\mathrm{j}\omega t} + c.c. \right]$$

$$= \frac{1}{2}\left[\tilde{\mathbf{E}}_{\mathrm{s}}(\mathbf{x}, \omega) \mathrm{e}^{-\mathrm{j}\omega t} + c.c. \right] \tag{8.178}$$

is, of course, also a solution. As the above equation shows, the phase conjugate wave is formally identical to the time reversed signal wave; a phase conjugate wave can therefore be visualized by running a movie of the signal wave backwards.

A device that can produce such a wave acts like a mirror that reflects the signal wave back into itself (while a conventional mirror just reverses the direction of propagation, Fig. 8.22) and is called phase conjugate mirror (PCM). A diverging wave, for example, after reflection at a PCM is converging again; while circularly polarized light changes its sense upon reflection at a conventional mirror, the phase conjugate wave maintains the sense of rotation. The most important application of such mirrors is the compensation of phase front aberrations of waves passing through an inhomogeneous medium, such as a thermally stressed laser crystal: if a wave is reflected at a conventional mirror and passes the same medium again, the phase distortion is doubled, while after the reflection at a PCM, the distortion is reversed by the reflection and compensated during the second pass (Fig. 8.23). Laser resonators that consist of a PCM and a conventional mirror are therefore insensitive to phase distortions in the gain medium.

The generation of a phase conjugate wave relies on a phase conjugate polarization in a (nonlinear) medium. A possible realization, based upon the cubic susceptibility $\chi^{(3)}$, is described in the following; for simplicity, all waves are assumed to be linearly polarized in the x direction. In addition to the signal $E_s(\omega)$, the medium is irradiated by two intense, counterpropagating pump waves

$$E_{p\pm} = \frac{1}{2}\left[\tilde{A}_{p\pm}e^{-j(\mathbf{k}_{p\pm}\mathbf{x}-\omega t)} + c.c.\right] \qquad (8.179)$$

with wave vectors $\mathbf{k}_{p+} = -\mathbf{k}_{p-}$ and the same frequency ω as the signal. The signal wave is described as a plane carrier wave with wave vector $\mathbf{k}_s$

$$E_s = \frac{1}{2}\left[\tilde{A}_s(\mathbf{x})e^{-j(\mathbf{k}_s\mathbf{x}-\omega t)} + c.c.\right] \qquad (8.180)$$

Fig. 8.22 Comparison of a conventional mirror (**a**), and a phase conjugate mirror (**b**), shown for the reflection of a spherical wave

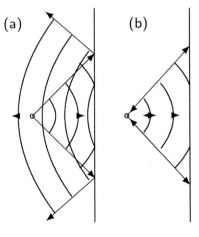

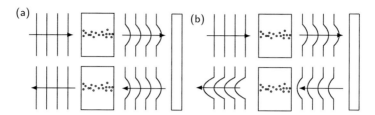

Fig. 8.23 Compensation of phase front distortions after a double pass through an inhomogeneous medium and reflection at a phase conjugate mirror (**a**); the same setup with reflection at a conventional mirror gives rise to doubled distortions (**b**)

and a spatially varying complex amplitude $\tilde{A}_s(\mathbf{x})$. The cubic polarization $\tilde{P}^{(3)}(\omega)$ includes a term proportional to

$$\tilde{A}_{p+}\tilde{A}_{p-}\tilde{A}_s^* e^{-j[(\mathbf{k}'_{p+}+\mathbf{k}'_{p-}-\mathbf{k}_s)\mathbf{x}-\omega t]}, \tag{8.181}$$

where $\mathbf{k}'_{p+}$ and $\mathbf{k}'_{p-}$ are the pump wave vectors modified by SPM and XPM. This polarization acts as a source for the phase conjugate wave, provided that $\mathbf{k}'_{p+} + \mathbf{k}'_{p-} = 0$, which requires that $|\tilde{A}_{p+}| = |\tilde{A}_{p-}|$.

Once the process gets started, the newly produced phase conjugate field

$$\tilde{E}_c(\mathbf{x}) = \tilde{A}_c(\mathbf{x})e^{+j\mathbf{k}_s\mathbf{x}} \tag{8.182}$$

also participates in the interaction and contributes to the signal wave, since the conjugate of the conjugate is the signal. The total field

$$\tilde{E} := \tilde{E}_{p+} + \tilde{E}_{p-} + \tilde{E}_s + \tilde{E}_c \tag{8.183}$$

generates, according to Eq. (8.91), a number of different terms, of which we collect only those with an $e^{\pm j\mathbf{k}_s\mathbf{x}}$-carrier that can act as source terms for the signal and phase conjugate waves, respectively:

$$\tilde{P}_s^{(3)}(\omega) = \frac{3}{2}\varepsilon_0\chi^{(3)}\left[|\tilde{A}_{p+}|^2\tilde{A}_s + |\tilde{A}_{p-}|^2\tilde{A}_s + \tilde{A}_{p+}\tilde{A}_{p-}\tilde{A}_c^*\right]e^{-j k_s z}$$

$$\tilde{P}_c^{(3)}(\omega) = \frac{3}{2}\varepsilon_0\chi^{(3)}\left[|\tilde{A}_{p+}|^2\tilde{A}_c + |\tilde{A}_{p-}|^2\tilde{A}_c + \tilde{A}_{p+}\tilde{A}_{p-}\tilde{A}_s^*\right]e^{-j k_c z}; \tag{8.184}$$

in the summation, terms that contain the factors $\tilde{A}_{s,c}$ or $\tilde{A}_{s,c}^*$ more than once have been neglected, since these amplitudes are assumed to be much smaller than the pump amplitudes.

Similar to Eq. (8.38), we obtain differential equations for the development of the fields $\tilde{A}_{s,c}$

$$\frac{\partial}{\partial z}\tilde{A}_s = -\frac{3j\omega}{4c_0 n_\omega}\chi^{(3)}\left[(|\tilde{A}_{p+}|^2 + |\tilde{A}_{p-}|^2)\tilde{A}_s + \tilde{A}_{p+}\tilde{A}_{p-}\tilde{A}_c^*\right]$$

$$\frac{\partial}{\partial z}\tilde{A}_c = \frac{3j\omega}{4c_0 n_\omega}\chi^{(3)}\left[(|\tilde{A}_{p+}|^2 + |\tilde{A}_{p-}|^2)\tilde{A}_c + \tilde{A}_{p+}\tilde{A}_{p-}\tilde{A}_s^*\right],\qquad(8.185)$$

where $k_s = -k_c = \omega n_\omega/c_0$ was used. Introducing the coupling factors

$$\kappa_{xpm} := \frac{3\omega}{4c_0 n_\omega}\chi^{(3)}\left(|\tilde{A}_{p+}|^2 + |\tilde{A}_{p-}|^2\right)$$

$$\kappa_{pcm} := \frac{3\omega}{4c_0 n_\omega}\chi^{(3)}\tilde{A}_{p+}\tilde{A}_{p-},\qquad(8.186)$$

these equations assume the form

$$\frac{\partial}{\partial z}\tilde{A}_s = -j\kappa_{xpm}\tilde{A}_s - j\kappa_{pcm}\tilde{A}_c^*$$

$$\frac{\partial}{\partial z}\tilde{A}_c = j\kappa_{xpm}\tilde{A}_c + j\kappa_{pcm}\tilde{A}_s^*.\qquad(8.187)$$

The XPM of the signal and phase conjugate, respectively, by the pump is represented by κ_{xpm} and can be included into the complex amplitudes by the transformations

$$\tilde{A}_s =: \tilde{A}'_s e^{-j\kappa_{xpm}z}, \quad \tilde{A}_c =: \tilde{A}'_c e^{j\kappa_{xpm}z}\qquad(8.188)$$

which is simply a modification of the wave vectors according to Eq. (8.104). In this way, (8.187) assumes the form

$$\frac{\partial}{\partial z}\tilde{A}'_s = -j\kappa_{pcm}\tilde{A}'^*_c$$

$$\frac{\partial}{\partial z}\tilde{A}'_c = j\kappa_{pcm}\tilde{A}'^*_s.\qquad(8.189)$$

Differentiation of the first equation and using $\frac{\partial}{\partial z}\tilde{A}'^*_c = -j\kappa^*_{pcm}\tilde{A}'_s$, we obtain

$$\frac{\partial^2}{\partial z^2}\tilde{A}'_s + |\kappa_{pcm}|^2\tilde{A}'_s = 0\qquad(8.190)$$

Fig. 8.24 Degenerate four wave mixing as a source for a phase conjugate wave

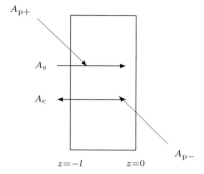

with the solution

$$\tilde{A}'_s(z) = C_1 \sin \kappa z + C_2 \cos \kappa z$$

$$\tilde{A}'_c(z) = -\frac{j}{\kappa}\frac{\partial}{\partial z}\tilde{A}'^*_s = -jC_1 \cos \kappa z + jC_2 \sin \kappa z, \qquad (8.191)$$

where $\kappa = |\kappa_{pcm}|$. In the configuration Fig. 8.24, the boundary conditions are $\tilde{A}_s(-l) = \tilde{A}_{s,in}$ and $\tilde{A}_c(0) = 0$, so that $C_1 = 0$ and $C_2 = \tilde{A}'_{s,in}/\cos \kappa l$. The output amplitudes are then

$$\tilde{A}_{s,out} = \tilde{A}_s(0) = \tilde{A}'_{s,in}\frac{1}{\cos \kappa l}$$

$$\tilde{A}_{c,out} = \tilde{A}_c(-l) = \tilde{A}'_{s,in} \tan \kappa l; \qquad (8.192)$$

the process not only produces a phase conjugate signal but can also amplify the incoming signal.

8.4 Electro-Optic Effects

In the framework of nonlinear optics, electro-optic effects (Sect. 2.3.4) can be understood as a mixing of electrostatic and optical fields. The Pockels effect, for example, is a manifestation of the quadratic susceptibility $\chi^{(2)}$; similarly, the quadratic electro-optic effect relies on the cubic susceptibility $\chi^{(3)}$. Other nonlinear optical effects that involve static fields are optical rectification (where a dc-field is generated by a nonlinear optical process) and electric field induced SHG, where a static electric field breaks the symmetry of a centrosymmetric medium (Sect. 8.4.3) and allows for quadratic nonlinear effects that are symmetry forbidden in the absence of the dc-field.

8.4.1   Linear Electro-Optic Effect

Consider a monochromatic optical wave in a quadratic nonlinear medium that is subject to an electrostatic field $\mathbf{E}^{\mathrm{dc}}$; the total electric field is then

$$\mathbf{E}(t) = \mathbf{E}^{\mathrm{dc}} + \frac{1}{2}\left[\tilde{\mathbf{E}}(\omega)e^{-j(k_\omega z - \omega t)} + c.c.\right] \qquad (8.193)$$

and contains frequency components at $\omega_1 = 0$ and $\omega_2 = \omega$. According to Eq. (8.6), the resulting polarization $\mathbf{P}^{(1)} + \mathbf{P}^{(2)}$ comprises frequency components at 0, ω, and 2ω. The SH-component was dealt with in Sect. 8.2.1; here, we concentrate on the polarization contributions at ω, with the complex amplitude

$$\tilde{\mathbf{P}}(\omega) = \varepsilon_0 \chi^{(1)}\tilde{\mathbf{E}}(\omega) + \varepsilon_0 \chi^{(2)}\mathbf{E}^{\mathrm{dc}}\tilde{\mathbf{E}}(\omega) = \varepsilon_0\left[\chi^{(1)} + \chi^{(2)}\mathbf{E}^{\mathrm{dc}}\right]\tilde{\mathbf{E}}(\omega). \qquad (8.194)$$

This is equivalent to a field dependent linear susceptibility $\chi(E^{\mathrm{dc}}) = \chi^{(1)} + \chi^{(2)}\mathbf{E}^{\mathrm{dc}}$ with the tensor components

$$\chi_{ij}(E^{\mathrm{dc}}) = \chi_{ij}^{(1)} + \chi_{ijk}^{(2)}E_k^{\mathrm{dc}}. \qquad (8.195)$$

The permittivity $\varepsilon = 1 + \chi$ therefore changes by

$$\Delta\varepsilon = \chi^{(2)}\mathbf{E}^{\mathrm{dc}}. \qquad (8.196)$$

Thus, the static field induces or changes the crystal anisotropy, depending on the direction and magnitude of the electrostatic field; in particular, an optically isotropic (but non-centrosymmetric) medium such as GaAs can become birefringent.

For historical reasons it is common to describe the electro-optic effect as a Taylor expansion of the impermeability tensor $\eta = \varepsilon^{-1}$ [Eq. (2.124)]

$$\eta_{ij}(\mathbf{E}^{\mathrm{dc}}) = \eta_{ij}^0 + r_{ijk}E_k^{\mathrm{dc}} + \ldots; \qquad (8.197)$$

or

$$\Delta\eta = r\mathbf{E}^{\mathrm{dc}}. \qquad (8.198)$$

As a third rank tensor, r_{ijk} vanishes in centrosymmetric media [Eq. (8.3)]. Table 8.4 shows electro-optic coefficients of important materials. Because of $\eta_{ij} = \eta_{ji}$, we can set $r_{ijk} = r_{jik}$, and the first two indices of r_{ijk} are usually contracted according to Table 8.2.

To relate r and $\chi^{(2)}$, we use the approximation

$$\eta(E^{\mathrm{dc}}) = \eta^0 + \Delta\eta = [\varepsilon^0 + \Delta\varepsilon]^{-1}$$

Table 8.4 Electro-optic coefficients of selected materials; also listed are the propagation index and the dc-permittivity

Material		Symmetry	$r_{\xi k}\ [10^{-12}\ \text{mV}^{-1}]$	n	ε^{dc}
KDP	KH$_2$PO$_4$	$\bar{4}$2m	$r_{41} = 8.6$	1.5	$\varepsilon_{11,22} = 42$
			$r_{63} = 10.6$		$\varepsilon_{33} = 21$
Lithium niobate	LiNbO$_3$	3m	$r_{13} = 9.6$	2.2	$\varepsilon_{11,22} = 78$
			$r_{22} = 6.8$		$\varepsilon_{33} = 2132$
			$r_{33} = 31$		
			$r_{51} = 33$		
Gallium arsenide	GaAs	$\bar{4}$3m	$r_{41} = 1.1$	3.3	$\varepsilon = 13$

$$\approx [\boldsymbol{\varepsilon}^0]^{-1} - [\boldsymbol{\varepsilon}^0]^{-1}\Delta\boldsymbol{\varepsilon}\,[\boldsymbol{\varepsilon}^0]^{-1}$$

$$= \boldsymbol{\eta}^0 - \boldsymbol{\eta}^0\Delta\boldsymbol{\varepsilon}\,\boldsymbol{\eta}^0, \tag{8.199}$$

so that

$$\Delta\boldsymbol{\eta} = -\boldsymbol{\eta}^0\Delta\boldsymbol{\varepsilon}\,\boldsymbol{\eta}^0 \tag{8.200}$$

Assuming that $\boldsymbol{\eta}^0$ is diagonal [Eq. (2.124)] with components $\eta_{ii} = 1/n_{(i)}^2$, we obtain, in linear approximation,

$$r_{ijk} = -\eta_{ij}\chi_{ijk}^{(2)}\eta_{kk} = -\frac{\chi_{ijk}^{(2)}}{n_{(j)}^2 n_{(k)}^2}. \tag{8.201}$$

Note that the relevant values for $\chi^{(2)}$ are not the same as for optical SFG, since $\chi^{(2)}$ strongly depends on the frequency of the fields involved (Sect. 8.1.2).

An important electro-optic material is KDP (KH$_2$PO$_4$), belonging to the symmetry class $\bar{4}$2m with the non-vanishing third rank tensor components $r_{231} = r_{321} = r_{41}$, $r_{132} = r_{312} = r_{52}$, and $r_{123} = r_{213} = r_{63}$, with $r_{52} = r_{41}$; in contracted form

$$r_{\xi k} = \begin{bmatrix} 0 & 0 & 0 \\ 0 & 0 & 0 \\ 0 & 0 & 0 \\ r_{41} & 0 & 0 \\ 0 & r_{52} & 0 \\ 0 & 0 & r_{63} \end{bmatrix}, \tag{8.202}$$

so that Eq. (8.197) has the form

$$\boldsymbol{\eta} = \begin{bmatrix} \eta_{11} & r_{123}E_3^{\text{dc}} & r_{132}E_2^{\text{dc}} \\ r_{213}E_3^{\text{dc}} & \eta_{22} & r_{231}E_1^{\text{dc}} \\ r_{312}E_2^{\text{dc}} & r_{321}E_1^{\text{dc}} & \eta_{33} \end{bmatrix} = \begin{bmatrix} \eta_1 & r_{63}E_3^{\text{dc}} & r_{41}E_2^{\text{dc}} \\ r_{63}E_3^{\text{dc}} & \eta_2 & r_{41}E_1^{\text{dc}} \\ r_{41}E_2^{\text{dc}} & r_{41}E_1^{\text{dc}} & \eta_3 \end{bmatrix}. \tag{8.203}$$

In the absence of an external field, KDP is uniaxial with $\eta_{11} = \eta_{22} = \eta_o$ and $\eta_{33} = \eta_e$.

In the following, we assume a dc-field $\mathbf{E}^{dc} = (0, 0, E_3^{dc})$, parallel to the z-axis (Fig. 2.35), so that

$$\boldsymbol{\eta} = \begin{bmatrix} \eta_o & r_{63}E_3^{dc} & 0 \\ r_{63}E_3^{dc} & \eta_o & 0 \\ 0 & 0 & \eta_e \end{bmatrix}. \tag{8.204}$$

This tensor can be diagonalized by a $45°$-rotation of the reference system around the z-axis [Eq. (2.3.1.1)]

$$\boldsymbol{\eta}' = \begin{bmatrix} \eta_o + r_{63}E_3^{dc} & 0 & 0 \\ 0 & \eta_o - r_{63}E_3^{dc} & 0 \\ 0 & 0 & \eta_e \end{bmatrix}; \tag{8.205}$$

in the presence of the dc-field, the originally uniaxial crystal becomes biaxial.

The diagonal form of $\boldsymbol{\eta}'$ allows for an immediate calculation of the field dependent propagation index, since $\eta'_{ii} = n_{(i)}^{-2}$. With the approximation $d\eta'/dn = -2/n^3$, we obtain $\Delta n \approx -(n^3/2)\Delta\eta'$ and finally

$$n_{(x',y')} = n_o \mp \frac{n_o^3}{2}r_{63}E_3^{dc}, \qquad n_{(z)} = n_e. \tag{8.206}$$

Applications of the electro-optic effect are discussed in Sects. 2.3.4 and 5.3.

8.4.2 Quadratic Electro-Optic Effect

While the linear electro-optic (Pockels) effect is restricted to crystals lacking a center of inversion, the quadratic electro-optic effect (also called electro-optic Kerr effect) relies on the cubic susceptibility and thus can be observed in all materials; the term "quadratic" refers to the fact that the propagation index changes with the square of the applied electrostatic field (Sect. 2.3.4).

For a brief discussion of this effect, we assume an isotropic medium exposed to a superposition of a linearly polarized optical wave and a dc-field, both fields oriented along the x-axis; the total field can then be expressed in a scalar form

$$E(t) = E^{dc} + \frac{1}{2}\left[\tilde{E}(\omega)e^{-j(k_\omega z - \omega t)} + c.c.\right]. \tag{8.207}$$

The resulting cubic nonlinear polarization contains a component at the frequency of the optical field

$$\tilde{P}^{(3)}(\omega) = 3\varepsilon_0\chi_{1111}^{(3)}\left(E^{dc}\right)^2\tilde{E}(\omega) \tag{8.208}$$

which can be treated as a change of the effective linear susceptibility by

$$\Delta\chi^{(1)} = 3\chi^{(3)}_{1111} \left(E^{dc}\right)^2 . \tag{8.209}$$

Similar to Eq. (8.98), this allows us to introduce a field dependent propagation index

$$n(E^{dc}) = n_0 + \frac{3\chi^{(3)}_{1111}}{2n_0} \left(E^{dc}\right)^2 . \tag{8.210}$$

While this effect is symmetry allowed in all materials, it is generally much smaller than the Pockels effect for fields below the dielectric breakdown limit.

8.4.3   Field Induced Second Harmonic Generation*

The cubic polarization resulting from the composite field Eq. (8.207) also contains a component at the second harmonic 2ω of the fundamental optical frequency

$$\tilde{P}^{(3)}(2\omega) = \frac{3}{2}\varepsilon_0\chi^{(3)}_{1111}E^{dc}\tilde{E}(\omega)\tilde{E}(\omega). \tag{8.211}$$

A comparison with Eq. (8.7) shows that this is equivalent to a field induced quadratic susceptibility

$$\chi^{(2)}_{111}(E^{dc}) = 3\chi^{(3)}_{1111}E^{dc}; \tag{8.212}$$

(the effect, shown here for parallel optical and dc-fields is, of course, not restricted to this simple configuration). A centrosymmetric medium such as glass, placed in an electrostatic field, can thus produce second harmonic or sum/difference frequency waves of incoming optical fields; a periodic alternation of the static field orientation can be used for QPM.

This effect is not in contradiction to the statement, made earlier, that centrosymmetry rules out effects such as SHG, because the electrostatic field breaks the centrosymmetry of the total system.

8.5   Acousto-Optics

8.5.1   Light Scattering at Sound Waves

As we have seen in Sect. 8.3.6, electromagnetic waves can also interact with acoustic waves. While Brillouin scattering deals with acoustic waves present as phonons in any medium, acousto-optics refers to the interaction of light with sound waves

Table 8.5 Acoustic and acousto-optic properties of selected materials (for longitudinal acoustic waves); acoustic phase velocity v_{ak}, mass density ρ, elasto-optic coefficient p, and acousto-optic figure of merit $M = p^2 n_0^6 / \rho v_{ak}^3$

Medium	v_{ak} [10^3 m s^{-1}]	ρ [10^3 kg m^{-3}]	n_0	p	M [10^{-15} m^{2}W^{-1}]
Water	1.5	1.0	1.3	0.31	137
Polystyrene	2.4	1.1	1.6	0.31	106
Silica	6.0	2.2	1.5	0.20	1.19
Flint glass	3.1	6.3	1.9	0.25	16
LiNbO$_3$	7.4	4.7	2.2	0.15	1.75
Gallium arsenide	5.2	5.3	3.5	0.41	104

externally excited by (piezoelectric) transducers. Acousto-optic effects rely on the dependence of the susceptibility on the acoustic strain. The strain S is defined as the relative deformation of a medium induced by a mechanical stress (force per area). Stress and strain can be longitudinal or transverse, i.e., the force (deformation) can be orthogonal or parallel to a given surface element; stress and strain are therefore tensors of second rank. Consequently, acoustic waves can be longitudinal as well as transverse.

The material property traditionally used to describe the acousto-optic interaction is the so-called elasto-optic coefficient p that relates the impermeability η to the stress

$$\Delta \eta = pS; \tag{8.213}$$

connecting two second rank tensors, the elasto-optic coefficient is a fourth rank tensor that has non-vanishing components in all symmetry classes (exactly like $\chi^{(3)}$). In anisotropic media, the description of the acousto-optic interaction can become very involved, because two electromagnetic modes and three acoustic modes have to be considered. Here, we restrict ourselves to a scalar description that is valid, for example, for the interaction of a longitudinal acoustic mode with light in an isotropic medium.

We start from $\eta = 1/\varepsilon = 1/(1 + \chi)$ to obtain

$$\Delta \chi = -(1 + \chi)^2 \Delta \eta = -n_0^4 p \Delta S \tag{8.214}$$

where n_0 is the propagation index of the unperturbed medium and $p = p_{1111}$ is the relevant elasto-optic coefficient. Table 8.5 shows this coefficient and other relevant properties of selected materials.

We assume an acoustic strain wave

$$S(\mathbf{x}, t) = \frac{1}{2} \left[\tilde{S} e^{-j(\mathbf{K} \cdot \mathbf{x} - \Omega t)} + c.c. \right]; \tag{8.215}$$

the angular frequency Ω and the wave vector $\mathbf{K}$ are related by the dispersion relation

$$|\mathbf{K}| = \frac{\Omega}{v_{ak}} \tag{8.216}$$

where v_{ak} is the acoustic phase velocity. The acoustic power density I_{ak} is related to the complex acoustic amplitude $\tilde{S}$ by

$$I_{ak} = \frac{|\tilde{S}|^2}{2} \rho v_{ak}^3 \tag{8.217}$$

where ρ [kg m^{-3}] is the density of the medium.

According to Eq. (8.214), the strain wave corresponds to a susceptibility wave

$$\Delta \chi(\mathbf{x}, t) = -\frac{p n_0^4}{2} \left[\tilde{S} e^{-j(\mathbf{K}\cdot\mathbf{x} - \Omega t)} + c.c. \right]. \tag{8.218}$$

The incoming light wave

$$E_i(\mathbf{x}, t) = \frac{1}{2} \left[\tilde{E}_i e^{-j(\mathbf{k}_i \cdot \mathbf{x} - \omega_i t)} + c.c. \right] \tag{8.219}$$

produces a polarization density wave with the stress induced component $\Delta P = \varepsilon_0 \Delta \chi E_i$

$$\Delta P(\mathbf{x}, t) = \frac{1}{2} \left[\Delta \tilde{P}(\mathbf{x}, t) + c.c. \right]$$

$$= -\frac{\varepsilon_0 p n_0^4}{4} \left[\tilde{S} e^{-j(\mathbf{K}\cdot\mathbf{x} - \Omega t)} + c.c. \right] \left[\tilde{E}_i e^{-j(\mathbf{k}_i \cdot \mathbf{x} - \omega_i t)} + c.c. \right]$$

$$= -\frac{\varepsilon_0 p n_0^4}{4} \left[\tilde{S} \tilde{E}_i e^{-j[(\mathbf{k}_i + \mathbf{K})\cdot\mathbf{x} - (\omega_i + \Omega)t]} \right.$$

$$\left. + \tilde{S}^* \tilde{E}_i e^{-j[(\mathbf{k}_i - \mathbf{K})\cdot\mathbf{x} - (\omega_i - \Omega)t]} + c.c. \right]. \tag{8.220}$$

The two side bands at $\omega_i \pm \Omega$ can serve as sources for two waves

$$E_d(\mathbf{x}, t) = \frac{1}{2} \left[\tilde{E}_d e^{-j(\mathbf{k}_d \cdot \mathbf{x} - \omega_d t)} + c.c. \right] \tag{8.221}$$

with frequency

$$\omega_d = \omega_i \pm \Omega, \tag{8.222}$$

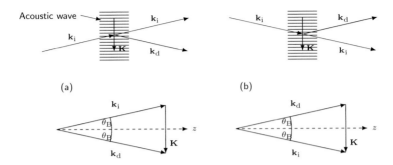

Fig. 8.25 Optical and acoustic wave vectors in an acousto-optic cell: (a) $\mathbf{k}_d = \mathbf{k}_i + \mathbf{K}$, (b) $\mathbf{k}_d = \mathbf{k}_i - \mathbf{K}$

provided that the phase matching condition

$$\mathbf{k}_d = \mathbf{k}_i \pm \mathbf{K} \tag{8.223}$$

is met. This scattering mechanism is called Bragg scattering, because it is essentially a diffraction of an electromagnetic wave at a moving Bragg grating.

Since the acoustic phase velocity is smaller than the optical by about five orders of magnitude, acoustic waves must have frequencies much smaller than that of optical waves to have wave numbers comparable to optical waves; therefore, $\omega_d \approx \omega_i$ and $|\mathbf{k}_d| \approx |\mathbf{k}_i|$; in other words, the three involved wave vectors Eq. (8.223) form an isosceles triangle (Fig. 8.25). In terms of the angle θ_B between the two optical wave vectors, the phase matching (or Bragg) condition can be written as

$$\sin\theta_B = \frac{|\mathbf{K}|}{2|\mathbf{k}_i|} = \frac{\lambda}{2\Lambda} = \frac{\lambda_0}{2n_0\Lambda}, \tag{8.224}$$

where

$$\Lambda = \frac{2\pi}{|\mathbf{K}|} = \frac{2\pi v_{ak}}{\Omega} \tag{8.225}$$

is the acoustic wavelength and λ_0 the vacuum wavelength of the optical waves. In a quantum picture, Eqs. (8.222) and (8.223) can be interpreted as energy and momentum conservation in an interaction between two photons and a phonon.

Once the scattered wave E_d with the wave vector $\mathbf{k}_i + \mathbf{K}$ is built up, it also interacts with the acoustic wave to produce a negative side band with the wave vector $\mathbf{k}_i + \mathbf{K} - \mathbf{K} = \mathbf{k}_i$. In this way, E_i is coupled to E_d and *vice versa*. To calculate the amplitudes of the two waves, we assume in the following that the Bragg condition Eq. (8.224) is met. We follow Eqs. (8.30) to (8.38) in the form (8.58), and replace

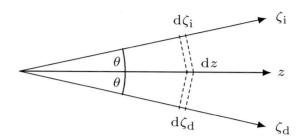

Fig. 8.26 Coordinate system for Eq. (8.227): the z-axis is the bisector of the isosceles triangle defined by the optical wave vectors

the source term Eq. (8.37) by

$$\Delta \tilde{P}_{d,i} = -\frac{\varepsilon_0 p n_0^4}{2} \tilde{S} \tilde{E}_{i,d} \tag{8.226}$$

[Eq. (8.220)], assuming, without loss of generality, $\tilde{S}$ to be real. Since the optical waves are not collinear in this case, we introduce separate propagation coordinates ζ_i, ζ_d (Fig. 8.26) and obtain the coupled amplitude equations

$$\frac{d\tilde{E}_i(\omega_i)}{d\zeta_i} = \frac{j\omega_i p n_0^3}{4c_0} \tilde{S} \tilde{E}_d$$

$$\frac{d\tilde{E}_d(\omega_d)}{d\zeta_d} = \frac{j\omega_d p n_0^3}{4c_0} \tilde{S} \tilde{E}_i. \tag{8.227}$$

Introducing a joint propagation coordinate z along the angular bisector, we set $d\zeta_i = d\zeta_d = \cos\theta\, dz$ and write Eq. (8.227) in the form

$$\frac{d\tilde{E}_i(\omega_i)}{dz} = j\kappa_{id} \tilde{E}_d$$

$$\frac{d\tilde{E}_d(\omega_d)}{dz} = j\kappa_{di} \tilde{E}_i, \tag{8.228}$$

where we have introduced the coupling coefficients

$$\kappa_{id} = \frac{\omega_i p n_0^3}{4c_0 \cos\theta} \tilde{S} \tag{8.229}$$

$$\kappa_{di} = \frac{\omega_d p n_0^3}{4c_0 \cos\theta} \tilde{S}; \tag{8.230}$$

because of $\omega_{\mathrm{i}} \approx \omega_{\mathrm{d}}$ we can set

$$\kappa := \kappa_{\mathrm{id}} \approx \kappa_{\mathrm{di}}. \tag{8.231}$$

With the boundary condition $\tilde{E}_{\mathrm{d}}(0) = 0$, the solution of Eq. (8.228) is

$$\tilde{E}_{\mathrm{i}}(z) = \tilde{E}_{\mathrm{i}}(0) \cos \kappa z$$

$$\tilde{E}_{\mathrm{d}}(z) = -\tilde{E}_{\mathrm{i}}(0) \sin \kappa z. \tag{8.232}$$

The acousto-optic coupling efficiency is then

$$\frac{I_{\mathrm{d}}(z)}{I_{\mathrm{i}}(0)} = \sin^2 \kappa z. \tag{8.233}$$

After the distance $z = \pi/2\kappa$, the energy transfer to the diffracted wave is complete. With Eq. (8.217) and using the material specific figure of merit

$$M := \frac{p^2 n_0^6}{\rho v_{\mathrm{ak}}^3} \tag{8.234}$$

(see Table 8.5), Eq. (8.233) has the form

$$\frac{I_{\mathrm{d}}(z)}{I_{\mathrm{i}}(0)} = \sin^2 \left(\frac{\pi}{\lambda_0} \sqrt{M I_{\mathrm{ak}} z} \right), \tag{8.235}$$

where $\theta \ll 1$ was assumed. For $z \ll \pi/2\kappa$, the diffraction efficiency

$$\frac{I_{\mathrm{d}}(z)}{I_{\mathrm{i}}(0)} \approx \frac{\pi^2}{\lambda_0^2} M I_{\mathrm{ak}} z^2 \tag{8.236}$$

is proportional to the acoustic power and M.

The above treatment relies on plane waves; in practice, the incoming light wave and the acoustic field are beam shaped and may actually be strongly focused (to obtain a sufficiently high acoustic intensity). As we have seen in Sect. 3.1.6, beams can be treated as a superposition of plane waves with different wave vectors. In the acousto-optic interaction of beam shaped acoustic and optical waves, the Bragg condition can be met by individual spatial Fourier components of the fields, even if the central wave vectors along the beam axes do not meet the phase matching condition. In particular, acousto-optic beam deflection can also work in a geometry where the optical beam crosses the acoustic beam orthogonally (Fig. 8.27). This is

Fig. 8.27 Acousto-optic interaction of an optical beam with a focused acoustic beam (Raman–Nath scattering)

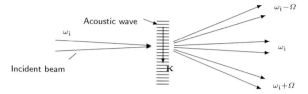

Fig. 8.28 Acousto-optic Bragg cell

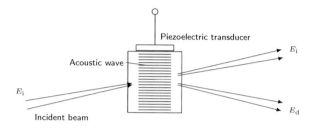

called Raman–Nath scattering and produces two diffracted beams at the frequency side bands $\omega_i + \Omega$ and $\omega_i - \Omega$.

8.5.2 Acousto-Optic Modulators

The acousto-optic effect is usually implemented in the form of a so-called Bragg cell, that is a piece of a suitable material to which a piezoelectric transducer is attached that is driven by an RF-source (Fig. 8.28). To get a feeling for typical operating parameters, we consider a Bragg cell made of flint glass, an acoustic transducer at 500 MHz delivering 1 W of acoustic power into a 1×1 mm^2 cross section beam; the optical wavelength is $\lambda_0 = 632$ nm (HeNe laser). Using the values in Table 8.5, we obtain an acoustic wavelength of $\Lambda = 6.2$ μm and a Bragg angle of $\theta_B = 26$ mrad (1.5°). Within an interaction length of 1 mm, Eq. (8.235) yields a diffraction efficiency of 34 %.

According to Eq. (8.236), the acoustic intensity can be varied to modulate the intensity of the diffracted beam. Varying the acoustic *frequency* allows selection of a certain optical wavelength from a broadband optical input. Bragg cell are also used to produce frequency shifted signals from a monochromatic laser for spectroscopic and interferometric applications.

For certain applications (such as actively mode locked lasers, Sect. 7.3.2), the cell is operated as an acoustic resonator, where the acoustic wave is reflected at the end facet of the cell to produce a standing acoustic wave of high intensity, similar to an optical Fabry–Perot resonator (Sect. 4.2.3). The two counterpropagating waves form the superposition

$$S(\mathbf{x}, t) = \frac{1}{2} \left[\tilde{S} e^{-j(\mathbf{K} \cdot \mathbf{x} - \Omega t)} + \tilde{S} e^{-j(-\mathbf{K} \cdot \mathbf{x} - \Omega t)} + c.c. \right]$$

$$= 2|\tilde{S}| \cos \mathbf{K} \cdot \mathbf{x} \cos \Omega t \tag{8.237}$$

(where $\tilde{S}$ is assumed to be real). As can be seen from this expression, the spatial modulation vanishes twice per acoustic period $2\pi/\Omega$ giving rise to a modulation of the transmitted optical power at twice the acoustic frequency.

8.6   Summary

Our treatment of nonlinear optics stays within the perturbative limit, valid for electric fields that are small in comparison to atomic fields. The source of any nonlinear optical effect is the nonlinear polarization induced by the participating electric fields. The material property relating the electric field to the nonlinear polarization density is the nonlinear susceptibility; it is derived from a series expansion of the polarization density and is a tensor of third or higher order. The linear oscillator model, supplemented by a nonlinear term in the restoring force, provides a valuable estimate of the relative magnitude of the nonlinear susceptibility of different materials; Miller's rule summarizes these results.

The nonlinear polarization density induced by monochromatic plane waves comprises components at the second, third, and higher harmonics, at the sum and difference frequencies, as well as rectified dc-components of the input fields. The symmetry properties of the nonlinear medium determine which nonlinear effects are possible: SHG and other second order nonlinear effects are symmetry forbidden in centrosymmetric media, for example, while third order effects are generally possible in all symmetry classes.

The nonlinear polarization density is the source term for new waves; applying the slowly varying amplitude approximation to the wave equation, we derive a first order differential equation for the amplitudes of these waves. This equation includes a term that takes the phase mismatch between the nonlinear polarization and the electric field into account; the amplitude of the wave radiated by the nonlinear polarization can only grow as long as the phase difference between the electromagnetic wave and the polarization is less than π. Means to achieve phase matching are crucial for the application of nonlinear optical effects. In addition to the exploitation of birefringence for this purpose, we analyze quasi phase matching that relies on the spatially periodic modification of the nonlinear medium.

Besides harmonic generation, optical parametric amplification (OPA) is of particular practical and theoretical interest. In the presence of an intense pump wave, a signal wave can be amplified, consuming energy from the pump; in addition to the amplified signal, a so-called idler wave is generated at the difference frequency between pump and signal frequency. Other nonlinear optical amplification schemes include Raman and Brillouin amplification, where the energy transfer from the pump to the signal wave is mediated by acoustic vibrations or waves, respectively, that are driven by the optical fields in the gain medium. Related acousto-optic effects are based on the interaction of the light field with acoustic waves that are launched in the medium by external acoustic transducers. The effect is used, for example, to modulate laser beams or to select a particular frequency component out of a polychromatic beam. Similar to sum and difference frequency generation, the scattered light is up or down shifted by the acoustic frequency.

A variety of interesting nonlinear optical effects results from the intensity dependence of the refractive index, a third order nonlinear effect that occurs in media of arbitrary symmetry. In combination with spatially or temporally varying fields, phenomena such as self-focusing, self phase modulation, and white light generation can be observed and exploited. The generation of phase conjugate waves is another manifestation of this third order nonlinearity.

The class of electro-optic effects, discovered long before the invention of the laser and treated in Chap. 2, are shown to be special cases of nonlinear optics, with one of the electric fields being a static electric field modifying the susceptibility. We link the conventional electro-optic tensor to the nonlinear susceptibility and show the qualitative validity of Miller's rule for this effect by comparing different electro-optic media.

8.7 Problems

1. Assuming that the linear and quadratic restoring force terms in Eq. (8.20) are of comparable magnitude if the displacement x is equal to the interatomic distance, and using Eq. (2.53) to estimate the linear "spring constant" a, calculate the nonlinear force coefficient D. Use this estimate to calculate $\chi^{(2)}$ from Miller's rule Eq. (8.28) ($N \approx 10^{22}\,\text{cm}^{-3} \approx 1/d^3$).

2. Use the nonlinear oscillator model Eq. (8.20) with a *cubic* restoring force term $\propto x^3$ to derive Miller's rule for the third order susceptibility $\chi^{(3)}(3\omega)$; restrict the calculation to a monochromatic input field. Use the arguments of problem 1 to estimate the value of $\chi^{(3)}(3\omega)$.

3. $BaTiO_3$ is an important quadratic nonlinear medium, belonging to point group 4mm; this means that it is invariant under rotations of 90° around the z-axis and reflection across the xz and yz plane, respectively. Find the non-zero elements of $\chi^{(2)}$ and compare with Table 8.1.

4. Consider an elliptically polarized 1 nJ, 100 fs pulse in a silica fiber with $20\,\mu\text{m}^2$ effective core area; which fiber length is required to rotate the polarization state by 90° (assume $\chi^{(3)}_{1122} = \chi^{(3)}_{1111}/3$ and $n_2 = 3.2 \times 10^{-20}\,\text{m}^2\text{W}^{-1}$). What is the extinction ratio between the original pulse and the rotated pulse if a linear polarizer is inserted after the fiber? What is the ellipticity that maximizes the extinction ratio?

5. Reproduce Fig. 8.10 and use the result to reproduce Fig. 8.9, lower panel.

6. Reproduce Fig. 8.14.

References and Suggested Reading

Agrawal, G. P. (2012). *Nonlinear fiber optics*. New York: Academic.

Bloembergen, N. (1982). *Nonlinear optics*. New York: Benjamin.

Boyd, R. W. (2008). *Nonlinear optics*. New York: Academic.

Butcher, P. N., & Cotter, D. (1991). *The elements of nonlinear optics*. New York: Cambridge University Press.

Cerullo, G., & De Silvestri, S. (2003). Ultrafast optical parametric amplifiers. *Review of Scientific Instruments, 74*(1), 1–18.

Gibbs, H. M., Khitrova, G., & Peyghambarian, N. (1990). *Nonlinear photonics*. New York: Springer.

Haussuehl, S. (2008). *Physical properties of crystals*. New York: Wiley.

Islam, M. N. (2004). *Raman amplifiers for telecommunications*. London: Springer.

Menzel, R. (2007). *Photonics*. Berlin: Springer.

New, G. (2011). *Introduction to nonlinear optics*. New York: Cambridge University Press.

Nye, J. F. (1985). *Physical properties of crystals*. New York: Oxford University Press.

Rottwitt, K., & Tidemand-Lichtenberg, P. (2015). *Nonlinear optics*. Boca Raton: CRC Press.

Shen, Y. R. (2002). *The principles of nonlinear optics*. New York: Wiley.

Stegeman, G. I., & Stegeman, R. A. (2012). *Nonlinear optics*. Hoboken, NJ: Wiley.

Starunov, V. S., & Fabelinskii, I. L. (1970). Stimulated mandel'shtam-brillouin scattering and stimulated entropy (temperature) scattering of light. *Soviet Physics Uspekhi 12*, 463. http://stacks.iop.org/0038-5670/12/i=4/a=R02

Suhara, T., & Fujimura, M. (2003). *Waveguide nonlinear-optic devices*. New York: Springer.

Trebino, R. (2000). *Frequency-resolved optical gating: The measurement of ultrashort laser pulses*. New York: Springer.

Tsai, C. S. (1990). *Guided-wave acoustooptics*. New York: Springer.

Xu, J., & Stroud, R. (1992). *Acousto-optic devices*. New York: Wiley.

Photodetection

9

Regarding detection, the optical frequency regime is quite distinct from the radio frequency (RF) range; while RF signals can be picked up by antennas, and the resulting current—which is essentially a replica of the electric field of the wave—can be amplified and processed electronically, detection of optical electric *fields* is extremely difficult; thus, practically all optical detectors rely on the excitation of electrons by absorption of photons, a process that scales with the signal *intensity* (Sect. 6.1) instead of its electric field.

The fact that the photoexcitation rate is proportional to $\tilde{E}\tilde{E}^*$ has important consequences: first of all, the output of a photodetector is a nonlinear (quadratic) function of the optical field amplitude; secondly, it does not contain information on the phase of the field. This does not imply, however, that the phase of the optical field is inaccessible to measurement with quantum detectors. Superposition of the signal field with a known reference field (e.g., from a local oscillator) produces, by interference, a photocurrent that contains information about the relative phase of the field and the local oscillator wave, allowing, for example, phase shift key modulation in optical communications.

The photosensitive component of optical detectors is usually a semiconductor or a metal layer. Electrons are photoexcited either from the valence band into the conduction band (semiconductors) or from the Fermi edge of a metal into a vacuum state. In both cases, a minimum quantum energy, i.e., a minimum frequency of the light, is required to induce a transition. This inherent high pass characteristics of the photoelectric effect is one of the outstanding advantages of quantum detectors: thermal background radiation and electric interference are practically irrelevant for the detection process.

An alternative detection scheme, used predominantly in the far infrared, relies on the conversion of electromagnetic radiation into thermal energy and measurement of the resulting temperature change. Such detectors can detect radiation of virtually any wavelength (which makes them, however, susceptible to thermal noise) and are very slow in comparison to quantum detectors.

© Springer International Publishing Switzerland 2016
G.A. Reider, *Photonics*, DOI 10.1007/978-3-319-26076-1_9

Of limited, but fundamental interest are so-called quantum-non-demolition detections schemes, where photons are detected without being absorbed; a possible implementation relies on the intensity dependence of the refractive index due to the Kerr effect (Sect. 8.3.2).

9.1 Photoelectric Detectors

The photoelectric effect relies on the transition of an electron from a bound state into a "mobile" state. Like any other transition, its probability is given by the Fermi rule Eq. (6.24). The external photoelectric effect, where the excited state is a freely propagating electron wave in vacuum, is schematically shown in Fig. 9.1. In metals, the energy threshold for this process is the work function Φ_m, i.e., the energy barrier between the Fermi level and the vacuum level. In semiconductors, the barrier is given by the sum of the band gap E_g and the so-called electron affinity E_A.

9.1.1 Photoelectron Multiplier Tubes

Although photoelectron multiplier tubes (PMTs) do not play a major role in photonics, they are of some practical interest because of their sensitivity, speed, and large photosensitive area. A PMT consists of a photocathode, usually made of a semiconductor layer, and a series of secondary electron multiplications stages, placed in a vacuum tube. A photon impinging on the photocathode produces, with a certain quantum efficiency, a photoelectron. This primary electron enters a cascade of so-called dynodes, i.e., electrodes that are optimized to emit secondary electrons when hit by an energetic electron. Starting with the photocathode, the dynodes are biased at increasingly positive potentials, so that the electrons are accelerated towards the following dynode; typical potential differences between successive electrodes are about 100 V (Fig. 9.2), giving rise to an impact energy of

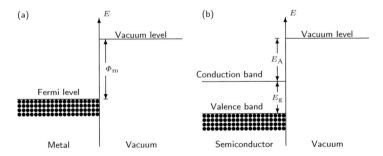

Fig. 9.1 External photoelectric effect: the barrier between the highest occupied electronic state and the vacuum must be overcome by the energy of absorbed photons; in metals (**a**) the barrier is the work function Φ_m (energy difference between Fermi and vacuum level); in intrinsic semiconductors (**b**) it is the sum of band gap energy E_g and electron affinity E_A

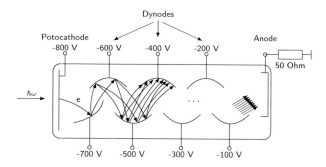

Fig. 9.2 Photoelectron multiplier tube (PMT)

the electrons of about 100 eV. Since this energy is several times the work function of the dynodes, every incoming electron produces several secondary electrons, resulting in an exponential growth of the electron package that has been initiated by a single photon. The final electrode (anode) is usually at ground potential and collects the electron package; the resulting current pulse produces a voltage spike at the output resistor which constitutes the output signal of the PMT.

The active material of the photocathode and dynodes is a thin layer either of an alkali-metal (Na, K, Cs) or a semiconductor (GaAs). The quantum efficiency depends on the absorption efficiency of the cathode and on the fraction of photoexcited electrons that escape into the vacuum. Semiconductor photocathodes are superior to metals in both respects, since their reflectance is lower than that of metals and the escape depth is much larger. In metals, only photoelectrons generated in the topmost atomic layer contribute to the photoemission, while in a semiconductor, electrons that are not immediately released into the vacuum populate the conduction band where they can propagate towards the vacuum interface during their relatively long recombination life time. To facilitate their escape into vacuum, the semiconductor is heavily p-doped and the surface is coated with highly electropositive atoms (Cs) that donate their valence electron to the semiconductor. The positively charged metal ions at the surface deform the semiconductor bands such that the bulk conduction band edge actually is above the vacuum level (Fig. 9.3)—a situation that is called negative electron affinity (NEA). Such photocathodes allow pushing the spectral sensitivity into the near infrared ($\approx 1.7 \, \mu$m). The UV response is limited by the transparency of the glass tube; dedicated UV-PMTs have a cutoff at ≈ 180 nm.

Due to the large electron escape depth, semiconductors are also the preferred dynode material. The secondary electron yield depends on the impact energy; at 100 eV, the gain per dynode is 3–5 so that a gain of $5^{10} \approx 10^7$ can be achieved with ten multiplication stages. Statistical variations of the electron trajectories result in a temporal spread of the electron package arriving at the anode. Optimized PMTs produce sub-ns current pulses, typical values of commercial PMTs are 1–2 ns (corresponding to a peak current of about 1 mA).

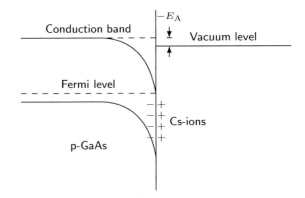

Fig. 9.3 Energy bands at the surface of a semiconductor with "negative" electron affinity: the band bending results from ionized Cs-atoms at the surface

PMTs can be operated in a photon counting mode; dark counts resulting from natural radioactivity of the tube materials can be suppressed efficiently, because most of these pulses do not experience the full gain and can be deselected by a pulse height discriminator; the residual dark count rate is on the order of 30 counts/s. Alternatively the average anode current is used as a measure of the incident optical power, a scheme that is linear over more than 6 orders of magnitude.

A related multiplier scheme is the so-called micro-channel plate (MCP, Fig. 9.4), a ceramic disc of about 2 mm thickness that is penetrated by millions of channels of typically $10\,\mu$m diameter. The inner wall of the channels is coated with a semiconductor or metal layer, and a voltage of ≈ 1 kV is applied to both ends. Photoelectrons that enter such a channel are accelerated in the strong axial electric field and produce a secondary electron avalanche by collisions with the walls, similar to the process in a PMT. The obvious advantage of this device is its spatial resolution: in combination with a fluorescent screen or a detector array (CCD) in the output plane, MCPs serve as image intensifiers with a gain of up to 10^6.

9.1.2 Semiconductor Photodetectors

9.1.2.1 pn-Photodiodes

By far the most important detectors in photonics are semiconductor photodiodes, i.e., pn-junctions that absorb photons in or close to the depletion zone (Fig. 9.5) and produce a photocurrent or a photovoltage, respectively, as output signal. A pn-junction consists of two sections of semiconductor material, one doped with electron donors (n-zone), the other with electron acceptors (p-zone). Driven by thermal motion, electrons from the n-zone diffuse into the p-zone and holes from the p-zone into the n-zone, leaving a space charge region of positively charged donor and negatively charged acceptor atoms behind. In equilibrium, the drift current induced by the electric field in this depletion zone balances the diffusive migration of charge carriers, and the bands in the two semiconductor zones are bent in such a way that

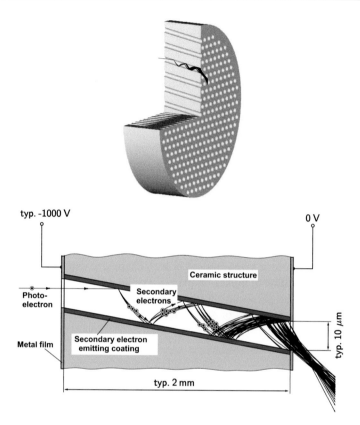

Fig. 9.4 Micro-channel plate (MCP); the μm-sized pores are coated with a secondary electron emitting material

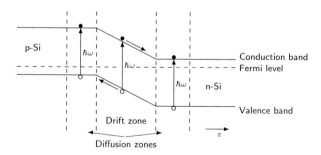

Fig. 9.5 Photon absorption in a pn-junction (photodiode)

there is a single Fermi level within the entire device. When a photon is absorbed in the drift zone, the resulting electron–hole pair is separated by the drift field: the electron is transported into the n-zone and the hole into the p-zone. This results in a potential difference between the two zones, the so-called photovoltage; the equilibrium situation can be re-established by recombination or an external current flow, the so-called photocurrent. It should be noted that despite the fact that two carriers are involved in the process, the photocurrent is constituted of a single electron charge, because at each point in the circuit, only one carrier contributes to the current. The process is completed once the hole recombines with an electron at the contact of the p-zone with the conductor.

In an ideal photodiode, every incident photon produces one electron/hole pair, so that the photocurrent is equal to the photon flux times the electron charge

$$J_{\text{ph}} = \eta_{\text{q}} \frac{e}{\hbar\omega} P_{\text{ph}}, \tag{9.1}$$

where the detector quantum efficiency is ideally $\eta_{\text{q}} = 1$. In practice, $\eta_{\text{q}} < 1$ for reasons that will be discussed below, but almost constant at a given wavelength so that the photocurrent is an extremely linear function of the incident optical power.

The photocurrent, at a given optical power, decreases with increasing optical frequency, since a photon of higher frequency, despite its higher energy, contributes only one carrier to the photocurrent (the excess energy is converted, via phonons, to heat). Expressed in terms of wavelength, Eq. (9.1) assumes the convenient form

$$J_{\text{ph}} = \eta_{\text{q}} \frac{\lambda}{1.240} P_{\text{ph}}, \tag{9.2}$$

where the current is given in [A], the optical power in [W], and the wavelength λ in [μm] (compare Table 1.1).

If the electrical circuit is open, every carrier pair, after being separated in the drift zone, reduces the space charge, resulting in a forward voltage across the two diode terminals. As a result, carrier diffusion is not fully compensated by the drift field and the excess carriers recombine within the diffusion time. If the incident light is not a single photon but a steady stream of photons, the photovoltaic forward voltage assumes a stationary value such that diffusion compensates the internal photocurrent. To calculate this voltage, we use Shockley's diode equation which relates the diode current J_{d} to the applied voltage U_{d} and supplement it with the photocurrent J_{ph}

$$J_{\text{d}} = J_{\text{s}} \left(e^{eU_{\text{d}}/k_{\text{B}}T} - 1 \right) - J_{\text{ph}}, \tag{9.3}$$

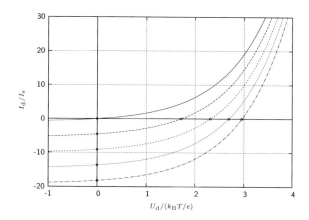

Fig. 9.6 Current–voltage diagram of a photodiode at different optical signal powers; the *dots* on the current axis show the short circuit photocurrent, those on the voltage axis indicate the open circuit photovoltage

where J_s is the so-called saturation current of the diode, typically on the order of 1 nA (Fig. 9.6). Setting $J_d = 0$ and using Eq. (9.1), we obtain

$$U_{ph} = U_{d,0} = \frac{k_B T}{e} \ln\left(\eta_q \frac{e P_{ph}}{\hbar \omega J_s} + 1\right). \tag{9.4}$$

For $J_{ph} \gg J_s$ ($P_{ph} \gg 1\,\text{nW}$), the photovoltage is a logarithmic measure of the incident optical power, and a photodiode in this mode of operation can be conveniently used in sensor applications with very large dynamic range.

Figure 9.7 shows the design of a typical photodiode; the optical signal is impinging on the pn-junction which is formed by a thin p-doped layer, contacted by a transparent electrode, on top of an n-doped substrate. To improve the yield, the detector face is usually antireflection coated. The photosensitive area of commercial diodes ranges from some $100\,\mu\text{m}^2$ to several $100\,\text{mm}^2$.

The responsivity R of a photodiode, that is the ratio of photocurrent to incident optical power, follows from Eq. (9.1) to be

$$R = \eta_q \frac{e}{\hbar \omega} = \eta_q \frac{\lambda[\mu\text{m}]}{1.240} [\text{A/W}], \tag{9.5}$$

where the quantum efficiency η_q is the fraction of incident photons that contribute to the photocurrent. If R is the reflectance of the detector, the fraction $1 - R$ of photons is absorbed, essentially within the absorption length $1/\alpha$ [Eq. (2.71)]; only those electron/hole pairs that are generated within the drift zone and the adjacent diffusion

Fig. 9.7 Cross section of a
commercial photodiode

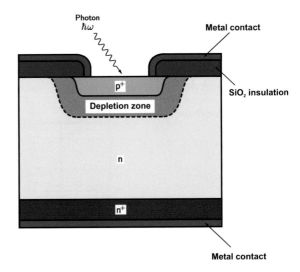

Fig. 9.8 Spectral
responsivity of a typical
Si-pin-photodiode: the
maximum quantum efficiency
is $\eta_q = 0.8$. The *dashed lines*
indicate the theoretical limits,
set by the ideal quantum
efficiency of $\eta_q = 1$ and the
cutoff due to the band gap

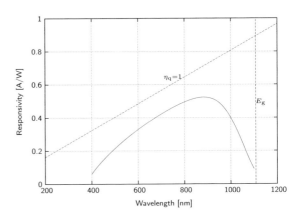

zones can participate in the photocurrent; if the total thickness of this range is given
by d,

$$\eta_q = (1 - R)\left(1 - e^{-\alpha d}\right)\eta_i, \tag{9.6}$$

where the internal quantum efficiency η_i is the fraction of carriers not lost by
recombination or traps. Figure 9.8 shows the responsivity of a typical commercial
Si-photodiode; the IR-sensitivity may extend slightly beyond the band gap of silicon
(1.12 eV) because of absorption by exciton states below the conduction band edge.
The cutoff in the UV will be discussed below.

The two contacts of a photodiode form a capacitor that limits the high frequency performance of the detector; fast photodiodes usually have a very small sensitive area to minimize the capacity. The capacity can be further reduced by operating the diode under a reverse bias voltage, increasing the thickness of the depletion zone that acts as a dielectric in the capacitor.

9.1.2.2 PIN-Photodiodes

The photodiode response can also be improved by placing an intrinsic (undoped) layer between the p- and n-zone (pin-structure). Apart from reducing the capacity, the intrinsic layer is part of the drift zone, improving the detector responsivity. The thickness of the intrinsic zone has to be carefully optimized, because the time the carriers need to get to the terminals is also increased, which is detrimental for the speed of the detector.

9.1.2.3 Avalanche Photodiodes

Unlike PMTs or a MCPs, the photodiodes discussed so far do not provide any gain. However, carrier multiplication by impact ionization is also possible within a semiconductor. To this purpose, a reverse voltage exceeding the band gap by a large factor is applied to the diode. Electrons in the conduction band and holes in the valence band can gain so much energy in the drift zone that they create new electron–hole pairs by collision (Fig. 9.9). In this way, a carrier avalanche can build up, similar to the electron avalanche in an electron multiplier tube; such diodes are known as avalanche photodiodes (APDs). An important difference to the multiplication process in PMTs is the fact that there may be actually two counterpropagating avalanches of electrons and holes, respectively, each of them producing new holes *and* electrons, so that the process is not self-terminating and can lead to catastrophic breakdown. Such APDs can be used as single photon

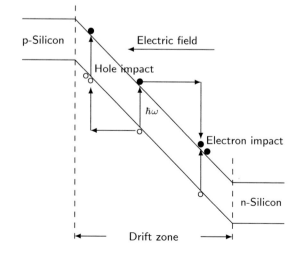

Fig. 9.9 Operation of an avalanche photodiode (APD): photogenerated electron/hole pairs are accelerated in the electric bias field and produce additional carriers by impact excitation, resulting in a carrier avalanche

detectors (see, e.g., Eisaman et al. 2011); catastrophic breakdown can be prevented, e.g., by a serial resistor in the bias supply that terminates the breakdown when the current exceeds a certain limit. APDs exhibit a dead time after each break down, so that the count rate is limited to about 10^7 counts per second.

For a single carrier species, the avalanche current density develops according to

$$\frac{dj_{e,h}}{dz} = \alpha_{e,h} j_e,$$

(9.7)

where the ionization coefficients $\alpha_{e,h}$ depend on the semiconductor material and the electric field, and usually differ significantly from each other. In silicon, for example, $\alpha_h \ll \alpha_e$, so that the hole-avalanche is negligible and the process comes to a halt when the electrons reach the end of the drift zone. In this operating regime, a Si-APD is stable and the gain factor along a drift zone of length l is given by

$$\frac{j_e(l)}{j_e(0)} = e^{\alpha_e l}.$$

(9.8)

In APDs, the light absorbing zone is usually spatially separated from the drift zone, so that all photoelectrons experience about the same gain and amplifier noise is kept low. APDs are also very fast; optimized APDs have cutoff frequencies of several 10 GHz.

9.1.2.4 Spectral Response

Depending on the signal wavelength, different materials are used for photodiodes (see Fig. 7.39). In the visible and near IR (400–1100 nm), Si-photodiodes are preferentially used, with a quantum efficiency of up to 0.9 (Fig. 9.8). In optical communications with frequency bands at 1.3 and 1.5 μm, Ge- and InGaAsP-pin-photodiodes are used as well as heterostructure APDs with an InGaAs photoexcitation zone and a (transparent) InP-avalanche zone. The ternary $In_{1-x}Ga_xAs$ system provides a wide range of band gaps, from InAs (0.35 eV/3.5 μm) to GaAs (1.43 eV/0.87 μm). For longer wavelengths, $Hg_{1-x}Cd_xTe$ is a widely used variable gap semiconductor, reaching up to 0.1 eV (12 μm). GaAs/AlGaAs multiple quantum well detectors reach 15 μm.

The low responsivity of Si (and other) photodiodes in the UV is due to the extremely short absorption length (about 10 nm for Si at a wavelength of 350 nm), so that the photocarriers are not generated in the depletion zone but have to diffuse to the pn-junction; close to the surface, however, the probability of defect-mediated recombination is very high, rendering the internal quantum efficiency low. For UV-applications, large gap photodiodes based on silicon carbide (4H-SiC, $E_g = 3.3$ eV) are used, with a spectral sensitivity ranging from 375 to 210 nm.

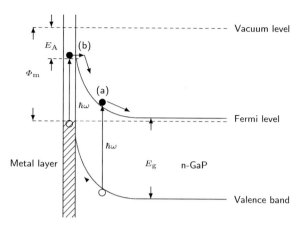

Fig. 9.10 Band diagram of a Schottky photodiode (without bias voltage); photoexcitation can occur (**a**) within the semiconductor for $\hbar\omega > E_g$, or (**b**) from the Fermi level of the metal across the Schottky barrier for $\hbar\omega > \Phi_m - E_A$

9.1.2.5 Schottky Photodiodes

The UV response can be improved by replacing the pn-junction with a metal-semiconductor junction (Schottky photodiode, Fig. 9.10). Since the work function of the (very thin and therefore transparent) metal is larger than the electron affinity of the (n-doped) semiconductor, electrons diffuse into the metal, leaving positively charged donor atoms behind. The depletion zone thus reaches up to the metal interface, so that practically all photons are absorbed within the drift zone; the metal layer also reduces the density of defects. Schottky photodiodes have the additional advantage of being extremely fast, with response times of several ps and a bandwidth above 100 GHz.

Schottky diodes can also be operated at photon energies that are below the band gap of the semiconductor; photons are absorbed by the metal electrode and the resulting photoelectron can migrate across the Schottky barrier into the semiconductor; the cutoff is then given by the height of the barrier, which can be adjusted by proper choice of the metal semiconductor system. While conventional photodiodes are illuminated from the front, these IR-photodiodes can be backside illuminated, since the semiconductor is transparent in the IR. Important examples of this group of detectors are PtSi Schottky barrier diodes that can also be readily integrated in Si CCD-arrays.

9.1.3 Detector Arrays

For imaging and related applications, photodetectors can be arranged in one- or two-dimensional arrays. Silicon-based photodiodes are particularly well suited for integration into arrays because of the high technological maturity of the supporting MOS electronics. For image acquisition, the individual photodiodes, called pixels, with a typical size of $10 \times 10\,\mu m^2$ or less, transfer the photoelectrons into an underlying capacitor; the readout is accomplished either by a charge coupled device

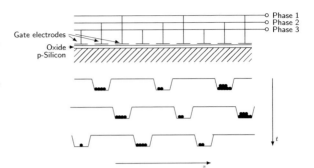

Fig. 9.11 Operation of a CCD as an analog shift register: the (photogenerated) carriers are shifted from one MOS-capacitor to the adjacent by cycles of positive bias of the transfer gates. In the scheme shown here, a complete transfer cycle consists of three steps

(CCD) or by individually addressable CMOS-amplifiers. The operation of a CCD is shown in Fig. 9.11: it consists of an array of microscopic metallic gate electrodes, contacted in three groups that form capacitors with an intermediate silicon oxide layer as dielectric. Electrons in the substrate are collected under the positively biased electrode and can be transferred to the adjacent capacitor by switching the corresponding gates to a positive bias. In this fashion, the charge can be shifted through the entire array in the way of a bucket chain. At the (serial) output, a transimpedance amplifier converts the charge-signal into a proportional voltage. The performance of such a shift register depends on the charge transfer efficiency (CTE), defined as the fraction of electrons that "survive" the transfer from one cell to the next; a CTE value exceeding 0.999999 is necessary to operate large CCDs without significant data deterioration. Alternatively, every pixel is supplied with an amplifier and can be read out individually (active pixel sensor).

Two-dimensional detector arrays are also known as focal plane arrays (FPAs). For IR-applications, PtSi/Si Schottky diodes can be integrated directly into the MOS-structure, hybrid structures of InSb-photodetectors and CMOS-readout electronics are also used.

9.1.4 Photoresistors

The internal photoeffect in semiconductors also changes the *conductivity* of a (homogeneous) semiconductor. Photodetectors that rely on this effect are known as photoresistors or photocells. They consist of a thin semiconductor film with metallic contact structures on top and can be realized with virtually any semiconductor, including those that are not suitable for the production of high quality diode junctions. The principle of operation is illustrated in Fig. 9.12: absorption of photons creates electron–hole pairs that migrate in the electric field $E = U/l$ produced by the applied external voltage U. To calculate the photocurrent, we assume that the electron mobility μ_e exceeds the hole mobility μ_h by far so that the hole usually recombines before it reaches the negative contact. Assuming the electron velocity to be $v_e = \mu_e E$, the average time that it takes an electron to arrive at the positive

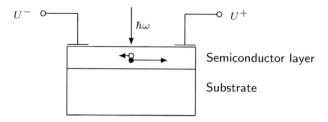

Fig. 9.12 Functional diagram of a photoconductor

contact is $t_e \approx l/2\mu_e E$. If the hole is still present in the layer, another electron from the negative contact is injected to conserve charge neutrality. This cycle is repeated until the hole recombines with an electron, which statistically takes the time τ_{rec} to happen. One absorbed photon therefore results in a transported charge of τ_{rec}/t_e electrons. A photoresistor thus produces a photocurrent

$$J_{ph}[A] = \eta_q \frac{\tau_{rec}}{t_e} \frac{P_{ph}}{\hbar\omega} e = \eta_q G \frac{\lambda[\mu m]}{1.240} P_{ph}[W], \qquad (9.9)$$

where $G = \tau_{rec}/t_e$ is the gain of the detector. A disadvantage of this device is its slow response time, limited by the lifetime τ_{rec} of the holes. Photoresistors are used primarily in the IR and produced from InAs, InSb, and $Hg_{1-x}Cd_xTe$, where the band gap can be chosen anywhere below $1.6\,eV$ by adjusting the stoichiometric parameter x. Extrinsic (doped) semiconductors are also used, where the photoexcitation happens from the dopant level to the conduction level. In this case, the hole is actually an immobile dopant atom and the cutoff wavelength is given by the energy difference between dopant level and band edge. If this spacing is comparable to the thermal energy $k_B T$, thermal excitation of dopant states contributes significantly to the dark current, increasing the noise. For operation at wavelength beyond $3\,\mu m$, cooling of the detector by liquid nitrogen or helium is required.

9.2   Characteristic Parameters of Detectors

Apart from their spectral and time response, detectors are characterized by various performance parameters. The responsivity has already been introduced [Eq. (9.5)] and is generally proportional to the wavelength for quantum detectors; PMTs, MCPs, APDs, and photoresistors have a responsivity that is enhanced by a gain factor that can be as large as 10^7 in comparison to a pn-photodiode.

For most applications, the signal-to-noise ratio is more relevant; its definition relies on the variance σ^2 of the photocurrent

$$(S/N) = \frac{J_{ph}^2}{\sigma^2(J_{ph})}; \qquad (9.10)$$

in addition to various electronic sources of noise, optical detectors show a fundamental noise contribution resulting from the quantum nature of photons and cannot be reduced by cooling or electronic means; we will discuss some aspects of photon statistics in Sect. 9.3.

Another important parameter is the "speed" of the detector, usually described by the dependence of the responsivity on the modulation frequency f_m of the optical signal. Most detectors show a simple low pass behavior

$$R(f_m) = R_0 \frac{1}{\sqrt{1 + 2\pi f_m \tau_d}}; \qquad (9.11)$$

the bandwidth of the detector is usually specified by the modulation frequency at which the detector power, which is proportional to J_{ph}^2, is reduced to one half ($-3\,dB$) of the low-frequency responsivity

$$f_{m,3\,dB} = \frac{1}{2\pi \tau_d}; \qquad (9.12)$$

τ_d is the characteristic time constant of the detector, determined by its capacitance and internal characteristic times such as the transit time or carrier life time. Well above this frequency, in particular for light pulses much shorter than τ_d, the detector operates as integrator, delivering an electrical pulse whose peak is proportional to the optical pulse energy and whose shape is the pulse response function of the detector, independent of the optical pulse shape.

9.3   Photon Statistics

The photodetection process usually relies on the excitation of electrons, and the discrete nature of the electric charge gives rise to a fundamental shot noise. Assuming that the number of photoelectrons is proportional to the number of photons impinging on the detector, the photoelectron statistics is a replica of the photon statistics.

Shot noise can often be described by a Poisson distribution valid for discrete, independent processes. A (stationary) optical signal with power P_{ph} can produce, in an ideal photodetector without gain, $P_{ph}/\hbar\omega$ electrons per second. If we count the number n of photoelectrons in many consecutive time intervals of duration t_M, the

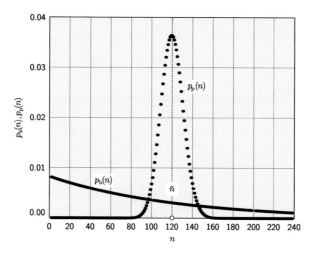

Fig. 9.13 Poisson distribution p_p and Bose–Einstein distribution p_b for the same average value $\bar{n}=120$

mean count number is $\bar{n} = (P_{ph}/\hbar\omega)t_M$. According to the Poisson distribution, the probability $p(n)$ to find n electrons in a randomly selected time interval is

$$p_p(n) = \frac{1}{n!}(\bar{n})^n e^{-\bar{n}}. \tag{9.13}$$

This distribution (shown in Fig. 9.13) exhibits a peak approximately at $\bar{n}$; the variance is equal to the mean value

$$\sigma_p^2 = \sum_n p_p(n)(n - \bar{n})^2 = \bar{n}. \tag{9.14}$$

According to Eq. (9.10), the signal-to-noise ratio due to Poisson distributed shot noise is therefore

$$(S/N)_p = \frac{(\bar{n})^2}{\sigma_p^2} = \bar{n}. \tag{9.15}$$

Poisson statistics describes the fluctuations of light from a single mode laser far above threshold. Thermal light, however, follows a different statistics. We assume a setup that allows us to measure the number of photons in a selected electromagnetic mode of a cavity at temperature T. According to Planck's theory of black body radiation, the energy of such a mode is an integer multiple of $\hbar\omega$; the probability that a given mode contains n photons, i.e., the energy $n\hbar\omega$, is given by Boltzmann's distribution

$$p(n) = p_0 e^{-n\hbar\omega/k_B T} = p_0 u^n, \tag{9.16}$$

where $u := e^{-\hbar\omega/k_B T}$ is introduced for convenience; p_0 follows from the condition $\sum p(n) \equiv p_0/(1-u) = 1$ to be $p_0 = 1-u$, so that Eq. (9.16) can be written as

$$p(n) = (1-u)u^n. \tag{9.17}$$

The mean value $\bar{n}$

$$\bar{n} = \sum_n n p(n) = \sum_n n(1-u)u^n \tag{9.18}$$

is calculated by taking the derivative of the relation $\sum u^n = 1/(1-u)$ in respect to u

$$\sum_n n u^{n-1} = \frac{1}{(1-u)^2} \tag{9.19}$$

and multiplying the result with $u(1-u)$:

$$\bar{n} = \frac{1}{u^{-1}-1} = \frac{1}{e^{\hbar\omega/k_B T}-1}. \tag{9.20}$$

We can now express Eq. (9.17) in terms of $\bar{n}$ by setting, according to Eq. (9.20), $u = \bar{n}/(\bar{n}+1)$:

$$p_b(n) = \frac{(\bar{n})^n}{(\bar{n}+1)^{n+1}}. \tag{9.21}$$

This is known as Bose–Einstein distribution; it denotes the probability to find n energy quanta in a mode under thermal equilibrium with matter, if the mean value is $\bar{n}$. As shown in Fig. 9.13, this distribution displays a maximum at $n=0$ and then falls off continuously; also, the width of the distribution is much larger than that of a Poisson distribution with identical mean value $\bar{n}$. The most likely number of detector counts is zero for any value of $\bar{n}$: many intervals with no or very view counts are followed by intervals with counts well above the average (Fig. 9.14). This behavior is known as photon bunching and is typical for thermal light.

To calculate the variance of this distribution, we use the identity $\sigma^2 = \overline{(n-\bar{n})^2} = \overline{n^2} - (\bar{n})^2$ and take the second derivative of $\sum u^n = 1/(1-u)$ to obtain

$$\overline{n^2} = \sum_n n^2 p_b(n) = \bar{n} + 2(\bar{n})^2, \tag{9.22}$$

Fig. 9.14 Count statistics of photons: (**a**) Poisson distribution, (**b**) Bose–Einstein distribution

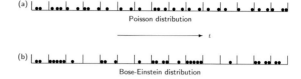

and finally

$$\sigma_{\mathrm{b}}^2 = \bar{n} + (\bar{n})^2. \tag{9.23}$$

Compared to Eq. (9.14), there is an additional term $(\bar{n})^2$ which can be attributed to interference effects between spontaneously emitted photons. The first term in Eq. (9.23) can be understood as representing the "particle" nature of photons while the second one is due to their wave character. The signal-to-noise ratio Eq. (9.10) of single mode thermal light is given by

$$(\mathrm{S/N})_{\mathrm{b}} = \frac{\bar{n}}{\bar{n}+1}, \tag{9.24}$$

and cannot exceed unity. If many thermal modes contribute to the signal, the counting statistics tends towards a Poisson distribution, however.

As can be shown, Poisson and Bose distributions are conserved if the events that are counted are selected randomly from the initial set of events; in photodetection, a beam splitter or a detection process with quantum efficiency < 1 gives rise to such a selection. For a Poisson distributed photon stream, this means that the photoelectron statistics from a detector with quantum efficiency η_{q} provides a signal-to-noise ratio of

$$(\mathrm{S/N})_{\mathrm{e}} = \eta_{\mathrm{q}}(\mathrm{S/N})_{\mathrm{ph}} = \eta_{\mathrm{q}}\bar{n}, \tag{9.25}$$

which is the S/N of an ideal detector, reduced by η_{q}.

In digital optical communications, the bit error rate (BER) is of particular importance, that is the probability to mistake an "1" for a "0" and *vice versa*. Assuming a simple encoding where "1" is represented by a package of $n \neq 0$ photons and "0" by $n = 0$, we find the probability to mistake a "1" for "0" to be $p_{\mathrm{p}}(0) = \mathrm{e}^{-\eta_{\mathrm{q}}\bar{n}}$; the reverse case is impossible in this case. Assuming an approximately equal number of "1" and "0" bits, a maximum permissible BER of 10^{-9} requires $\bar{n} \approx 40$ per "1" bit, if a quantum efficiency of 0.5 is assumed. At a photon energy of 0.8 eV (1.55 μm) and data rate of 10 Gbit/s, this corresponds to an optical average power at the detector of 26 nW. Further assuming a length of the fiber link of 100 km and losses of 0.2 dB/km, a power of 1 μW must be launched into the fiber. This estimate takes only quantum noise into account and disregards all other sources of noise; in practice, the launched power is about 1 mW, limited by nonlinear optical processes in the fiber.

9.4   Photometry and Colorimetry

9.4.1   Photometry

Photometry deals with the measurement of light levels as perceived by the human eye. It uses its own set of units (Table 9.1) that are related to the radiometric (photonic) units by the so-called luminosity function. For physiological reasons, the luminosity itself depends on the light level; photopic vision requires "bright" illumination and allows color vision; the relevant luminosity function $V(\lambda)$ is shown in Fig. 9.15, selected values are given in Table 9.2. Physiologically, cone cells in the retina are responsible for this type of vision. Low level light vision is called scotopic vision and is mediated by rod cells; it is characterized by the luminosity function $V'(\lambda)$. At intermediate light levels occurs mesopic vision with ill-defined color perception and luminosity.

For the conversion of radiometric data into photometric (photopic), the power spectrum is multiplied with the luminosity function V, integrated over the wave-

Table 9.1 Correspondence between photometric and radiometric quantities; important photometric units are Lux [lx], Lumen [lm], and Candela [cd]

Photonic quantity	Units	Photometric quantity	Units
Radiant flux/power	[W]	Luminous flux	[lm]
Radiant energy	[Ws = J]	Luminous energy	[lm s]
Irradiance/intensity/flux density	[W m^{-2}]	Illuminance	[lm m^{-2} = lx]
Radiant exposure/fluence	[J m^{-2}]	Luminous exposure	[lx s]
Radiant intensity	[W sr^{-1}]	Luminous intensity	[lm sr^{-1} = cd]
Radiance/brightness	[W sr^{-1} m^{-2}]	Luminance	[lm sr^{-1} m^{-2}]

Fig. 9.15 Luminosity functions $V(\lambda)$ (for bright light) and $V'(\lambda)$ (for very low light levels) according to ISO/CIE 10527 (1991)

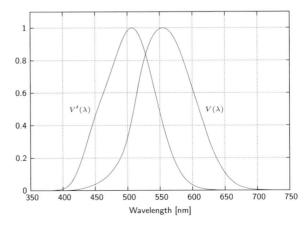

Wavelength [nm]

Table 9.2 Luminosity function for photopic $[V(\lambda)]$ and scotopic $[V'(\lambda)]$ vision according to ISO/CIE 10527 (1991) standard; $V(\lambda)$ is identical with the colorimetric $\bar{y}(\lambda)$-function (Fig. 9.16)

λ [nm]	$V(\lambda)$	$V'(\lambda)$	λ [nm]	$V(\lambda)$	$V'(\lambda)$	λ [nm]	$V(\lambda)$	$V'(\lambda)$
380	0.0000	0.0006	510	0.5030	0.9970	610	0.5030	0.0159
400	0.0004	0.0093	520	0.7100	0.9350	620	0.3810	0.0074
420	0.0040	0.0966	540	0.9540	0.6500	640	0.1750	0.0015
440	0.0230	0.3281	550	0.9950	0.4810	660	0.0610	0.0003
460	0.0600	0.5670	560	0.9950	0.3288	680	0.0170	0.0001
480	0.1390	0.7930	580	0.8700	0.1212	700	0.0041	0.0000
500	0.3230	0.9820	600	0.6310	0.0332	720	0.0011	0.0000

length and multiplied with the conversion factor 683 [lm/W]. For scotopic vision, the weight function is $V'(\lambda) \times 1700$ [lm/W]. If the light is practically monochromatic, the respective radiometric value is simply multiplied with $V(\lambda) \times 683$ [lm/W] at the respective wavelength. Photometric measurements do not require spectral resolution, however, if the spectral detector sensitivity matches the luminosity function (eye response photodiodes).

A 1 mW-HeNe laser at $\lambda = 632$ nm $[V(632) = 0.247]$, for example, has a luminous flux of $683 \times 0.247 \times 10^{-3}$ lm $= 0.17$ lm; assuming a beam cross section of 1 mm^2, the illuminance is 170,000 lx. The sun, for comparison, provides an illuminance of about 70,000 lx, the full moon 0.2 lx. At the sensitivity peak of the human eye, at 555 nm, 1 lx $\hat{=} 0.1464\ \mu$W/cm^2 and 1 lm $\hat{=} 1.464$ mW.

The luminous intensity of a light source, given in candela [cd], is the luminous flux per solid angle [sr] and takes the degree of collimation of the emitted light into account; an isotropic emitter with a luminous flux of 1 lm has a luminous intensity of $1/4\pi$ cd, while the aforementioned HeNe laser, with a divergence angle Eq. (3.19) of, say, 1 mrad [corresponding to $(\pi/4) \times 10^{-6}$ sr] produces a luminous intensity of 214,000 cd.

9.4.2 Colorimetry

Color vision relies on the existence of three kinds of cone cells with differing spectral sensitivity. Based upon the empirical Grassmann's laws, color perception can be described in a three dimension linear vector space, called color space. The direction of a (position) vector in this space determines the chromaticity, while its length is proportional to its luminance. Accordingly, three so called color matching functions are required and sufficient to determine the chromaticity.

9.4.2.1 Color Matching Functions
Because of the linearity of the color space, the choice of color matching functions is not unique, and color coordinates referring to one set of functions can be transformed into any other base by a linear transformation. Among the many

Fig. 9.16 Color matching functions $\bar{x}(\lambda), \bar{y}(\lambda), \bar{z}(\lambda)$ of the CIE (1931) XYZ color space (for numerical values, see CIE 1931)

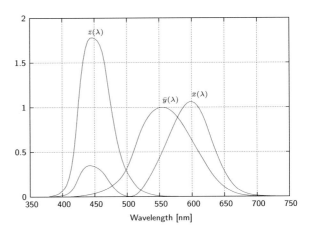

different sets of color matching functions, the CIE (1931) XYZ-metric is the most popular; the three weight functions $\bar{x}(\lambda), \bar{y}(\lambda), \bar{z}(\lambda)$ are shown in Fig. 9.16. The location of a signal with the power spectrum $S(\lambda)$ in the XYZ color space is given by the coordinates

$$X = \int S(\lambda)\bar{x}(\lambda)\,d\lambda \qquad (9.26)$$

$$Y = \int S(\lambda)\bar{y}(\lambda)\,d\lambda \qquad (9.27)$$

$$Z = \int S(\lambda)\bar{z}(\lambda)\,d\lambda, \qquad (9.28)$$

also called tristimulus values.

Mathematically, Eqs. (9.26)–(9.28) are inner products in Hilbert space and projection of the infinite-dimensional spectrum on the base vectors $\bar{x}(\lambda), \bar{y}(\lambda), \bar{z}(\lambda)$. Note that such a projection is not isomorphic: while every possible spectrum is mapped onto exactly one point in color space, a given color point corresponds to an infinite number of different, so-called metameric spectral distributions; exceptions are only the color points that correspond to monochromatic signals.

Unlike most other sets of color matching functions, the $\bar{x}\bar{y}\bar{z}$- functions are non-negative, implying that the tristimulus values of any signal are also positive. A convenient consequence is that the XYZ-coordinates can be directly measured, without the requirement of spectral analysis, by three photodetectors with spectral sensitivity $\bar{x}(\lambda), \bar{y}(\lambda)$, and $\bar{z}(\lambda)$. Another convenient feature of this color space is that the Y component is identical to the illuminance, because $\bar{y}(\lambda) = V(\lambda)$. On the other hand, the three base vectors do not correspond to any existing color, because there exists no possible spectral distributions for which only one of the coordinates is non-zero.

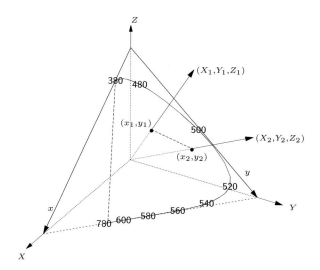

Fig. 9.17 In three-dimensional *XYZ*-space, every given power spectrum corresponds to a vector; its chromaticity can be identified by the intersection of the vector with the $X+Y+Z = 1$ plane (*xy*-coordinates). Additive mixing of two colors produces a new color that is represented by the vector sum of the two input colors, with a chromaticity that lies on a straight line between the input chromaticity points

Since chromaticity is independent of luminance, it can be characterized by normalized coordinates

$$x, y, z := X, Y, Z/(X + Y + Z) \qquad (9.29)$$

that can be localized in a *two*-dimensional map, because $x + y + z = 1$; it is common to use an orthogonal *xy*-system. In the three-dimensional *XYZ*-space, the resulting map (called chromaticity diagram) lies in the plane defined by $X+Y+Z = 1$; the *xy* coordinates denote the point where the *XYZ*-vector (or its extension) intersects this plane (Fig. 9.17).

9.4.2.2 Additive Color Mixing

As a consequence of the linearity of the color space, the superposition of two or more color signals (called primary colors) produces a signal that is represented by the vector sum of the input signals. If the luminance of the primaries is varied, the resulting chromaticity lies within the polygon formed by the chromaticity coordinates of the primaries, since the luminance cannot assume negative values; this polygon is called gamut. Obviously, the gamut is a sub-space of the color space. No finite set of primaries allows producing all colors, i.e., the entire gamut of human vision. In particular, the choice of three primary colors produces a triangular gamut that is significantly smaller than the complete color space.

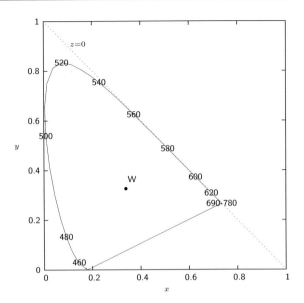

Fig. 9.18 Chromaticity diagram according to CIE (1931): the set of visible colors lies within the locus of spectrally pure colors and the line of purples; W denotes the white point $x = y = z = 1/3$

The locus of spectrally pure colors, (i.e., the chromaticity of monochromatic light) forms a concave, horseshoe-shaped curve in the chromaticity diagram, stretching from 780 nm (deep red) to 400 nm (extreme blue); the two end points are connected by the so called line of purples. Since any spectrum can be understood as linear combination of spectrally pure colors, all possible colors lie within this curve (see Fig. 9.18). In the center of the chromaticity diagram there is the achromatic white point W, defined by $x = y = z = 1/3$. It can be realized by an infinite variety of spectra, for example, two (so-called complementary) monochromatic colors lying on a line through W, or by the very broad and smooth spectrum of a filament of appropriate temperature. Since the chromaticity of the light scattered by an object depends on the spectral properties of the object itself (reflectance, scattering) *and* the power spectrum of the illuminant, it will usually display very different chromaticities when illuminated by different illuminants (day light, incandescent bulbs, white LEDs), even if the chromaticity of the illuminants is identical.

9.5   Summary

Practically all photonic detectors are quantum detectors: a photon is absorbed and the excited electron contributes to an electrical current or serves to build up a voltage. We describe the underlying photoelectric effect and possible internal gain mechanisms. PMT and MCP rely on the external photoelectric effect, require vacuum environment and a high voltage supply and provide very high gain and speed; both detectors are capable of single photon detection. The photodiode, a semiconductor pn-junction constructed in such way that light is absorbed in the drift zone of the junction, delivers a closed loop current that is proportional to the absorbed optical power, equivalent to ideally one electron per absorbed photon. The responsivity (the ratio of photo current to detected optical power) of quantum detectors decreases, for fundamental reasons, with the light frequency, because a high frequency photon carries more energy, but still produces only one photoelectron.

Photodiodes can also provide gain by internal excited carrier multiplication; such avalanche diodes are also capable of single photon detection and can outperform PMTs in terms of speed.

For many applications, the signal-to-noise ratio is the most important feature of a detector; we discuss Poisson distributed photon streams provided by lasers and compare them to Bose–Einstein distributed photon streams from thermal light sources. A comparison of the respective BER shows the statistical advantage of laser sources for optical communications.

Display applications of photonic light sources require the understanding of human color vision. Color measurement is based on Grassmann's laws and can be understood as a projection of the spectral space on a three-dimensional vector space. The inverse process, additive color mixing, is discussed and its limits in terms of color reproduction is analyzed.

9.6 Problems

1. What is the photon flow (photons/s) of a 1 mW HeNe-laser beam ($\lambda = 632$ nm)? What is the photon current induced by this laser beam in an ideal photo diode? What is the photocurrent induced by a blue ($\lambda = 310$ nm) 1 mW beam? What is the maximum photovoltaic power in either case [use Shockley's diode equation (9.3) to maximize the product of photovoltage and corresponding photocurrent]?

2. How many photons impinge on a CCD-camera pixel (area $10\,\mu m \times 10\,\mu m$) at 300 lux (office illumination level) during a 1/100 s exposure time (assume monochromatic light of wavelength 550 nm)? What is the statistical pixel to pixel variation of the number of photons, assuming Poisson statistics? Compare this to sunlight (70,000 lux) and moonlight (0.2 lux).

3. Show that the mean value $\sum_n n\, p_P(n)$ of the Poisson distribution Eq. (9.13) is indeed $\bar{n}$; prove that the variance of the Poisson distribution is equal to the mean value Eq. (9.14).

4. We want to communicate with our friends on the moon and use a 1 W Gaussian laser beam, wavelength $1\,\mu m$, with a telescope expanding the beam to 1 m diameter. The detector area on the moon is $1\,cm^2$, the quantum efficiency is 1. What is the maximum bit rate (on/off keying), if the BER is supposed to be less than 10^{-6} (neglect light scattered from the earth or coming from stars)? How do things change if our friends also use a telescope of 1 m input aperture?

5. Calculate the color coordinates of "white" light (a) with $S(\lambda_0) = $ const., (b) with $S(\omega) = $ const., and compare it to the white point W.

References and Suggested Reading

Boyd, R. W. (1983). *Radiometry and the detection of optical radiation*. New York: Wiley
Burns, G. (1985). *Solid state physics*. Orlando: Academic
CIE. (1931). *1931 color observer*. http://files.cie.co.at/204.xls
Eisaman, M. D., Fan, J., Migdall, A., & Polyakov, S. V. (2011). Single-photon sources and detectors. *Review of scientific instruments, 82*(7), 071101
Haus, H. A. (2002). *Electromagnetic noise and quantum mechanical measurements*. Berlin: Springer
Jenkins, T. E. (1983). *Optical sensing techniques and signal processing*. Englewood Cliffs: Prentice-Hall
MacAdam, D. L. (1985). *Color measurement*. New York: Springer
Nabet, B. (2015). *Photodetectors*. Oxford: Elsevier
Saleh, B. E. (1977). *Photoelectronic statistics*. New York: Springer
Schneider, H., & Liu, H. C. (2007). *Quantum well infrared photodetectors*. New York: Springer
Seitz, P., & Theuwissen, A. J. P. (2011). *Single-photon imaging* (Vol. 160). New York: Springer
Trishenkov, M. A. (1997). *Detection of low-level optical signals*. New York: Springer

Index

ABCD-transformation, 117, 217
Absorption, 60, 253
  coefficient, 60
  length, 60
  saturation, 264, 272
  two-photon, 387
Acousto-optic modulator, 320, 406
Active zone. *See* Laser diode
Additive color mixing, 433
Amplification
  Brillouin, 392
  free electron laser, 348
  parametric
    second order, 369
    third order, 384
  Raman, 388
  stimulated emission, 256
Amplitude, complex, 6
Anisotropic media, 69
  energy transport, 83
  Poynting vector, 83
  wave equation, 75
Anti-reflection coating, 176
Anti-Stokes line, 389
APD. *See* Avalanche photodiode
Aperture, numerical, 198
AR coating. *See* Anti-reflection coating
Autocorrelation, 191
  intensity, 377
  interferometric, 377
  second order, 377
Avalanche photodiode (APD), 422
Average power, 14
Axis, optical, 78

Babinet-Soleil compensator, 35
Band edge. *See* Semiconductor
Band gap. *See* Semiconductor
Base, 20

circular, 29
  linear, 21
Base transformation, 29
Beam propagation, 101
Beam splitter, 158
  scattering matrix, 163
  waveguides, 223
Beam velocity, 12
Bessel functions, 208
Biaxial media. *See* Anisotropic media
Birefringence, 82
Bit error rate, 429
Blazing. *See* Optical grating
Bloch wave, 276
Boltzmann distribution, 262
Boltzmann's constant, 2
Bose-Einstein distribution, 428
Boundary conditions, 40, 45, 208
  conducting resonator, 187
  fiber waveguide, 206
  multilayer, 170
  periodic, 276
Bragg
  cell, 409
  condition, 228
  grating, 228
  reflection, 228
  reflector, 339
  scattering, 406
Brewster angle, 50
Brillouin
  amplification, 392
  zone, 276

c_0. *See* Vacuum speed of light
Causality, 67
CCD. *See* Charge coupled device
Charge coupled device (CCD), 424
$\chi^{(2)}$-. *See* Susceptibility, second order

© Springer International Publishing Switzerland 2016
G.A. Reider, *Photonics*, DOI 10.1007/978-3-319-26076-1

$\chi^{(3)}$-. *See* Susceptibility, third order
Chirp, 146
Chromaticity, 433
Circularly polarized base, 29
Coherence
   complex degree, 191
   length, 191
   length, nonlinear optical, 363
   spatial, 193
   stimulated emission, 254
   temporal, 189
   time, 191
Collisions, 63
Color matching functions, 432
Color space, 431
Colorimetry, 431
Complex analytic signal, 189
Confocal parameter. *See* Gaussian beam
Coulomb potential, 246
Coupled modes formalism, 219
Coupling, 219, 300, 371, 398, 408
   optimum, 303
Critical angle. *See* Total reflection
Cross correlation, 191
Cross phase modulation (XPM), 382
Cutoff frequency. *See* Waveguide

dB. *See* Decibel
DeBroglie wave, 44, 246
Decibel (dB), 213
Density of modes
   3D resonator, 187
   electromagnetic field, 187
Density of states
   joint, 284, 285
   quantum dot, 292
   quantum well, 291
Dephasing time, 260
Difference frequency generation, 353
Diffraction, 43, 168
Dipole interaction, 250
Dipole matrix element, 251
Dipole moment, 57
Dirac distribution, 250
Dispersion
   group velocity
      anomalous, 141
      normal, 141
   phase velocity
      anomalous, 11
      normal, 11
   pulse broadening, 137
   waveguide, 204

Dispersion coefficient, 141
Dispersion flattened, 216
Dispersion length, 145
Dispersion relation
   free electron, 246
   light, 7
Dispersion shifted, 216
Dispersive media, 56
Doping, 280
Doppler effect, 97, 161
Drag coefficient, 98
Drude-Lorentz model, 56
Dynode, 415

EDFA. *See* Erbium doped fiber amplifier
Efficiency
   differential, 302
   quantum, 415
Eigenbase, 30
Eigenfrequencies. *See* Resonators
Eigenmode. *See* Mode
Einstein's convention, 69
Electro-optic effect, 86
   linear, 86, 400
   quadratic, 402
Electron affinity, 415
   negative, 415
Electron momentum, 283
Electronvolt, 9
Emission
   spontaneous, 256
   stimulated, 253
Energy flux, 13
   density, 13
Energy transport, 12
Envelope, 142
Epitaxial growth, 335
ε_0. *See* Vacuum permittivity
Erbium, 241
Erbium doped fiber amplifier (EDFA), 241
Escape depth, 415
Etalon, 179
eV. *See* Electronvolt
Evanescent wave. *See* Total reflection
Excimer, 329
   laser, 329

Fabry-Perot interferometer. *See* Interferometer
Faraday
   effect, 92
   isolator, 93
   rotation, 93

FEL. *See* Free electron laser
Fermi-Dirac distribution, 278
Fermi factor, 286
Fermi level, 278
  quasi-, 278
Fermi's golden rule, 250
Fiber gyroscope, 238
Fiber laser, 240
Fiber waveguide, 205
Finesse. See Interferometer
Focal plane array (FPA), 424
Four-level system, 269, 297
Four-vector, 95
Four wave mixing, 399
Fourier limit, 139
Fourier transformation
  2-*f* system, 134
  far field, 133
FPA. *See* Focal plane array
Free electron gas model, 62
Free electron laser (FEL), 340
Free spectral range. *See* Interferometer
Frequency mixing, 353
Frequency, normalized, 200
Fresnel coefficients, 47
Fresnel rhomb, 54
Füchtbauer-Ladenburg equation, 259
Full width at half maximum (FWHM)
  beam diameter, 107
  line width, 260
FWHM. *See* Full width at half maximum

Gain
  avalanche photodiode, 421
  Brillouin, 395
  coefficient, 268
  condition, 268, 287
  free electron laser, 347
  modulation, 309
  parametric
    second order, 372
    third order, 386
  photoelectron multiplier tube, 415
  photoresistor, 422
  Raman, 391
Gamut, 433
Gaussian beam, 102
  ABCD-transformation, 117
  beam divergence, 106
  beam radius, 106
  confocal parameter, 106
  focusing, 122
  phase front curvature, 107

q-parameter, 109
SHG, 363
Gaussian modes, 182
Glass fibers. *See* Waveguide
Gouy phase, 105
Gradient index fiber, 206
Gradient index lens (GRIN-lens), 113
Grating. *See* Optical grating
Grassmann's laws, 433
GRIN-lens. *See* Gradient index lens
Group delay, 141
Group velocity, 9
Group velocity dispersion (GVD), 141
GVD. *See* Group velocity dispersion

$\hbar$. *See* Planck's constant
Hamilton operator, 245
Helmholtz equation, 6
  paraxial, 102
  scalar, 6
Hermite-Gaussian beams, 126
Hilbert space, 36, 432
Hilbert transformation, 98
Hole burning
  spatial, 308
  spectral, 266

Idler wave. *See* Optical parametric amplifier
Impermeability, 74, 400, 404
Index ellipsoid, 81
Index profile, parabolic, 216
Indicatrix, 81
Integrated optics, 218, 237, 326
Intensity, 17
Interaction cross section, 251
Interband transitions, 278
Interference, 157
  multiple beam, 167
  two field-, 157
  visibility, 191
Interferometer
  Fabry-Perot, 177
    finesse, 179
    etalon, 179
    free spectral range, 179
  Mach-Zehnder, 162
  Michelson, 158
  Sagnac, 162, 165
  waveguide, 237
Intraband transitions, 278
Inversion, 264, 267
Irradiance. *See* Intensity
Isotropic media, 5

Joint density of states, 284
Jones matrix, 21
   eigenvectors, 25
   Pockels cell, 87
   reflection, 52
   transformation, 27
   transmission, 52
Jones vector, 19

k_B. *See* Boltzmann's constant
KDP, 86, 360, 401, 402
Kerr effect, 380
   electrooptic, 86
Kerr lens, 114
Kramers-Kronig relations, 66, 293
k-surface, 76

Laplace operator, 5
Laser
   diode, 331
      active zone, 335
      heterostructure, 334
      quantum well, 336
   efficiency, 302
   fiber-, 240
   free electron-, 340
   gain modulation, 312
   gas-, 328
   Helium-Neon-, 329
   linewidth, 309
   mode locking, 318
   mode selection, 307
   optimum coupling, 303
   q-switched, 314
   relaxation oscilllations, 309
   semiconductor-, 331
   solid state-, 325
   threshold, 302
   Ti:sapphire, 323
   velocimeter, 161
Laser materials, 324
LCD. *See* Liquid crystals
Lens, 112
   GRIN-lens, 113
   Kerr, 114
Light amplification. *See* Amplification
Line broadening
   collisions, 260
   crystal field, 263
   Doppler, 262

   homogeneous, 262
   inhomogeneous, 262
   natural, 259
Line shape
   Lorentzian, 58, 260
   saturated, 266
Liquid crystals (LCD), 90
Lithium niobate, 88, 224, 360, 368, 401
Lorentz force, 56, 341
Lorentz transformation, 96
Luminescence, 333
Luminosity, 430

μ_0. *See* vacuum permeability
Mach-Zehnder interferometer. *See*
      Interferometer
Magnetic constant. *See* Vacuum permeability
Magneto-optic effect, 92
Manley-Rowe relations, 357, 375
Mass
   effective, 277
   reduced, 285
Matrix
   ABCD-, 118
   dipole, 250
   Jones, 21
   perturbation, 248
   S- (*see* scattering matrix)
Maxwell's equations, 2
MCP. *See* Micro-channel-plate
Metals, optical properties, 62
Michelson interferometer. *See* Interferometer
Micro-channel-plate (MCP), 416
Miller's rule, 359
Mirror
   concave, 183
   convex, 183
   dielectric, 170
      bandwidth, 174
      reflection coefficient, 173
      transmission coefficient, 173
   metal, 173
   spherical, 115
Mode
   index, 181
   longitudinal, 183
   resonator, 181
   transverse, 186
   waveguide, 197
Mode condition. *See* Waveguide
Mode locking, 318
Momentary frequency. *See* Pulse propagation

Momentum
    electron, 283
    photon, 283
Monomode condition. *See* Waveguide
M^2-parameter, 107
Multilayer system, dielectric, 170

NA. *See* Numerical aperture
Nonlinear length, 151
Nonlinear polarization rotation (NPE), 382
Nonlocal effects, 4
NPE. *See* Nonlinear polarization rotation
Numerical aperture (NA), 198

OPA. *See* Optical parametric amplifier
Optical activity, 90
Optical grating, 168
    blazing, 170
Optical parametric amplifier (OPA), 358, 369,
        384
    idler wave, 369
Optical tunneling effect, 56
Oscillator
    harmonic, 56
    linear, 56
    parametric, 375

Parametric fluorescence, 375
Parametric frequency conversion, 376
Parametric process, 351
Parity, 253
Parseval's theorem, 138
Permittivity
    complex, 59
    relative, 4
    tensor, 69
    vacuum-, 3
Perturbation theory, 249
Phase, 6
Phase conjugation, 395
Phase matching, 82, 356, 365
    90°, 366
    birefringence, 365
    boundary, 40
    parametric amplification
        second order, 373
        third order, 386
    quasi-, 368
    second harmonic generation, 362
Phase velocity, 9
    in dispersive media, 60

Phonon, 278, 392
Photocathode, 414
Photodetector, 413
    array, 423
    bandwidth, 426
    integrator, 426
    responsivity, 425
Photodiode, 416
    avalanche, 421
    photocurrent, 418
    photovoltage, 418
    Schottky, 423
Photoelectric effect, 414
Photoelectron multiplier tube (PMT), 415
Photometry, 430
    units, 430
Photon, 252
    counting, 416
    energy, 252
    momentum, 283, 357
    statistics, 426
Photonic band gap, 236
Photonic crystals, 241
Photopic vision, 430
Photoresistor, 425
π polarization. *See* Polarization state
Planck's constant, 2, 245
Plasma frequency, 64
PMT. *See* Photoelectron multiplier tube
pn Junction, 419
Pockels effect. *See* Electro-optic effect, linear
Poincaré sphere, 32
Point groups, 352
Point spread function, 131
Poisson distribution, 427
Polarizability, 57
Polarization. *See* Polarization state
Polarization density, 3
    nonlinear, 351
Polarization maintaining fiber, 216
Polarization rotator, 22
Polarization state, 18, 75
    circular, 18
    eigenstate, 25
    elliptic, 18
    ellipticity, 31
    linear, 18
    orthogonal, 20
    π-, 45
    σ-, 45
    $\sigma^{\pm}$-, 20, 45
Polarizer, 22
Power density, 192
    spectral, 192

Poynting
  theorem, 13
  vector, 13
    anisotropic media, 83
    inhomogeneous fields, 35
Principal value, 73
Propagation constant, 6
  waveguide, 199
Propagation index, 11
  complex, 59
  effective, 200
Pulse
  chirp, 146
  compression, 146
  dispersion length, 145
  envelope, 137
  Gaussian, 138, 144
  group velocity, 141
  intensity autocorrelation, 377
  momentary frequency, 146
  nonlinear length, 151
  propagation, 141
  solitons, 152
Pumping, 268, 301, 358, 369

q-Parameter, 109, 217
Q-switching, 314
QPM. *See* Quasi-phase matching
Quantum cascade laser, 293
Quantum dots, 292
Quantum efficiency, 415
Quantum well, 291
  laser, 336
Quasi phase matching (QPM), 366

Raman amplification, 388
Raman effect, 388
Raman-Nath scattering, 409
Rate equations, 255
  four-level system, 297
  relaxation oscillations, 311
  three-level system, 304
  two-level system, 255
Rayleigh range. *See* Gaussian beam, confocal
    parameter
Recombination, 415
Rectification, optical, 355
Reflectance, 49
  absorbing media, 62
  dielectric multilayer, 173
Reflection, 40
  Brewster angle, 50
  coefficient, 47

Refraction
  anisotropic media, 82
  isotropic media, 41
Refractive index, 7
  extraordinary, 73
  frequency dependence, 61
  ordinary, 73
  principal values, 73
Relaxation oscillations, 309
Resonators, 180, 297
  confocal, 185
  eigenfrequencies, 182
  Hermite-Gaussian modes, 186
  mode sparation, 181
  modes, 181
  stability condition, 184, 187
Responsivity, 425
Retarder, 21
  achromatic, 54
  circular, 24, 90
  general, 33
  linear, 21, 75
  variable, 35

Sagnac interferometer. *See* Interferometer
Saturation, 264
  absorption, 264
  amplification, 269
  fluence, 273
  intensity, 265, 270
Scattering matrix, 26, 165
  beam splitter, 163
  interferometer, 237
  waveguide coupler, 226
Schawlow-Townes limit, 309
Schrödinger equation, 245
  eigenfunctions, 246
  eigenvalues, 246
  nonlinear, 153
  time independent, 246
Scotopic vision, 430
Second harmonic generation (SHG), 358
  Gaussian beam, 364
  field induced, 403
Selection rules, 253, 283
Self-consistency, 182, 199
  condition, 207
Self-focusing, 115
Self-phase modulation (SPM), 149, 381
  spectral broadening, 152
  white light generation, 152
Semiconductor, 275, 415
  band edge, 277

band gap, 275
bands, 275
  density of states, 276
  doping, 280
  gain bandwidth, 288
  gain condition, 287
  intrinsic, 278
  laser, 331
  optical transitions, 283
  transparency carrier density, 332
Sensitivity
  human eye, 431
  photodiode, 420
  PMT, 415
SFG. *See* Sum frequency generation
SHG. *See* Second harmonic generation
Shockley's diode equation, 418
Shot noise, 426
σ polarization. *See* Polarization state
Signal-to-noise ratio, 426
  Bose-Einstein statistics, 429
  Poisson statistics, 427
Silica glass fibers, 213
Single photon detector, 416, 422
Slowly varying envelope approximation, 102,
      361, 370
S-matrix. *See* Scattering matrix
Snell's law. *See* Refraction
Solitons, 4, 152
Space charge region, 416
Spectrometer, 135
Spectrum, 192
SPM. *See* Self-phase modulation
Spontaneous emission, 256
Stability condition. *See* Resonators
Step index fiber, 206
Stimulated emission, 253
Stimulated Raman scattering, 389
Stokes line, 389
Stokes's theorem, 4
Stop band, 175, 231
Sub-k space, 289
Sub-band, 290
  density of states, 292
Sum frequency generation (SFG), 353
Susceptibility, 3
  complex, 57
  linear oscillator model, 56
  nonlinear
    second order, 351
    third order, 351
      anharmonic oscillator, 358

scalar, 5
tensor, 69
as transfer function, 67

Tensor, 70
  diagonal, 74
  symmetry, 71, 352
  transformation, 70
THG. *See* third harmonic generation
Third harmonic generation (THG), 379
3 dB-coupler. *See* Waveguide
Three-level system, 271
Three wave mixing, 358, 370
Total reflection, 41, 48, 197
  amplitude, 55
  critical angle, 42
  evanescent wave, 53
  optical tunneling effect, 56
Transfer function, 67
  spatial, 128
Transition
  direct, 283
  forbidden, 253
  indirect, 283
  probability, 250
Transmission coefficient, 47
Transmittance, 49
  dielectric multilayer, 173
Transparency carrier density, 332
Two-level systems, 245
Two-photon absorption, 387

Ultraviolet (UV), 66
Undulator, 341
Uniaxial media. *See* Anisotropic media
UV. *See* Ultraviolet

Vacuum
  energy density, 13
  fluctuations, 257, 375
  impedance, 16
  permeability, 3
  permittivity, 3
  speed of light, 1, 5
VCSEL. *See* Vertical cavity surface emitting
      laser
Verdet constant, 92
Vertical cavity surface emitting laser (VCSEL),
      339

Wave
  Bloch, 276
  DeBroglie, 44, 246
  diffracted, 43
  evanescent, 53
  plane, 8
Wave equation, 5
  anisotropic media, 75
  dispersive media, 59
  nonlinear media, 360
  optically active media, 91
  paraxial, 101
Wave number, 6
Wave plate, 22, 54
Wave vector, 6
Waveguide, 197
  active, 240
  amplifier, 240
  coupler, 220
    3 dB, 223
    eigenmodes, 226
    S-matrix, 226
  cutoff frequency, 202, 210
  dispersion, 204
  eigenmodes, 199
  electro-optic, 198
  filter, 223
  gradient index, 206
  grating, 228
  gyroscope, 238

  implementation, 198
  interferometer, 237
  laser, 240
  loss, 213
  mode condition, 200
  monomode condition, 202
  normalized frequency, 200
  planar, 197
  self-consistency condition, 207
  step index, 206
  TE mode, 202
  TEM mode, 208
  weak guiding, 202, 208, 209
Wavelength, 8
Wavelength division multiplexing (WDM),
      216, 382
WDM. *See* Wavelength division multiplexing
White light generation. *See* Self-phase
      modulation
Wiener-Khinchin theorem, 192

XPM. *See* Cross phase modulation

Young's double slit, 165

Z_0. *See* Vacuum impedance